Thomas Muench

# 3D Inversion for Seismic Structure and Hypocenters in Germany

Thomas Muench

# 3D Inversion for Seismic Structure and Hypocenters in Germany

3D Simultaneous Inversion for Seismic Structure and Local Hypocenters in Germany under Consideration of Anisotropy in the Upper Mantel

**Südwestdeutscher Verlag für Hochschulschriften**

**Imprint**
Any brand names and product names mentioned in this book are subject to trademark, brand or patent protection and are trademarks or registered trademarks of their respective holders. The use of brand names, product names, common names, trade names, product descriptions etc. even without a particular marking in this work is in no way to be construed to mean that such names may be regarded as unrestricted in respect of trademark and brand protection legislation and could thus be used by anyone.

Publisher:
Südwestdeutscher Verlag für Hochschulschriften
is a trademark of
Dodo Books Indian Ocean Ltd., member of the OmniScriptum S.R.L Publishing group
str. A.Russo 15, of. 61, Chisinau-2068, Republic of Moldova Europe
Printed at: see last page
**ISBN: 978-3-8381-2490-2**

Zugl. / Approved by: Kassel, Universität Kassel, Diss., 2009

Copyright © Thomas Muench
Copyright © 2011 Dodo Books Indian Ocean Ltd., member of the OmniScriptum S.R.L Publishing group

# Contents

**1 Introduction**   **7**
- 1.1 Historical development . . . . . . . . . . . . . . . . . . . . . . . . . . . . . . . . . 7
- 1.2 Introduction to the problem . . . . . . . . . . . . . . . . . . . . . . . . . . . . . . 9
- 1.3 Geology and Tectonics of Germany . . . . . . . . . . . . . . . . . . . . . . . . . . 9
- 1.4 Simultaneous Inversion for structure and hypocenters . . . . . . . . . . . . . . . . 15
- 1.5 Previous regional seismological studies within Germany . . . . . . . . . . . . . . 15
- 1.6 Objectives and goals of the thesis . . . . . . . . . . . . . . . . . . . . . . . . . . 17

**2 Theory of seismic tomography**   **19**
- 2.1 Wave theory . . . . . . . . . . . . . . . . . . . . . . . . . . . . . . . . . . . . . . . 19
  - 2.1.1 Stress, Strain and the Wave equation . . . . . . . . . . . . . . . . . . . . . 20
  - 2.1.2 Seismic Velocity . . . . . . . . . . . . . . . . . . . . . . . . . . . . . . . . 21
- 2.2 Ray Theory . . . . . . . . . . . . . . . . . . . . . . . . . . . . . . . . . . . . . . . 22
  - 2.2.1 Basic concepts of Ray Theory . . . . . . . . . . . . . . . . . . . . . . . . . 22
  - 2.2.2 Eikonal equation . . . . . . . . . . . . . . . . . . . . . . . . . . . . . . . . 23
  - 2.2.3 Application to tomography . . . . . . . . . . . . . . . . . . . . . . . . . . 23
- 2.3 Seismic inverse theory . . . . . . . . . . . . . . . . . . . . . . . . . . . . . . . . . 24
  - 2.3.1 The forward problem . . . . . . . . . . . . . . . . . . . . . . . . . . . . . . 24
  - 2.3.2 The inverse problem . . . . . . . . . . . . . . . . . . . . . . . . . . . . . . 26
  - 2.3.3 Validation of the inversion results . . . . . . . . . . . . . . . . . . . . . . . 30
  - 2.3.4 Limitations and inherent sources of errors in the SSH traveltime inversion . . . 33
  - 2.3.5 Practical considerations of the SSH traveltime inversion . . . . . . . . . . . 35

**3 Anisotropy**   **37**
- 3.1 Origins of anisotropy . . . . . . . . . . . . . . . . . . . . . . . . . . . . . . . . . . 37
  - 3.1.1 Mineralogical sources of anisotropy . . . . . . . . . . . . . . . . . . . . . . 37
  - 3.1.2 Anisotropy caused by structural complexity . . . . . . . . . . . . . . . . . 39
- 3.2 Mathematical description of anisotropy . . . . . . . . . . . . . . . . . . . . . . . . 39
- 3.3 Occurrence of seismic anisotropy in the Earth's interior . . . . . . . . . . . . . . 41
  - 3.3.1 Anisotropy in the mantle . . . . . . . . . . . . . . . . . . . . . . . . . . . . 41
  - 3.3.2 Anisotropy in the crust . . . . . . . . . . . . . . . . . . . . . . . . . . . . 42
- 3.4 Shear wave splitting due to seismic anisotropy . . . . . . . . . . . . . . . . . . . . 42

**4 Data Set**   **45**
- 4.1 General aspects of data requirements in seismic inversion . . . . . . . . . . . . . 45
- 4.2 Data used in the present study . . . . . . . . . . . . . . . . . . . . . . . . . . . . 46
- 4.3 Initial data preprocessing and selection criteria . . . . . . . . . . . . . . . . . . . 47
- 4.4 Tests of the impacts of the various selection criteria . . . . . . . . . . . . . . . . 51
  - 4.4.1 Effect of maximal GAP . . . . . . . . . . . . . . . . . . . . . . . . . . . . 51
  - 4.4.2 Influence of the number of observations per event (NOBS) . . . . . . . . . 52
  - 4.4.3 Influence of the maximally allowed residuals . . . . . . . . . . . . . . . . . 56

|     |       |                                                                     |     |
| --- | ----- | ------------------------------------------------------------------- | --- |
|     | 4.5   | Testing for ambiguous $P_n$-phases                                  | 60  |
|     | 4.6   | Testing for anisotropy                                              | 60  |
|     | 4.7   | Summary of the data preprocessing analysis                          | 65  |

# 5 SSH-inversions for 1D velocity model — 69
    5.1 General approach of SSH — 69
    5.2 Control of important input parameters in the SSH-program — 70
    5.3 Analysis of the azimuthal $P_n$-anisotropy — 72
        5.3.1 Previous evidence of $P_n$-anisotropy across Germany — 72
        5.3.2 Control of the $P_n$-anisotropy ellipse — 72
    5.4 Finding the optimal average depth of the Moho — 74
        5.4.1 Information on the Moho depths across Germany — 74
        5.4.2 SSH-control of the Moho depth — 78
        5.4.3 Effects of different initial models on the inversion results — 80
    5.5 Sensitivity tests for the 1D velocity models — 81
        5.5.1 Test of the convergence of various isotropic and anisotropic input models — 81
        5.5.2 1D velocity models with finer layer discretization — 92
    5.6 Selection of the optimal 1D velocity model — 94
    5.7 Inversions with $S$-phases and tests of the $V_P/V_S$-ratio — 98
    5.8 Final considerations about the optimal 1D velocity model — 100

# 6 SSH-inversions for 3D velocity models — 101
    6.1 Introduction, general approach and model-setup — 101
    6.2 Synthetic random resolution tests — 103
        6.2.1 General approach — 103
        6.2.2 Synthetic test with 15x15 blocs — 104
        6.2.3 Synthetic test with 25x25 blocs — 118
        6.2.4 Synthetic test with 35x35 blocs — 123
        6.2.5 Final remarks on synthetic random tests — 130
    6.3 Trade-off characteristics of the synthetic models — 130
    6.4 Testing for resolution — 133
        6.4.1 Resolution matrix — 133
        6.4.2 Covariance matrix — 134
    6.5 Checkerboard tests — 138
        6.5.1 Checkerboard test for $15 \times 15$ bloc models — 139
        6.5.2 Checkerboard test for $25 \times 25$ bloc models — 148
        6.5.3 Checkerboard test for $35 \times 35$ bloc models — 149
        6.5.4 Conclusive remarks on the checkerboard tests — 149
    6.6 3D laterally heterogeneous seismic velocity models — 149
        6.6.1 $15 \times 15$ bloc models — 150
        6.6.2 $25 \times 25$ blocs models — 170
        6.6.3 $35 \times 35$ bloc models — 173
    6.7 F-Test statistics for the 3D Models — 174
    6.8 Calculation and interpretation of $V_P/V_S$-ratios — 177
        6.8.1 $V_P/V_S$-ratio and its relation to the composition of the lithosphere — 177
        6.8.2 $15 \times 15$ $V_P/V_S$-models — 181
    6.9 Geological and petrological interpretation, 3D models — 181

## 7 Relocation of hypocenters — 191
- 7.1 General approach — 191
- 7.2 Hypocentral relocations, optimal 1D model — 192
- 7.3 Relocations with former 1D velocity models — 201
- 7.4 Hypocenter relocations, optimal 3D velocity models — 201
- 7.5 Sensitivity tests — 209
  - 7.5.1 Influence of anisotropically generated traveltime data — 209
  - 7.5.2 Sensitivity test with randomly shifted hypocenters — 210
  - 7.5.3 Sensitivity tests with randomly perturbed arrivaltimes — 213
- 7.6 Summary of the major results of the hypocentral relocations — 217

## 8 Summary — 219
- 8.1 Theory — 220
- 8.2 Anisotropy — 221
- 8.3 Initial data analysis — 221
- 8.4 Analysis and interpretation of 1D-velocity models — 223
- 8.5 Analysis and interpretation of 3D velocity models — 224
  - 8.5.1 Synthetic tests — 225
  - 8.5.2 Structural interpretation of the 3D seismic models — 226
- 8.6 Effects of the seismic structure on relocations — 228
- 8.7 Concluding remarks — 229
- 8.8 Further developments and outlook — 230

## A Details of 1D sensitivity studies — 233
- A.1 Results of different 1D velocity model computations — 233
- A.2 Traveltime variations for anisotropy over distances — 237
- A.3 Traveltime variations for dipping Moho over distances — 237

## B Material properties — 239

# Kurzzusammenfassung

Das Hauptziel dieser Arbeit war die Untersuchung von anisotropen Effekten auf die Ausbreitung von seismischen Wellen entlang des oberen Mantels unterhalb Deutschlands und angrenzenden Gebieten. Bisherige refraktions- und reflexionsseismische Untersuchungen lieferten Ergebnisse für die Existenz von Anisotropie im oberen Mantel und ihren Einfluss auf die Ausbreitungsgeschwindigkeit von $P_n$-Wellen.

Für die Untersuchungen der Anisotropie in Europa wurde nun durch 3D-Untersuchungen für die Kruste und den oberen Mantel unterhalb Deutschlands unter Berücksichtigung des Einflusses der Anisotropie auf $P_n$-Phasen eine Lücke geschlossen. Diese Untersuchung wurde mit dem SSH-Inversionsprogramm von Dr. M. Koch durchgeführt, welches in der Lage ist die seismische Struktur und Hypozentren simultan zu berechnen.

Für die Untersuchung steht ein Datensatz seismischer Aufzeichnungen zwischen den Jahren 1975 und 2003 zur Verfügung mit Aufzeichnungen von 60249 P- und 54212 S-Phasen von insgesamt 10028 seismischen Ereignissen.

Die zu Beginn durchgeführte Analyse der Residuen (RES, Differenz zwischen berechneter und beobachteter Ankunftszeit) bestätigte das Vorhandensein von Anisotropie für $P_n$-Wellen. Die sinusförmige Verteilung wurde durch die Erweiterung des SSH-Programms um eine elliptische Korrektur mit einer schnellen und dazu senkrechten, langsamen Geschwindigkeitsvariation korrigiert.

Der Azimut der schnellen Achse wurde durch die Anwendung der simultanen Inversion für Hypozentren und Struktur bei $25-27°$ mit einer Variation der Geschwindigkeiten zu $\pm 2.5\%$ um einen mittleren Wert von $8\,km/s$ ermittelt. Dieser neue Wert unterscheidet sich vom bisherigen, der bei circa $35°$ lag, aufgrund der neuen Position der Hypozentren und der neu ermittelten Geschwindigkeitsstruktur.

Die Anwendung der elliptischen Korrektur für die Anisotropie ergab im weiteren eine bessere Anpassung eines vertikal geschichteten 1D-Modells im Vergleich zu den Ergebnissen der vorausgegangenen seismologischen Experimente und 1D und 2D Untersuchungen.

Das optimale Ergebnis der 1D-Inversion wurde als Startmodell für die 3D-Inversion für die dreidimensionale Abbildung der seismischen Struktur der Kruste und des oberen Mantels verwendet. Die simultane Inversion zeigte dabei eine Optimierung der Relokalisierung der Hypozentren und der Rekonstruktion der seismischen Struktur im Vergleich zu bisher ermittelten geologischen und tektonischen Strukturen.

Die Untersuchungen für die seismische Struktur und die Relokalisierung der Hypozentren wurden durch umfangreiche Tests abgesichert. In einem ersten Test wurden synthetische Laufzeitdaten mit anisotroper Variation erzeugt, womit nach der isotropen und anisotropen Relokalisierung der Einfluss auf die Genauigkeit abgeschätzt werden konnte. Des weiteren wurden Tests mit zufallsverteilten Hypozentren und Laufzeitdaten durchgeführt, um deren Einfluss auf die Relokalisierungsgenauigkeit und wiederum den Einfluss der Anisotropie zu ermitteln.

Abschliessend wurden die Ergebnisse auf die Relokalisierung des Waldkirchbebens 2004 angewandt, um die anisotrope Relokalisierung mit der isotropen und der ursprünglichen zu vergleichen, um festzustellen ob die anisotrope Korrektur zu einer Verbesserung führt.

## Abstract

The main task of this work has been to investigate the effects of anisotropy onto the propagation of seismic waves along the Upper Mantle below Germany and adjacent areas. Refraction- and reflexion seismic experiments proved the existence of Upper Mantle anisotropy and its influence onto the propagation of $P_n$-waves.

By the 3D tomographic investigations that have been done here for the crust and the upper mantle, considering the influence of anisotropy, a gap for the investigations in Europe has been closed. These investigations have been done with the SSH-Inversionprogram of Prof. Dr. M. Koch, which is able to compute simultaneously the seismic structure and hypocenters.

For the investigation, a dataset has been available with recordings between the years 1975 to 2003 with a total of 60249 P- and 54212 S-phase records of 10028 seismic events.

At the beginning, a precise analysis of the residuals (RES, the difference between calculated and observed arrivaltime) has been done which confirmed the existence of anisotropy for $P_n$-phases. The recognized sinusoidal distribution has been compensated by an extension of the SSH-program by an ellipse with a slow and rectangular fast axis with azimuth to correct the $P_n$-velocities.

The azimuth of the fast axis has been fixed by the application of the simultaneous inversion at $25-27°$ with a variation of the velocities at $\pm 2.5\%$ about an average value at $8\,km/s$. This new value differes from the old one at $35°$, recognized in the initial residual analysis. This depends on the new computed hypocenters together with the structur.

The application of the elliptical correction has resulted in a better fit of the vertical layered 1D-Model, compared to the results of preciding seismological experiments and 1D and 2D investigations.

The optimal result of the 1D-inversion has been used as initial starting model for the 3D-inversions to compute the three dimensional picture of the seismic structure of the Crust and Upper Mantle. The simultaneous inversion has showed an optimization of the relocalization of the hypocenters and the reconstruction of the seismic structure in comparison to the geology and tectonic, as described by other investigations.

The investigations for the seismic structure and the relocalization have been confirmed by several different tests. First, synthetic traveltime data are computed with an anisotropic variation and inverted with and without anisotropic correction.

Further, tests with randomly disturbed hypocenters and traveltime data have been proceeded to verify the influence of the initial values onto the relocalization accuracy and onto the seismic structure and to test for a further improvement by the application of the anisotropic correction.

Finally, the results of the work have been applied onto the Waldkirch earthquake in 2004 to compare the isotropic and the anisotropic relocalization with the initial optimal one to verify whether there is some improvement.

# Chapter 1

# Introduction

From outer space, the earth looks like a homogeneous sphere, but when approaching it, more and more of its details on the surface become present. Thus, first the major features, such as oceans and mountains on the surface of the sphere will appear, then valleys, plains and rivers, and eventually, even small grains of sand laying on the beach become visible. To understand the physics of the earth in more detail, one has to dig into the deep interior of earth. There, one will encounter a complex structure, consisting of many zones of different materials with different physical and chemical properties.

Applying the methods of seismology, one will be able to probe the earth's deep interior in detail by measuring its seismic properties, starting from the earth's surface down to its center at a depth of about $6378\,km$. Knowing these seismic properties at various depths, other chemical and physical properties of the layers there can be deduced. To reach this goal, many passive and active seismic experiments have been set up during the approximately 100-years-long history of seismology, all with the purpose to record the seismic waves, emanating from either a natural or an artificial seismic source and traveling through the earth's interior, at a seismometer at the earth's surface. Then, knowing the traveltime of this seismic wave, the velocity structure of the earth along the wave path can, in principle, be determined, by a process which is called seismic inversion.

The ultimate goal of seismic inversion is then the determination of the three-dimensional structure of the earth's interior from a set of two-dimensional seismic observations at its surface. In spite of more than 100 years of seismological research, especially the so called "seismic tomography" problem, is far from being solved, not only because of its intrinsic challenging mathematical and computational intricacy, but also due to the fact that the observed seismic signals often elude a precise and complete analysis as one wishes to extract the maximum of information hidden in there.

## 1.1 Historical development of seismological methods to probe the earth's interior

The principal aim of seismology is to measure and interpret arrivaltimes of seismic waves, generated by seismic events like earthquakes, nuclear explosions and other abrupt movements within the earth, and to establish a seismological model to explain these arrivaltimes. Especially tomographic methods like teleseismic and local earthquake tomography are powerful tools to compute these models.

Since the introduction of inversion methodology (Wiechert, 1910) and it's application to geophysical problems in teleseismics by Aki and Lee (1976), there was strong development of the method, the computation and the application. For example the works of Koch (1983a, 1985a); Koch and Kalata (1992), Eberhart-Phillips (1995), Thurber and Aki (1987), Tarantola et al. (Tarantola and Valette, 1982b,a; Tarantola, 1987) or Kissling (1988) have lead to an improvement in the tomographic methodology.

**A brief history of the development of seismic probing and inversion (highlights):**

- 1879 Start of continuous recording of earthquakes ins Switzerland

- 1899 First measurement of earthquake activity in Göttingen by Emil Wiechert

- 1971 HYPO71 to determine hypocenter, magnitude and first motion pattern of local earthquakes

- 1976 Aki, Christofferson and Husebye introduced a simple and approximate, yet elegant technique for using body wave arrivaltimes from teleseismic earthquakes to infer three dimensional (3D) seismic velocity heterogeneities (ACH method)

- 1976 HYPO2D for seismic tomography studies by W.L. Ellsworth and S. Roecker

- 1980 HYPOELLIPSE by Lahr (1999)

- 1981 layered-model ray-tracer (Thurber, 1981, 1983; Thurber and Aki, 1987)

- 1984 Minimum 1D Model with Velest (Kissling et al., 1994)

- 1985 SSH Program for simultaneous inversion of earthquake location and velocity structure including a non-linear ray tracing (Koch, 1983b)

- 1985 Improvement of the ACH-method by including a non-linear ray tracing. (Koch, 1985a)

- 1987 Three dimensional seismic imaging by Thurber (1981) and Thurber and Aki (1987)

Although "normal" tomography is nowadays "state of the art", there is still development to include new aspects, like the azimuthal traveltime correction for $P_n$-Phases to include anisotropic correction for the upper mantle, as done in this work, with the aim to refine the structural model of the Earth and to obtain more precise hypocenter locations.

The principal aim of geophysical inversion in general is to get a reasonable model of a particular physical property of the earth's interior from measured associated data at the surface. In seismology, the registrations of seismic signals, the traveltime arrivals at distinct stations, are used to deduce the seismic structure, i.e. the seismic properties of the earth material in general which can be those of rocks, magmas, subduction zones, rift zones, volcanoes, etc.

One part of seismology is the application of inversion calculus to the near surface of the earth which is commonly known as crust and upper mantle. Near surface is understood as the continental crust, which ranges from zero to about $30 - 40\,km$, sometimes deeper to about $70\,km$, and the oceanic crust, which only has a thickness of about $5 - 20\,km$ and the first few kilometers of the upper mantle.

One particular interesting application of seismology is the determination of the seismic structure near the surface of the earth which is commonly known as crust and upper mantle. The utmost upper layer is called the crust which, under continents, ranges from zero to about $30 - 40\,km$, sometimes deeper to about $70\,km$, whereas the oceanic crust only has a thickness of about $5 - 20\,km$.

The crust is ending at the so-called Mohorovîcic discontinuity, commonly known as Moho. This discontinuity is characterized by a sudden change of physical and chemical properties of the rock material, that leads to a significant increase of the propagation speed of seismic waves. Moreover, the abrupt change in seismic velocity at the Moho results in refraction and reflection of the incoming seismic wave field which are additional indicators of that vertical seismic inhomogeneity. But not only refraction or reflection occurs at such a seismic discontinuity, but a so-called Head- or Mintrop Wave develops that travels along the boundary, that, due to Huygens principle, is continuously sending back seismic energy to the surface, with an angle that is equal to the critical angle of incidence.

## 1.2 Introduction to the problem

The central aim of this work is to investigate the crust and upper mantle under Germany with an improved version of the method of "Simultaneous inversion of seismic Structure and local Hypocenters (SSH)". Several recent studies showed evidence for strong seismic anisotropy in the upper mantle in this region that, therefore, should be included in the numerical approach of the SSH method.

So, for example, the work of Enderle et al. (1996a) indicates that traveltime observations for $P_n$-phases have strong azimuthal dependence over many areas of Germany. Other investigations, like those done by Okaya et al. (1995); Gledhill (1991) for New Zealand, by Boness and Zoback (2004) for stress induced anisotropy in Parkfield, CA, USA, or for anisotropic effects in Bohemia by Plomerova (1997); Plomerová et al. (1998) or basic investigations done by Wu and Lees (1999); Anderson (1989) and Song et al. (2001b,a); Song and Koch (2002); Song et al. (2004) show that there is strong evidence of the occurrence of seismic anisotropy at the crust mantle boundary and the upper mantle. Hawkins et al. (2001) studied areas north of Germany, where they are applying anisotropy to achieve accurate imaging of the geological environment and to optimize the localization of gas fields in the Rotliegendes.

As one knows, a deficit (lack of resolution, ambiguous trade off between various parameters of the model) arises in the hypocenter relocation of earthquakes with the present isotropic SSH method using the existing isotropic velocity models. Improvements in the exactness of the hypocentral determinations and, moreover, in the fit of the model's predictions to the measured data can only be achieved if other seismic properties of the crust and the upper mantle which have an influence on the propagation of seismic signals, namely, seismic anisotropy, are taken into consideration in the analysis. Another factor is the Moho topography that has an essential influence on the exactness of the earthquake relocation or of the computed velocity models. In a first approximation this influence can be grasped by assuming an inclined Moho boundary, as has been proposed by Koch and Kalata (1992).

In this thesis a comparison between "normal", isotropic velocity models and improved velocity models with an anisotropic correction for the $P_n$ phases is done to show the increase in the quality of the model fit to the observed data and of the structure and hypocenter determination by incorporating an anisotropic correction technique into the original isotropic SSH method of Koch (1983a, 1985a); Koch and Kalata (1992)

## 1.3 Geology and Tectonics of Germany

The apparent topography (Figure 1.1) of Germany is directly related to its tectonic development and geological structure. The mountainous topography mainly shows a southwest northeast orientation, as epitomized by the Schwäbische Alb, the Frankish Jura, and the Fichtelgebirge. Perpendicular to the Fichtelgebirge, the Bohemian Massif forms the eastern border of Germany to Czechia. The Rhinegraben in the southwest of Germany extends from Basel northward up to Frankfurt (Main) and is a major crustal fracture (Graben or rift), and fault zone in Germany, dividing one geological ancient unit into two parts, the Vosges (France) in the west and the Black Forest (Germany) in the east. North of the Harz, the surface is rather flat due to soil abrasion from glaciers of Pleistocene Ice Age.

Following Berndt (1999) and Krumbiegel (1981) which describe in detail the geological history of Central Europe, the named topographic features are formed by the convergence of the African plate, the Eurasian plate and the American plate that resulted in a rather complex tectonic and geological structure, as can be seen in Figures 1.2 and 1.3. (Berthelsen, 1992)

The tectonic processes that occurred in central Europe during the Tertiary age had the most effective influence on today's geomorphology and, as consequence, the development of the Alps and the upper Rhinegraben. Tertiary rocks are found over large areas of central and northern Germany, the Voralps and in the upper Rhinegraben. During that time numerous reservoirs were formed that extend to

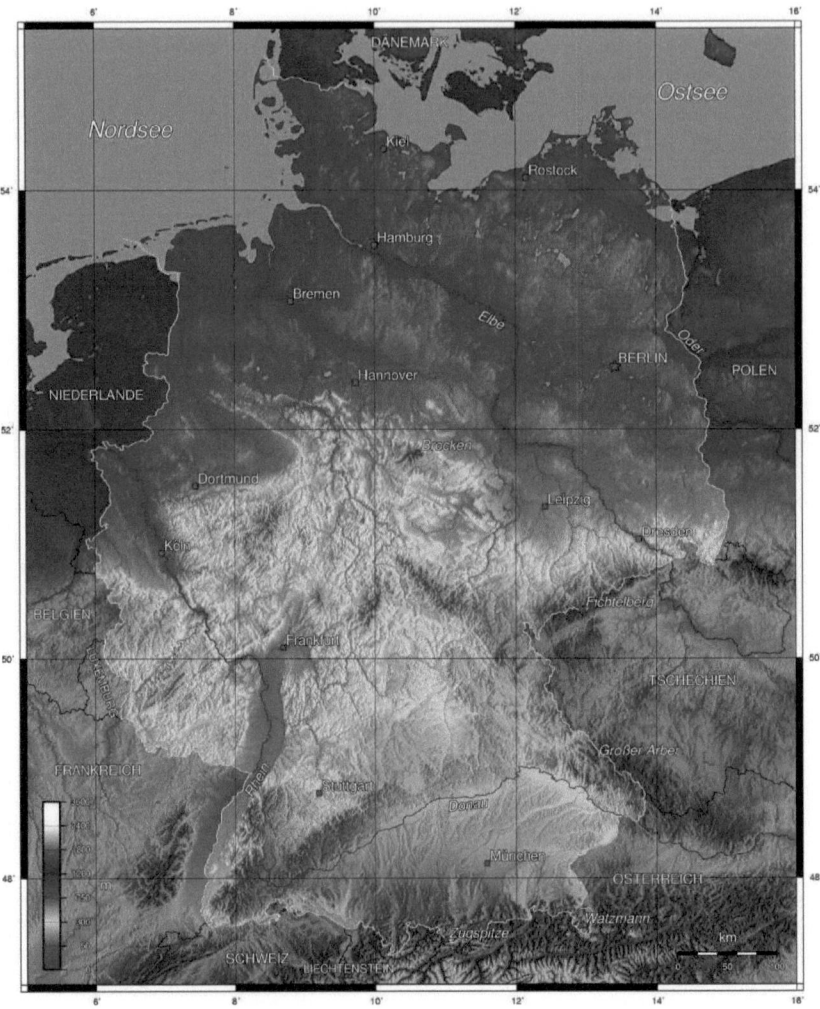

Figure 1.1: Topography of Germany and adjacent countries. The Alps in the south form the highest areas (up to 4809 meter for the Mont Blanc (France)), followed by the Black Forest, the Vosges (France) in the southwest and the Bohemian Massif in the east. The Harz is the most northwards located high mountain, followed by the "Norddeutsches Becken" with its rather flat topography extending north up to the North Sea and the Baltic Sea. This figure applies to the GNU Free Documentation License, where the complete text can be found here.

## 1.3. GEOLOGY AND TECTONICS OF GERMANY

depths of more than $10\,km$, especially lignite[1], gas and oil fields in tertiary layered structures.
The major geotectonic features of Germany and adjacent regions consist of the following geographical areas, ordered from south to north:

- The Alps
  They form the southern border of Germany, and are geologically separated into the Western and Eastern Alps. The Eastern Alps are separated again into three parts, the Northern Alps, the Central Alps and the Southern Alps. The Northern Alps mainly consist of the Subalpin Molasse, the Flyschzone[2], the northern "Kalkalpen", "Schieferalpen" and "Grauwackenzone". In the north, the bavarian and austrian Alpen foreland affiliates. The Central Alps consist of granite, gneiss and mica schist building warping structures. Because of their age, they are the highest mountains in Europe owing their orogeny to the tectonic collision of the African and Eurasian plates. The South Alps are mainly represented by the Southern Kalkalps.

  The Western Alps are again separated into three parts, which are several crystalline massifs in the center along the strike of the mountain, a zone with gneiss in the south and a zone made of Chalk in the west and north. The tectonic upfolding of the Alps started as a multistage process about 135 million years ago at the edge of the Jurassic to the Cretaceous period, terminating the biggest movements in the Tertiary about 30 - 35 million years ago.

- Alpine Foreland and the Molassebecken
  The Molassebecken consists of a massif sedimentary band (sandstones, shales and conglomerates) formed as terrestrial deposit in front of the rising alpine mountain chains (Stanley, 1999). Erosion especially took place in the former ice age, when alpine rocks were eroded by glaciers with the abrasive debris left in the region.

- The mid european Variscides
  The variscian[3] orogeny started in the early Devonian (410 Mio. years) during until the end of the Perm (250 Mio. years), when numerous mountains in middle and western Europe were build. It is divided into three parts: The Subvariszikum, the Rhenoherzynikum, the Saxothuringikum and the Moldanubikum. The variscian mountains represent a complex fault structure, where the enormous crustal movements lead to strong faulting, thrusting of sediments in the Devonian and early Carboniferous. In Germany, the Rhenish Massif, the Harz and the Erzgebirge result out of this tectonic process.
  http://www.geoglossar.de/tektonik/variszikum/variszikum.html

- The Moldanubikum
  The Moldanubikum is named after the rivers Moldau and Danube. It is the southern zone of the Variscides Mountains stretching from the Vosges over the Black Forest to the Bohemian Massif and is the oldest of the three zones of these mountains.

- The Vogtland (Bohemian Massif)
  Its border to Czechia is build by the Bohemian Massif, which is, like the Black Forest, one of the oldest mountains in Germany. The Bohemian Massif is the largest coherent surface outcrop of the Variscan basement in central Europe.

- The Saxothuringikum
  Characteristic for the Saxothuringikum are clastic rocks, mixed with carbonates and syn- to postorogenic vulcanites, undergoing strong metamorphic processes by strong folding with SE

---
[1]What is Braunkohle in German, and wherefore the Tertiary is also called Braunkohlezeit
[2]Flysch is a relatively archaic term describing synorogenic (occurring contemporaneously with mountain building) clastic sedimentation within marine depositional facies.
[3]A phase of orogeny all over the Earth starting in the mid Paläozoic.

convergence. In the Carboniferous, a lot of granitic intrusions arose caused by igneous activity. (M. Powell, 2005)

- Swabian Alb
  The Swabian Alb is a mountainous region in front north of the Alps, formed mostly by limestone, which formed the seabed during the Jurassic era. The Swabian Alb (German: Schwäbische Alb) is a plateau in Baden-Württemberg, Germany, extending 220 km from southwest to northeast and 40 to 70 km in width.
  http://en.wikipedia.org/wiki/Swabian_Alb

- The Rhinegraben
  This is one of the major tectonic zones in Germany extending from the Lake Constance in the south to Amsterdam in the north. Most of the seismic activity can be observed in the south between Basel and Karlsruhe, which is a part of the Upper Rhine Graben between Basel and south of Frankfurt. South of Frankfurt, the Rhine changes its direction, now being named the Lower Rhine Graben. The Upper Rhine Graben is the central segment of a rift valley extending from the north to the western see. It is assumed, that tensional stress in the Earth crust and mantle caused thinning of the crust, hence the surface along the rift zone dropped down. Opposite to this, the Crust Mantle Boundary was lifted up and raised up the Vosges and the Black Forest, building nowadays the shoulders of the Upper Rhine Graben.

- The Lower Rhine Depression
  It is the northern part of the geological structure of the Lower Rhine bay, that has its origin by subsidence, happening for the last 30 million years. The subsidence was compensated by sedimentary refill with vertical extension up to $1.3\,km$.

- Norddeutsches Becken
  As with Hoffmann et al. (1996) described the position of the Moho with her striking WNW ESE stroking depression in the area of the Ostelbischen massif probably a result of isostatic balance movements in the period by the higher "Oberrotliegenden" up to the middle "Buntsandstein".

- The "Norddeutsches Tiefland"
  It is also called "Norddeutsche Tiefebene" and is situated between the coasts of North Sea and the Baltic Sea and the "mitteleuropäischen Mittelgebirgsschwelle". It was formed during the last glacial period in the Quaternary, when the massive ice shields eroded the topography to a rather planar surface. It consists of massif sedimentary layers.

- The Mitteldeutsche Kristallin-Schwelle
  Also called the "mitteleuropäische Mittelgebirgsschwelle", a component of the variszic orogeny, in the middle of Germany with strike from North East to South West. Its main parts are the Harz, the Thüringer Wald, the Kellerwald, the Frankenwald, the Taunus, the Odenwald, the Spessart and the Rhön. (Henningsen and Katzung, 2002; Walter, 1995)
  http://www.tu-dresden.de/biw/geotechnik/geologie/

Over all, this is only a very small and incomplete review of the tectonics and geology in Germany and adjacent regions. The tectonic processes in detail are much more complex, causing a huge variety of geological structures.

## 1.3. GEOLOGY AND TECTONICS OF GERMANY

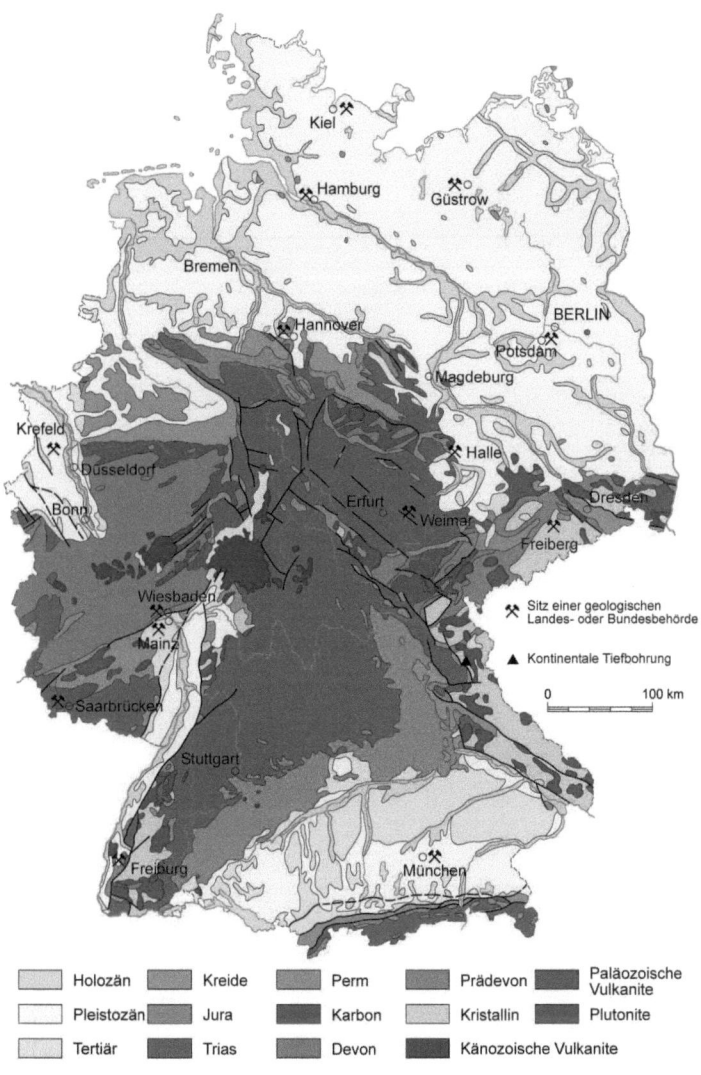

Figure 1.2: Geological overview of Germany. After "Die deutsche Subkommission für Tertiärstratigraphie" (http://www.palaeontologische-gesellschaft.de/palges/tertiaer/).

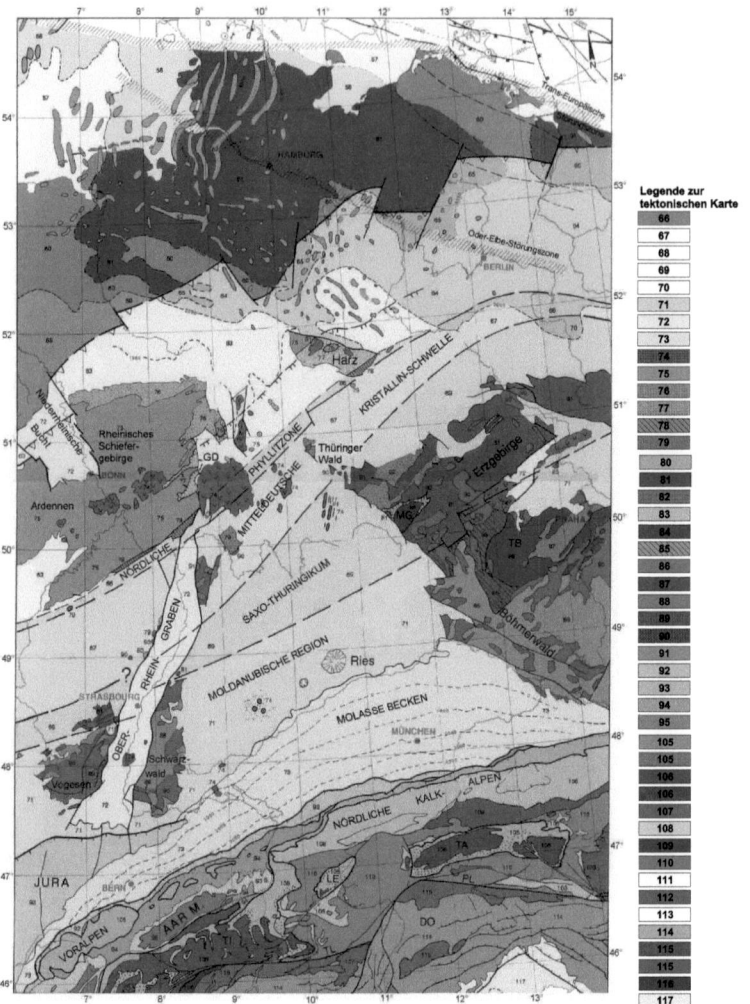

Figure 1.3: The tectonic situation in Germany
From: The European Science Foundation (1992): A Continent Revealed, The European Geotraverse,
Legend to tectonics south sheet (Berthelsen, 1992)
Abbreviations:
Aar M. Aar-Massiv
GD Gießen-Harz-Decke
MG Münchberger Klippe (und vergleichbare)
TA Tauern-Fenster
TI Ticino-Region (lepontinische Region)

DO Dolomiten
LE Unterengadin-Fenster
PL Pustertal-Linie
TB Tepla-Barrandium
TL Tonale-Linie (insubrische Linie)

## 1.4 Simultaneous Inversion for structure and hypocenters: SSH method

The inversion method used here is called "Simultaneous Inversion for Structure and Hypocenters", heretofore abbreviated "SSH". The advantage of this method is to compute together (simultaneously) both the hypocenters and the seismic velocity structure and, because the solutions for each of the parameter sets are usually not independent from each other, to find the best solution for both. SSH inversion is basically a technique which combines the methods of pure structure tomography using teleseismic data and of pure hypocenter relocation using local data.

In teleseismic tomographic techniques only the velocity structure is investigated, i.e. the hypocenters are fixed. Due to the large distance between the seismic sources and the regional velocity structure in question (usually the crust and the upper mantle are investigated), only steep rays travel through the model space, resulting in high horizontal, but low vertical resolution. Furthermore, the final velocity structure has strong dependence of the rest of the velocity model [4] of the earth. Therefore the computed velocity structure is more like a velocity contrast than absolute values. (Aki and Lee, 1976)

On the other side there are pure hypocenter relocation methods, such as HYPO71 (Lee and Valdes, 1985) or HYPOELLIPSE (Lahr, 1999), which use a fixed 1D velocity model. The relocations are usually accurate with regard to the epicenters, but the hypocentral depths are often poorly determined, particularly if there is a lack of recording stations in the epicentral vicinity of the earthquake. Moreover the hypocenters often show strong dependency on the velocity model assumed. Using a velocity model previously computed with the SSH Method will give the most accurate hypocenter localizations, especially using a 3D model.

Combining these two, completely different, methods (pure velocity structure computation and hypocenter relocation) in one is the salient feature of the SSH method, the latter then having the ability to fit the model, now consisting of both, hypocentral and velocity parameters, best to the available data (arrivaltime registrations). Further mathematical details of the SSH method will be provided in the "Chapter Theory".

## 1.5 Previous regional seismological studies within Germany

In the investigation area a huge number of different seismic experiments have been carried out over the years. To name a few, the refraction and reflection experiments from Enderle et al. (1996a) or Kummerow (2004) are to mentioned. These experiments delivered on one hand valuable additional information about the position of the reflectors and, on the other hand, the velocity structure. Another interesting experiment was the Eifel Plume project, a passive measurement campaign lasting from 1997 to 1998 which covered a surface of $500 \times 500\,km$ square kilometers, with the center in the Vulkaneifel, and which included about 140 portable and 80 permanent stations for the recording of, namely, teleseismic events.
(http://www.uni-geophys.gwdg.de/~eifel/Publications/ritter.dgg.html)
(Ritter et al., 2000, 2001) The analysis of the $P-S$ conversions with the method of receiver function analysis (RFA) proved an approximate depth of the Moho of $30 km$ with relatively small variations. The initial goal of that experiment was to find the sources of volcanism and their traces, the Maare[5], in the Eifel, which is expected to be a plume build out of magma burning through the lithosphere.

For Germany, until now, only investigations for small areas of the 3D crustal structure have been proceeded, compared to a huge number of 2D profiles and local 1D models for routine localization of earthquakes.

---
[4]Often used is PREM, a standard model for seismic velocities in the earth. Improved one's are IASP91 or AK135.
[5]Water filled holes, resulting from volcanic explosions, where the occurrence of typical volcanic gases can be observed

- **Refraction and reflexion seismics**
  1. Parts of the crust and the mantle have been surveyed by several refraction and reflexion seismic experiments. Here is to mention the DEKORP experiment (Deutsches Kontinentales Reflexionsseismisches Programm) from 1984 until 1989 with a length of 1700 km for all the profiles together.

  2. Oberrheingraben
  This area was investigated by Koch and Kalata (1992); Koch (1993a,b,c) applying the method of simultaneous inversion for hypocenters and crustal structure. The results of Koch and Kalata (1992); Koch (1993a) showed that the use of refined 3D models instead of vertical layered 1D models gives optimized relocation for earthquakes with optimizations for epicenters up to $5\,km$ and $2\,km$ for hypocenters

- **1D Model Schlittenhardt**
  During the study of Schlittenhardt (1999), he derived a new 1D model for the crust and upper mantle from first arrivals at GRSN stations together with epicenters localized with local networks. This new model led to an improvement for the localization of seismic events with GRSN stations. Additionally, he could prove, that the $P_n$-phases in the dataset suffer from anisotropy. Subsets of the observed $P_n$-phases for six azimuthal sectors of 30 degree showed an obvious dependence of the direction of the $P_n$-phases and their velocity. These results are in agreement with the results of refraction seismic experiments (Enderle et al., 1996a).

- **1D Analysis (Time Term Method)**
  Starting from this investigation, Song et al. (2001a) used the time term method Bamford (1973); Hearn and Clayton (1986a,b); Enderle et al. (1996a) to proceed a 1D tomographic inversion for $P_n$-phases. Using the normal time term method, the traveltime $t_{ji}$ from the event $E_i$ with the term $a_i$ to the station $B_j$ with the station correction $b_j$ is $t_{ji} = a_i + b_j + D_{ji}S$. Here, $D_{ji}$ is the epicentral distance and $S = 1/V_{ref}$ the refractor slowness.

  Here, 1D is equivalent to the estimation of the fast axis of anisotropy in the upper mantle, represented by a 2 dimensional plain.

  The results for the azimuth of the fast axis is about $N25°E$ with a $P_n$-velocity between maximal $8.29\,km/s$ and minimal $7.95\,km/s$.

  Another result of this study is the improved fit of the calculated arrivaltimes to the recorded arrivaltimes of about 64% and additional 20% after the application of the modified time term method.

- **2D lateral velocity anomalies**
  After modifying the 1D analysis to 2D velocity variations Song et al. (2004) could prove the influence of anisotropic correction onto the computed structure. With this procedure, the 2D lateral velocity variations and the 2D anisotropic structure of the upper mantle were examined. The result of including anisotropy gave reduced, more reasonable velocity contrasts without extreme values compared to those of isotropic computations, which are in good agreement with other studies like that of Enderle et al. (1996a).

  The following velocity anomalies could be found:
  One high velocity anomaly in the region of Ingolstadt, which just occurred in the study of Enderle et al. (1996a), but was not interpreted. Another high velocity anomaly is proposed to be in between the Harz in the North and the Vogelsberg in the South in the region around Kassel, but is not well resolved due to the position at the northern border of the model area. In the region of the Vosges, higher velocity values can be found and lower ones in the area south of the Eifel, which are in good agreement with the study of Judenherc et al. (1999).

## 1.6 Objectives and goals of the thesis

According to the topics mentioned above, the main goal of this thesis is to get an improved model of the structure for Germany and adjacent regions from the surface down to the Moho discontinuity. The improvement consists in applying anisotropic correction for $P_n$-phases traveling along the Moho boundary. Anisotropy is a major effect, that has to be included in the ray tracing part of the SSH program. Comparing the results of isotropic and anisotropic inversions will show the necessity to include this type of correction. Further, a lot of registered arrivaltimes now can be explained, whereas in the isotropic case, these arrivaltimes might be interpreted as errors. The deviations, as will be shown later have perfect fit in the anisotropic case. The effect on the velocity structure will be compared to isotropic inversions, and, in addition, the effect on precise hypocenter localization.

# Chapter 2

# Theory of seismic tomography

Seismology and, therefore, seismic tomography are based to a large extent on the propagation of seismic energy which occurs in the form of elastic seismic waves. Seismic waves have, unlike acoustic waves, two components, compressional and shear waves. The seismic waves emanating from a seismic source are released into the elastic medium, the earth, by different mechanisms. One of these, for example, can be due to a spontaneous nuclear explosion in a deep cave which leads to a radial compression of the surrounding sphere which then travels outwards, away from the center of the sphere. Another mechanism is triggered by an earthquake dislocation process that also represents a change in the local structure in the vicinity of the source. This change is called a fracture process (Lay and Wallace, 1995).

Simply expressed, earthquake mechanisms are represented as a dislocation between two fault planes which causes a sudden deformation of the structure once a threshold state of stress is reached, leading then to a spontaneous release of the energy accumulated in the asperities of the fault planes. The fracture process takes place within a limited timespan $T$ (where $T$ may range from some $\mu s$ to some s, depending on the size and complexity of the fracture). The strain or deformation accompanying the fracture process triggers the generation of seismic waves which then travel through the earth, following the rules of elasticity theory, i.e. seismic wave theory. With increasing distance from the initial fracture plain, the fracture elongation appears as a single point, a so called point source, the concept of which is often used in practical seismology and that helps to simplify the further mathematical analysis.

Starting from this point source, the wave field initially propagates with radial symmetry but, with increasing distance, is more and more influenced by the elastic properties of the environment. As such the observed wave field reveals integral information about the earth material along its path from the source to the receiver. This propagation of seismic energy is described mathematically by the wave equation which is shortly presented in Section 2.1.

## 2.1 Wave theory

The propagation of seismic energy is mathematically described by elastic wave theory which, in turn, is based on the relationships between strain and stress and Newton's second law. A more complete discussion can be found, for example, in Lay and Wallace (1995), Thurber (1981), Thurber and Aki (1987), Aki and Lee (1976), Studer et al. (2007), Koch (1983a) or Koch (1989). Later, in Section 2.2, the reduction of wave theory to ray theory by replacing the wave equation by the eikonal equation, and the ensuing restrictions which result from this simplification of wave theory for tomographic uses, are discussed.

## 2.1.1 Stress, Strain and the Wave equation

The propagation of a seismic wave in a linear elastic three-dimensional medium is completely described by the equation 2.3, which relates the force $F$ (stress and other forces) acting onto a volume to the resulting deformation (strain) $u$ (see for example Thurber and Aki (1987); Aki and Lee (1976); Lay and Wallace (1995). This equation is derived using

(1) the generalized form of Hooke's law (2.1):

$$\sigma_{ij} = c_{ijkl} u_{kl} \qquad (2.1)$$

for the relationship between stress $\sigma_{ij}$ and strain $u_{kl}$ for linear three dimensional media, with the elasticity tensor $c_{ijkl}$. Here, $c_{ijkl}$ completely describes the reaction of a small volume to compression and distortion. Combining Hookes law [1] and

(2) Newton's second law (2.2) gives:

$$\rho \ddot{u}(x,y,z,t) = \nabla \sigma_{ij} + f(x,y,z,t) \qquad (2.2)$$

Here, $\ddot{u}(x,y,z,t)$ is the acceleration to a small volume, caused by the differential internal forces (stresses) $\nabla \sigma_{ij}$, which do not lead to an overall movement of the volume, and external forces $f(x,y,z,t)$, like gravity which could lead to a movement. $\rho$ is the density.

Combining Newton's and Hooke's law leads to the general form of the equation of motion in an elastic solid:

$$\rho \ddot{u}_i = \frac{c_{ijkl} \partial u_{kl}}{\partial x_j} + f_i, \qquad i = 1,2,3 \qquad (2.3)$$

After Aki and Richards (2002) or Studer et al. (2007), the wave equation can be written as:

$$\frac{1}{v^2} \ddot{u} = \frac{\partial^2 u_{(x,t)}}{\partial x^2} \qquad (2.4)$$

Here, the velocity $v$ is equal to $\sqrt{\frac{c_{ijkl}}{\rho}}$, so it only depends on the density $\rho$ and the material properties $c_{ijkl}$, represented by the elasticity tensor. The latter has theoretically 81 independent components. Due to various symmetry conditions these can be reduced to a minimum of 21 for a general anisotropic medium. For a completely isotropic medium the number of independent coefficients can be reduced further, down to two. Various combinations of these two independent elasticity variables are used in elasticity theory, such as Young's modulus $E$, Poisson's ratio $\nu$, the Bulk modulus $k$, or the shear modulus $\mu$. In seismology it is common to use the two Lamé constants $\lambda$ and $\mu$. The named other variables can then be derived from the latter two.

Using the two Lamé constants Hooke's law can be written as

$$c_{ijkl} = \lambda \delta_{ij} \delta_{kl} + \mu (\delta_{ik} \delta_{jl} + \delta_{il} \delta_{jk}) \qquad (2.5)$$

whereby $\delta_{ij}$ is the Kronecker symbol, with $\delta_{ij} = 1$ for $i = j$ and $\delta_{ij} = 0$ for $i \neq j$.

With only two independent components of the elasticity tensor $C_{ijkl}$, depending on the boundary conditions (free surface or infinite volume), the solutions of the wave equation lead at least to four typical wave types, the body waves $P$ and $S$ and the surface waves $L$ and $R$ [2]. Other solutions of the wave equation lead to additional wave types, for example $T$ waves.

Further on, these wave types can be differentiated by the way they are traveling, how many reflections and conversion's they undergo, which leads to the term "phase". Over the history of seismology, a nomenclature of numerous different phases has been derived taking into account the structure of the earth. For example, subsets of $P$ are $P_g$, $P_n$, $P_m P$, and so on. Some examples of the traveltimes for some of these phases are illustrated in Figure 2.1.

---
[1] Hookes Law is a linearization of the stress-strain relationship and only valid for pure elastic media
[2] $P$ for undae primae or Compressional waves ($P$ for Pressure), $S$ for undae secundae or shear waves, $L$ for Love waves and $R$ for Rayleigh waves called by the names of their discoverers.

## 2.1. WAVE THEORY

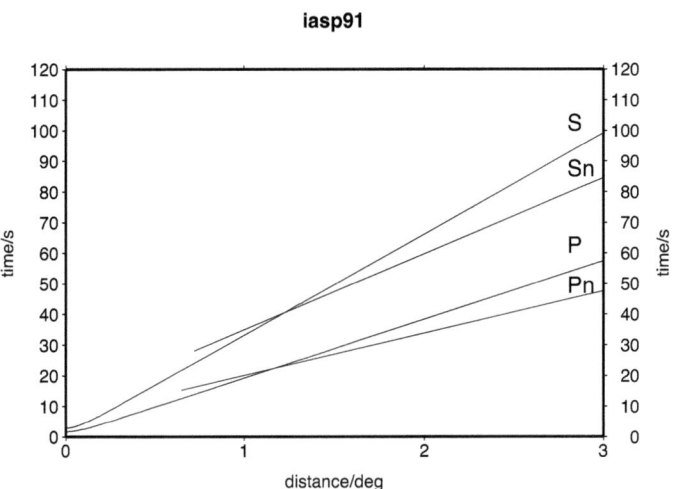

Figure 2.1: Traveltime - distance plot for regional P and S waves.

### 2.1.2 Seismic Velocity

The wave equation describes the propagation of small elastic deformations through an elastic body. These small deformations result in the so called body waves, because they travel through a material volume[3]. These deformations can be in the form of compressions and dilations or in the form of shearing leading to the two kinds of waves, $P$ and $S$. Using the isotropic Hooke's law 2.5 in the wave equation, the latter can be split into two parts, each describing then the motion corresponding to these two kinds of waves, with wave velocities

$$\alpha = \sqrt{\frac{\lambda + 2\mu}{\rho}} \quad \text{for the } P \text{ wave velocity and} \tag{2.6}$$

$$\beta = \sqrt{\frac{\mu}{\rho}} \quad \text{for the } S \text{ wave velocity.} \tag{2.7}$$

The equations derived so far are only valid in an isotropic medium, which is a sufficient approximation of first order to handle a lot of seismic investigations. For more precise investigations one has to extend the formulas of the isotropic case, for example, when studying anisotropy in the upper mantle. In such a case, the anisotropy is assumed to be horizontal and two-dimensional (elliptic anisotropy) and a simple way of treating this phenomenon is to add a correction term to the original isotropic values of $\alpha$ or $\beta$, which depends on the direction the ray is traveling from the source to the receiver.

---

[3]At the surface, Rayleigh ($R$) and Love ($L$) waves travel with different behavior due to the boundary condition at the surface.

# CHAPTER 2. THEORY OF SEISMIC TOMOGRAPHY

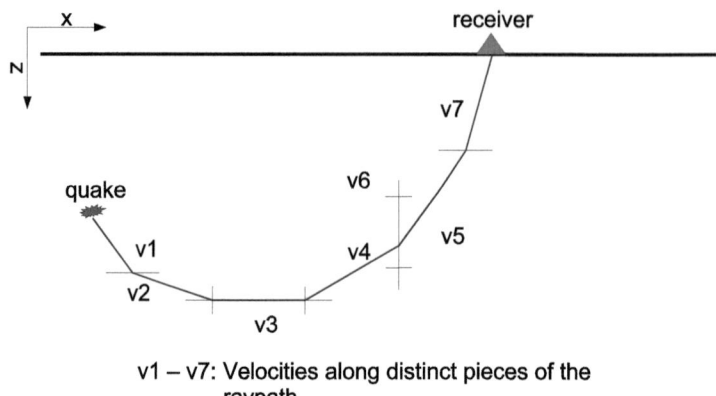

Figure 2.2: An exemplary ray path through the model space. In detail, the ray paths might be rather complicated, but due to the computer algorithms, they are more or less simpler. In a first approximation the ray path is described as a part of a circle for which the traveltime is calculated. The more exact ray paths include real reflection and refraction at the block-boundaries, representing the model space.

## 2.2 Ray Theory

Ray theory is until now one of the most powerful techniques for the investigation of seismic traveltime problems without huge computational efforts, compared to solve the wave propagation by FD or FE methods.

### 2.2.1 Basic concepts of Ray Theory

Ray theory is derived from wave theory by neglecting several aspects of how waves travel through space which, mathematically, is described by the Eikonal- instead of the Wave-equation. As ray theory is at the heart of seismic traveltime inversion, one needs to understand its theoretical basis as a simplification of wave theory. In fact, wave theoretical aspects, e.g. interferences or dispersion, are neglected.

Without taking into consideration the details of the mathematics, ray theory describes the direction of the propagation of the wave front which, by definition, is perpendicular to the ray. The direction vector linked with it only describes a small part of the wave front that runs through the space. Reflections and refractions (see Figure 2.2 for an exemplarily ray path) are then described by taking into account the continuity of horizontal and vertical components of the displacement- and stress vector at the boundary which, after using Fermat's principle of the shortest traveltime, eventually results in the classical formula of geometrical optics (i.e. Snell's law, etc.)[4] and not after Huygens principle of elementary waves.

---

[4]In fact, not only boundary layers exist but also continuous changes of the elastic properties. This already shows an essential restriction of resolution of the SSH method and the division of the model space into blocks and fixed, sharp borders and layers

## 2.2.2 Eikonal equation

Following Fermat's principle, the optimal ray is the one which gives the smallest traveltime between source and receiver. For a layered, or in blocks inhomogeneous earth this results in Snell's law at the layer/block interfaces. To find that optimal ray, one uses the high-frequency approximation of the wave equation which leads to the Eikonal equation for the traveltime $T(x, y, z)$ (the Eikonal) (2.8).

$$(\nabla T(x,y,z))^2 = s^2(x,y,z) \tag{2.8}$$

or written out

$$\left(\frac{\partial T}{\partial x}\right)^2 + \left(\frac{\partial T}{\partial y}\right)^2 + \left(\frac{\partial T}{\partial z}\right)^2 = s^2(x,y,z) \tag{2.9}$$

where $s$, the reciprocal of $v$ is called the slowness.

The Eikonal equation 2.8 is a non-linear partial differential equation for the seismic travel time $T$. The approaches that are used to solve the equation can generally be categorized as ray methods on the one hand, and eikonal solvers on the other. Ray methods, based on the method of characteristics, represent the classical approach. Here the partial differential equation is replaced by a set of ordinary differential equations, which are generally easier to solve. Nevertheless, a disadvantage of the ray methods is that they involve a change of coordinates and the evaluation of the medium properties, i.e. the slowness $n$ which has to be done at arbitrary positions in the medium, which, in practice, requires interpolation. As this interpolation has to be smooth, it is relatively expensive. In fact, in ray methods a considerable amount of computation time is spent on interpolating the medium properties. To avoid some of these problems, another option -used in the present thesis- consists in using a classical shooting ray-tracing through a block-discretized medium.

Eikonal solvers evaluate the eikonal equation directly in terms of the spatial coordinates, incorporating mostly finite difference (FD) methods. By evaluating the eikonal equation directly in terms of the spatial coordinates these methods bypass the change of coordinates inherent to ray methods and thus avoid the interpolation of medium properties. Notwithstanding, eikonal solvers face a number of other theoretical and practical difficulties, primarily associated with stability and accuracy and the inherent non-linearity of the eikonal equation. Extension to general anisotropic media is problematic. In spite of ongoing research related to these issues the eikonal solvers do not yet provide an alternative to ray methods. For further details of the mathematics of wave theory, I refer the reader to the excellent works of Aki and Lee (1976); Anderson (1984); Granet (1998); Spakman (1988); Williamson and Worthington (1993); Trampert (1998); Twoomey (1977); Hearn (1996). Also the work of G. Müller (Weber et al., 2007) here has to be mentioned which can be downloaded at http://bib.gfz-potsdam.de/pub/str0703/0703.htm. As for ray tracing methods, see the works of Pratt et al. (2002); Wielandt (1987); Cerveny (2001).

## 2.2.3 Application to tomography

There are several procedures to do seismic tomography, i.e. to determine the 3D seismic structure of the earth's interior. One technique, not considered further here, consists in the analysis of harmonic oscillations of the earth, which are influenced by its structural properties, though mostly on the global scale, as such normal mode inversion usually does not have the resolving power of traveltime tomography, which is the focus of the present thesis. Depending on the kind of traveltime data used in the inversion, one basically distinguishes two methods of tomography: teleseismic and local.

In teleseismic tomography only the velocity structure is investigated, i.e. the hypocenters are assumed to be known and fixed. Due to the large distance between the seismic sources and the regional velocity structure in question (usually the crust and the upper mantle are investigated), only steep rays travel through the model space, resulting in high horizontal, but low vertical resolution. Starting with the development of the so-called ACH-method by Aki et al. (Aki and Lee, 1976; Pujol, 1988), variants of this techniques have been used in hundreds of regional teleseismic tomography studies around the world.

In local or regional seismic tomography one uses traveltime data from local earthquakes. Here, in addition to the local seismic velocity structure, the exact locations of the hypocenters themselves are not known, so they have to be determined simultaneously. This is the reason that techniques of local traveltime tomography are commonly known as inversion methods of "Simultaneous Inversion for Structure and Hypocenters", here abbreviated "SSH" (Koch, 1983a, 1985a). As any other tomographic inversion method to find the model parameters, the SSH-method consists also of two parts, namely, (1) the forward modeling by ray-tracing and (2) the traveltime inversion.

In order to solve this inversion problem, a plausible starting solution $M_0 = (H_0, v_0)$ for both the hypocenters and the seismic velocity is set up. In the simplest case, $v_0$ may consist of just one value representing an average seismic velocity for the whole earth volume under investigation. A more realistic starting model will use a set of vertically stratified velocities or, even better, a full 3D heterogeneous velocity structure. For most tomographic investigations it has been proved that a simple vertically stratified optimal 1D starting model $v_0$ [5] is sufficient.

Now, using this starting model $M_0$, theoretical arrivaltime terms ($t_{calc}$) are computed by ray-tracing and compared to the measured ones ($t_{obs}$); the differences between these two are called the residuals ($t_{res}$). In the second step of the inversion process the model parameter improvements $\delta M$ with respect to the starting model $M_0$ are computed for several different increasing damping factors which are used to stabilize the inversion. After each inversion step, ray-tracing is done again to calculate the new arrivaltimes ($t_{calc}$). The optimal model solution $M_0 + \delta M$, which is measured by the data fit, the total sum of the squared residuals (TSS) or the root mean square (RMS) of TSS, is used as a new starting vector in the next iteration step.

## 2.3 Seismic inverse theory

Seismic inverse theory is the base to solve multiple seismological problems. It consists at least of two parts, the forward and the inverse problem.

The forward problem consists of the generation of new data, which can be "compared" to the original, observed data to calculate the so called residuals $t_{res}$.

The inverse problem mainly consists of a matrix coupling the data space with the model space. The solution of this is of intrinsic challenge.

### 2.3.1 The forward problem

The forward problem arises when calculating traveltimes through the earth volume under question and arrivaltimes at the earth's surface for an existing model $M$, that consists of the seismic velocity structure $v_i$ and the hypocenters $H_j$ with the three spatial $(x_j, y_j, z_j)$ and one temporal $(t_j)$ coordinates. The traveltimes can be calculated by the so-called ray-tracing, whereby a specific hypocenter is connected to the receiver at the earth's surface by a line, called the ray. The simplest ray is a straight line, but using it as such would not be sufficiently exact in practical applications. A comparatively better approximation is to describe the ray path as part of a circle connecting the two endpoints and assuming that the seismic velocity increases with depth. The best approximation of ray tracing is to include refraction and reflection at the layer boundaries or the vertical boundaries of the blocs which represent the geometrical model. (see Koch (1985a,b, 1989)) The different types of approximations are shown in Figure 2.3.

---

[5] For the 1D modeling the model space becomes vertically parametrized, the structure is accepted as laterally homogeneous. With this parametrization one can get a first impression of the model space. Now, above all, the vertical velocity structure can be used as a starting solution for the 3D model. Within the 3D model the 1D model is laterally and vertically varied so that the adaptation of the traveltime terms is further optimized.

## 2.3. SEISMIC INVERSE THEORY

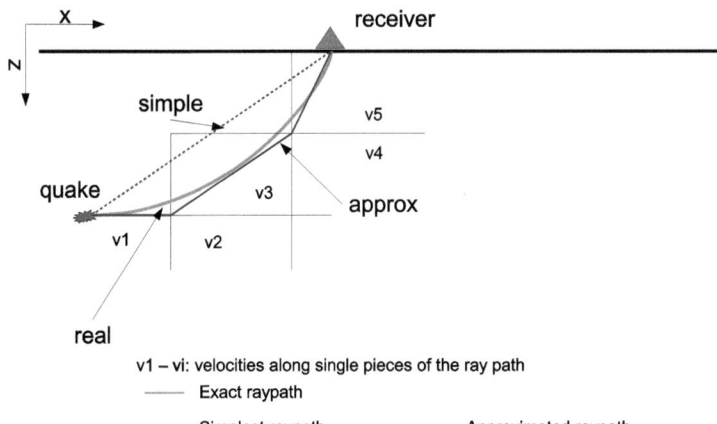

Figure 2.3: Some examples of approximations between the calculated and true rays with different complexity from left to right. Overall, even the best approximation is still only an approximation and will never represent the exact ray.

The most common form to represent the relationship between traveltime and velocity structure is provided in equation 2.10

$$T = \int_S \frac{1}{v(x,y,z)} ds \tag{2.10}$$

The integral 2.10 is nonlinear, because the ray path S itself is a function $S(v)$ of the unknown local velocity $v(x,y,z)$. For a number of hypocenters $i$ and stations $j$ the traveltime can also be written as:

$$t_{ij} = \int_{x_i;y_i;z_i}^{x_j;y_j;z_j} u(\vec{x}) ds \tag{2.11}$$

where $u(\vec{x}) = \frac{1}{v(\vec{x})}$ is the slowness along the ray path elements $ds$ from the hypo-center with coordinates $x_j; y_j; z_j$ to the station with coordinates $x_i; y_i; z_i$. This equation is the basis of nonlinear ray tracing and is discretized (whereby the integral is replaced by a sum) in actual calculations, such as in the SSH-method Koch (1985b). For the nonlinear inversion an iterative linear least-squares method, to be discussed later, is used. For that purpose the nonlinear traveltime equation is linearized which requires the calculation of the so-called Frèchet derivatives of the traveltime $T$ with respect to the model unknowns (hypocenters and velocity, i.e. $(H, v)$). With regard to the Frèchet derivatives with respect to the velocity parameter, these depend on the seismic phase considered, i.e. direct, refracted or reflected one. The ray tracing for direct phases is based mainly on the procedure explained in Koch (1985b). In Koch (1993a) the ray tracing has been extended to incorporate refracted and reflected phases as well.

Using Fermat's principle, the Frèchet derivatives $\partial T/\partial v_i$ of the traveltime $T$ with respect to the velocity unknowns are computed piecewise for each block from Equation 2.22 as $\partial T/\partial v_i = -L_i/v_i^2$ where $L_i$ is the path length of the ray in the $i_{\text{th}}$ block. Note, that this expression is also valid for refracted and reflected phases, once the path length $L_i$ has been evaluated (see below) for a particular block $i$.

The Frèchet derivatives $\partial T/\partial H_j$, $(j=1,q)$ for the hypocentral parameters $H_j = (x_j, y_j, z_j, t_j)$ can be found analytically using a variational approach (Lee and Stewart, 1981). This results in the following three well-known expressions for the x-,y-,and z-components, respectively:

$$\partial T/\partial x_j = -p*\sin\varphi,\ \partial T/\partial y_j = -p*\cos\varphi,\ \partial T/\partial z_j = \cos\theta/v \tag{2.12}$$

Here, $p = \sin\theta/v$ is the ray parameter of the ray, emanating from the source embedded in the layer or block with velocity $v$, and $\varphi$ is the azimuth between the source and the station. Doing the calculations for each ray between a particular seismic source and a station the linearized problem can be written in matrix notation as shown in the next section.

### 2.3.2 The inverse problem

#### 2.3.2.1 General approach

Once the forward problem is solved, the newly calculated traveltimes with the model from the previous iteration are available and can be compared with the observed ones, i.e. residuals can be computed to be used in the subsequent inversion iteration. If one only has a small data set with a small number of model parameters, it would be possible to do this manually, as Sir H. Jeffreys did it originally in the late 1890's, then coming up with a first analysis of the earth's vertically inhomogeneous structure. To do this in the age of computer technology would be awful, instead, the application of matrix based inversion computation has nowadays become the state of the art (See also Aki and Lee (1976); Tarantola and Valette (1982b); Koch (1983a, 1985b); Thurber and Aki (1987); Kissling (1988); Koch (1989); Pujol (1992); Koch (1993a); Kaschenz (2006)).

The starting point of the mathematical formulation of the inverse problem is equation 2.10 for the nonlinear integral relationship between the local velocities and the traveltime or, equivalently, equation 2.11 for the local slowness and the traveltime. In practical seismic tomography data are always measured as arrivaltimes $t_j$ of a particular seismic wave (phase) at a station $j$, the latter resulting from either an underground explosion or a natural earthquake $i$. Knowing then the origin time $t_i$ of the hypocenter $i$, the traveltime $t_{ij}$ of the wave is then computed as

$$t_{ij} = t_j - t_i \tag{2.13}$$

#### 2.3.2.2 Linearization

To solve the nonlinear integral equation 2.11, the problem has to be linearized, i.e one is looking how small changes $\delta m$ of the model vector produce changes $\delta t$ of the traveltime $t_{ij}$. For this one needs to linearize 2.11 around a reference model $m_0$ which results in a homogeneous Fredholm's integral equation (Fredholm, 1903; Atkinson, 1976) of the second kind (Courant and Hilbert, 1962) 2.14

$$\delta t = \int_{S_0} \delta m ds \tag{2.14}$$

The important point in the equation above is that the integration path is taken along the unperturbed ray path $S_0$. This is a consequence of Fermat's principle which states that the ray path $S$ chosen is always such that the traveltime $t$ is a minimum, i.e. changes of the traveltime $T$ with respect to a change in the ray path are zero to first order. This means that equation 2.14 is correct only for small changes $\delta m$ of the model. For larger variations $\delta m$, one must take, in principle, an incremental approach, whereby the total change of the model vector is split into smaller pieces and the corresponding traveltime change $\delta t$ is recalculated along the updated ray path obtained from the previous change $\delta t$.

For a number $n$ of rays, equation 2.14 is in fact a system of linear integral equations for the $m$ model unknowns which, in discrete form, can be cast as a linear system of equations

$$y = Ax + e, \tag{2.15}$$

## 2.3. SEISMIC INVERSE THEORY

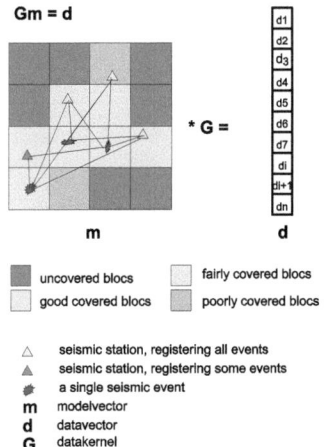

Figure 2.4: The model space **m**, the data kernel **G** and the data vector **d**. These are the fundamental elements in an inversion process. The model space most generally consists of multiple covered, well covered, poorly covered and uncovered areas. These areas correspond to the equations in the data kernel **G**, which correlates the model space to the data vector. These equations are poorly determined for uncovered areas, well determined for areas with exact coverage, and overdetermined for areas with multiple coverage.

where $y = (\delta T_1, \delta T_2, .., \delta T_n)^T \in \mathbf{R}$, is the data vector or, more precisely, the vector of the residuals of the observed minus the calculated arrivaltimes; $A = (\delta m_1, \delta m_2, .., \delta m_m) \in \mathbb{R}^m$, the matrix of the Frèchet derivatives (see previous section), $x$ is the parameter vector and $e$ is a general error vector that comprises errors in the data as well as those due to the model discretization.

Figure 2.4 shows the linear relationship between the model and the data as stated by the linear system 2.15 whereby, as common in the geophysical literature, the vector $d$ corresponds to $y$, $G$ to $A$, and $x$ to $m$ in equation 2.15. The goal of the inversion is to determine the model vector $x = m$ via measured data $y = d$, i.e. the solution of linear system 2.15 in the least squares sense, as will be discussed in the following sections.

### 2.3.2.3 Least-squares inversion

The least squares method is the basis of all calculations in inverse theory. Depending on whether one considers the inverse problem is being a linear one (as obtained from the linearization, discussed above) or straightforward as a nonlinear one (as it is originally by default), different approaches ensue. For the most widely used gradient methods, to be presented in the following, the nonlinear least squares problem is essentially being cut down to a series of linear least squares problems so that, at the end, one deals with the intricacies of the latter.

Starting directly with the nonlinear traveltime equation 2.10, one may write the relationship between the model vector $x$ and the observed traveltime data vector $y$ in functional form as

$$y = f(x) + e \qquad (2.16)$$

where f(x) is the theoretical traveltime function as defined by equation 2.10, and $e$ the error vector discussed above.

The objective of least squares inversion is to minimize the error $e$ in the least squares sense, i.e. to minimize the nonlinear residual function $F$, defined as

$$F = ||y - f(x)||^2 \mapsto Minimum \qquad (2.17)$$

Equation 2.17 is basically an optimization problem and, in this case, a minimization problem for the nonlinear function $F(x)$. Numerous minimization techniques for its solution exist and among those one essentially distinguishes between global and local minimization methods. In the following we discuss only the latter ones and, in particular, the gradient methods usually work well, when the starting solution is sufficiently close to the minimum of the function $F(x)$.

**2.3.2.3.1 Gauss-Newton method** The Gauss-Newton (GN) method is an extension of the conventional Newton method to the nonlinear function $F$ (equation 2.17). The GN-method is also known as the "tangent method", since tangents or gradients of the objective function $F$ are repeatedly calculated. Thus, starting with an initial value $x_0$ close to the expected solution, i.e. the minimum, the derivative (the gradient) $g = \nabla F$ of $F$ at $x_0$ is first calculated. Using essentially this gradient direction (see below) a step length vector $\Delta x_1$ is then computed to find a value $x_1 = x_0 + \Delta x_1$ which comes closer to the minimum. Repeating the procedure $i_{max}$ times, a final solution will be produced which, hopefully, is close to the true minimum of the objective function $F$.

Formally, the Gauss-Newton Method can be derived from the expansion of the functional $\mathbf{F}$ into a Taylor series up to order 2, neglecting terms $O(x^3)$:

$$F(x_0 + \Delta x) = F(x_0) + \Delta x^T g + \frac{1}{2}\Delta x^T \mathbf{H} \Delta x + O(x^3) \qquad (2.18)$$

In this equation, $g$ is the gradient $\nabla F$ and $\mathbf{H}$ is the Hessian $\nabla^2 F$, the second derivative in matrix formulation, at $x_0$. From the condition for the minimum $\frac{\partial \mathbf{F}}{\partial \Delta x} = 0$ [6] one gets the most general form of the correction vector $\Delta x_i$ in the $i^{th}$ iteration step

$$H_i \Delta x_i = -G_i \qquad (2.19)$$

For the quadratic function $F$ (Equation 2.17) the gradient $g_i$ is given by $g_i = -\mathbf{A}_i^T r_i$, where $\mathbf{A}_i = \nabla f(x_i)$ is the Jacobian or the matrix of the Frèchet derivatives of $f$ with respect to the model vector $x_i$, and $r_i = y - f(x_i)$ is the traveltime residual for the model $x_i$. For the Hessian, $H_i$, one gets the expression

$$H_i = 2(\nabla f)^T \nabla f + 2\nabla(\nabla f)^T r_i \qquad (2.20)$$

which, after neglecting the second quadratic term in Equation 2.18 (Koch, 1993a), for a discussion, and use of equation 2.19, results in the Gauss-Newton (GN) method

$$\mathbf{A}_i^T \mathbf{A}_i \Delta x_i = \mathbf{A}_i^T r_i \qquad (2.21)$$

which are also the normal equations for the linear least squares problem 2.15, with $\Delta x_i$ replacing $x$ and the $r$ replacing $y$ (Beck and Arnold, 1977; Koch, 1993a).

**2.3.2.3.2 Levenberg-Marquardt method** The Levenberg-Marquardt (LM) method, so-called after Kenneth Levenberg and Donald Marquardt (Levenberg, 1944; Marquardt, 1963) combines the Gauss-Newton algorithm with methods of regularization or damping. Regularization means, that the solution $\Delta x_i$ of the Gauss-Newton Method is stabilized by adding a damping factor $k$ on the diagonal of the normal matrix $\mathbf{A}_i^T \mathbf{A}_i$, resulting in

$$(\mathbf{A}^T \mathbf{A} + kI)\Delta x = \mathbf{A}^T r \qquad (2.22)$$

---

[6] the first derivative equals zero

## 2.3. SEISMIC INVERSE THEORY

where $I$ is the identity matrix, which might be replaced by an appropriate scaling matrix $D$ that allows damping of particular components of the solution vector $\Delta x_i$. An equivalent effect of the damping factor $k$ is to prevent the singular (Eigen) values of $A_i^T A_i$ from becoming zero, which would lead to an infinite step length $\Delta x_i$, i.e. to possible overshooting beyond the minimum of $F$, with no chance to return back to the latter.

The LM algorithm is more robust than the Gauss-Newton algorithm and may still converge even with bad initial values when GN fails. Nevertheless, convergence to the local minimum is not always guaranteed and may depend on the wise choice of the damping factor $k$. Since its original conception by Levenberg (1944) and Marquardt (1963), numerous search strategies for acceptable $k$ have been proposed. Nevertheless, it appears that the original strategy of Marquardt, which consists basically in increasing the damping parameter $k$, until $F(x_{i+1}) < F(x_i)$, still works best in most applications. Experience shows that, as the solution approaches the minimum of the objective function $F$, it is recommendable to reduce the size of $k$ to zero, so that the LM-method behaves more than a pure Newton method with its property of quadratic convergence close to the minimum.

### 2.3.2.4 Generalized Inverse

The least squares solution $x_s$ of the linear system of equations 2.15, with $n$ observations and $m$ unknowns, i.e. with an $m \times n$ matrix $A$, can also be written using the concept of the generalized inverse, also called the pseudo inverse, or Moore-Penrose pseudo inverse $A^+$ as

$$x_s = A^+ y \qquad (2.23)$$

Unlike the ordinary inverse $A^{-1}$ of a square ($m = n$) matrix $A$, which is only defined for full rank $p = m = n$, $A^+$ is defined for arbitrary $p$, $m$, and $n$, i.e. for the underdetermined ($p < n < m$) as well as for the overdetermined ($p < m < n$) case, the latter representing the classical least squares problem if $p = m$. In fact for this case, one gets from the least squares minimization of $||e|| = e^2 = (y - Ax)^2$ in equation 2.15 the well-known normal equations

$$(A^T A) x_s = A^T y \qquad (2.24)$$

which is formally identical to the iterative solution with the Gauss-Newton method 2.21. Solving for the least squares solution $x_s$ one gets

$$x_s = (A^T A)^{-1} A^T y \qquad (2.25)$$

Comparison of this equation with 2.23 shows that the generalized inverse for the overdetermined least squares problem is

$$A^+ = (A^T A)^{-1} A^T \qquad (2.26)$$

The same procedure can be used for the damped, rank deficient ($p < m < n$) least squares problem. One then gets for the generalized inverse $A^+$ the expression

$$A^+ = (A^T A + kI)^{-1} A^T \qquad (2.27)$$

which is formally identical to the expression one gets from the Levenberg-Marquardt equation 2.22. However, when applied to the linear least squares problem, is also called the Ridge estimator (Hoerl and Kennard, 1970).

Summarizing, one can state that the general solution of the linear (or of the nonlinear, linearized) inverse problem is given by 2.23, whereby the G-inverse $A^+$ takes the form of either 2.26, for the full rank, stable case, or 2.27, for the rank deficient, unstable case.

### 2.3.3  Validation of the inversion results

The validation of the results of the inversion procedure is a fundamental aspect of seismic inverse theory, especially in seismic tomography. One may ask, for example, what is the use of an inversion technique good for, when the solution bears no relation to reality? Although the inverse problem is in itself always non-unique, since the number of model parameters is in principle always larger than the number of observations, the power of inverse theory lies in the fact that it allows to somehow quantify this degree of non-uniqueness and to define the plausibility of a result or a model. This can be achieved by a skillful evaluation of the fundamental equations of inverse theory and various statistical properties of the inversion solution.

#### 2.3.3.1  General approach

The most elementary way to reduce, for example, the inherent uncertainty in the model space and improve the plausibility of a model is to use "a priori" data for the latter. Suitable for these are any sources, such as other seismic experiments, that can deliver general information about model parameters. For example, refraction seismic experiments supply extra data for seismic velocities which is particularly useful in the case, studied in the present thesis, of the simultaneous inversion for seismic velocities and hypocenters (SSH), which is especially plagued by problems of instability and non-uniqueness. Suitable as "a priori" data for hypocentral parameters are quarry blasts, also called ground truth events. The latter can be relocated by a new velocity model determined from a previous inversion run and, from the mislocations, one can draw conclusions about the quality of the velocity model in that region. For the quantitative evaluation of the reliability of the tomographic inversion models the following quantities pertinent to the data and model space are investigated throughout the SSH computations.

#### 2.3.3.2  Ray Density

One simple criterion for a first quality check is the ray density in the study area, respectively, the number of rays through the different blocks of the 3D model. However, the ray density plot has to be taken with a grain of salt and may be misleading since a high number of rays through a particular block may suggest rather good resolution here which, in fact, is not the case if the rays run parallel through a series of blocks, thus providing only redundant information. Therefore, a better way is to check the direction dependent ray density where the number of directions, from which the rays cross the distinct blocks, is considered.

#### 2.3.3.3  Goodness of fit (TSS)

The goodness of fit is computed from the total sum of residuals squared (TSS) $||r_s||^2$, as obtained a posteriori from the linearly estimated solution $x_s$ in 2.25 (and using the appropriate G-inverse $A^+$, defined earlier, when dealing with the linearized inverse problem by means of the GN- or LM-method $x_s$ is in fact $\Delta x_i$. $||r||^2$ is computed as

$$||r_s||^2 = (y - Ax_s)^T(y - Ax_s) \tag{2.28}$$

$||r_s||^2$ can be used to compute the percentile coefficient of multiple determination, also called explanation of variance, $R^2$:

$$R^2 = \left(1 - \frac{||r_s||^2}{s_y^2}\right) \cdot 100\% \tag{2.29}$$

where $s_y^2$ is the variance of the data $y$. Obviously, $0 \leq R^2 \leq 100\%$, and one has $R^2 = 100\%$, for a model which perfectly fits the data.

The total sum of residuals squared (TSS) can also be used in a classical F-test (Song et al., 2004)

## 2.3. SEISMIC INVERSE THEORY

to better optimize the number of model parameters and, especially, to follow the general principle of parsimony of an inverse model. Since it is always possible to fit an observed data set arbitrarily close using a sufficient large number of model parameters, though, at the sake of possibly fitting data noise, parsimony means that one should try to explain the data with the least number of model parameters (Occam's razor principle). Basically, the F-test answers the question whether a decrease in the TSS (or RMS) of the more complicated model with $m_2$ number of model parameters is worth the cost of the additional variables as compared with the simpler model $m_1$ ($m_2 > m_1$).

The F-test relies on the following test statistics:

$$\frac{TSS_{m_1}/(n-p1) - TSS_{m_2}/(n-p2)}{TSS_{m_2}/(n-p1)} = F(p_2 - p_1, n - p_2) \quad (2.30)$$

The F-Test is based on the statistical theorem that $TSS(m)/\sigma^2$ has approximately a $X^2$-distribution with $n-p$ degrees of freedom. The value $F(p_2-p_1, n-p_2)$ can be compared to the appropriate $F$-value, which can be taken from tables using the number of the model parameters $p_1$ and $p_2$ and the number of data $n$ or better be calculated using the Statistic program package R (http://www.r-project.org/).

### 2.3.3.4 Information density

The information density is a measure how well data are able to provide information on the model. It is quantified by the information density matrix $S$ defined as $S = AA^+$ and which acts in the data space. Using this definition of $S$, the residual vector $r_s$ can be written as (Koch, 1989)

$$r_s = (I - S)y \quad (2.31)$$

From this it follows for the case that the model fits perfectly well the data ($r_s$=0), that $S = I$, the identity matrix. Moreover one can show that the quantity $y_p = Sy$ is that part of the data which is exactly predicted by the inverted model $x_s$.

### 2.3.3.5 The resolution

The resolution of the model is computed from the resolution matrix $R = A^+A$ which, using 2.27 results in

$$R = (A^T A + kI)^{-1} A^T A \quad (2.32)$$

From this equation one notes that for $k$=0, i.e. no damping, $R = I$, i.e. optimal resolution is achieved, or the model is perfectly well resolved since, from the definition of $R$ and the fundamental property $E(x_s) = Rx$, the estimated solution model parameters $x_s$ represent, on average (as specified by the expectation $E$), exactly the true model $x$ as all off diagonal elements of $R$ are zero (Koch and Kalata, 1992).

### 2.3.3.6 The covariance

The covariance of the solution is specified by means of the covariance matrix

$$cov(x_s) = A^+ A^{+T} \sigma^2 \quad (2.33)$$

where $\sigma^2$ is the variance of the data which is either estimated a priori or is evaluated from the a posteriori residual sum squared defined earlier. For computational purposes the covariance matrix $cov(x_s)$ is computed over the resolution matrix $R$ (which is highly advantageous since these two parameters are usually calculated together) using the expression (Koch and Kalata, 1992)

$$cov(x_s) = R(A^T A + kI)^{-1} \sigma^2 \quad (2.34)$$

Because of the extreme computational efforts involved in the evaluation of the full expressions for $cov(x_s)$ and $R$ for very large scale problems (as is the case for the fully coupled SSH problem here) these quantities are usually only computed column wise for a number of selected model parameters.

### 2.3.3.7 Checkerboard test

Because of the large computational burden for the full resolution and covariance matrices and their somewhat ominous interpretation, a more intuitive and practical test to investigate the resolving power of the data and the model error in a tomographic problem consists in the use of the so-called checkerboard test. In this test a discontinuous velocity distribution is defined on a checkerboard like grid of given grid size. According to the given geometrical distribution of hypocenters and recording seismic stations, synthetic traveltimes are computed through this structure, like the registrations in the original data set. These theoretical traveltimes, possibly corrupted by noise, are then used as input in the subsequent inversion and the model output is compared with the input checkerboard structure. Ideally, the inverted structure should be very similar to the input checkerboard configuration. However, in reality the deficiency of the data, namely, a poor event and station distribution will show up in sections of the model as a "blurred" picture, i.e. there the model space will not be well resolved. One may then increase the size of the checkerboard grid and repeat the test on this coarser grid. Eventually one should find a minimal checkerboard size where all areas of interest are sufficiently well resolved. Otherwise, those model areas with poor resolution must be critically looked upon in the subsequent tomographic inversion with the real data. For more precise test, the checkerboard test can be changed to a "real structure" test, where the areas with coarser resolution are fit together, and the ones with good resolution can be refined.

Although the checkerboard test is more a qualitative than a quantitative test, its value cannot be understated, particularly, when dealing with a nonlinear inverse problem, in the case of which an unambiguous specification of both the resolution- and covariance matrices is not always possible, as this depends somewhat on the iteration sequence during the minimization of the objective function $F$, equation 2.17. In fact, the statistical properties of the nonlinear inverse solution appear to be valid only close to the true minimum of $F$ which, for obvious reasons, may not be reached in a real application (cf.Koch (1993a), for a discussion).

### 2.3.3.8 Regularization and Trade-off

One of the most intriguing issues in inverse theory is the regularization of the highly ill-posed, or unstable SSH inverse problem, i.e. the damping or stabilization of the inverse solution. Regularization is the family name for a variety of general techniques to restore the well posedness of an inverse problem. Since each damping, or regularization parameter $k$ in the g-inverse $A^+$ 2.27, or in the LM-procedure 2.22, results in a particular solution $x_s$ $(= \Delta x_{is})$, arbitrary regularization generates a whole manifold of models. Because of the inherent trade off between the various inversion solution characteristics in the model space as well as in the data space (see Table 2.1), on the one hand, regularization will have a positive effect of reducing (1) oscillations in the model vector and, (2) its statistical covariances, but on the other hand will (1) deteriorate the goodness of data fit, (2) decrease the resolution (as embodied by emerging off diagonal elements in the resolution matrix $R$) or, equivalently induce a widening of the resolution kernels as is conceptualized in the Backus and Gilbert formalism (Backus and Gilbert, 1968) and, (3) introduce a statistical bias, as discussed by Koch (1989). The ultimate issue in general inverse theory is then the determination of an optimal solution $x_{opt}$ using a particular regularization parameter $k_{opt}$ which possesses trade off characteristics that are somehow a best compromise compared with all other solutions $x_s$.

In the last decades several optimal regularization techniques (ORT's) have been developed under different names, and often independently, for the stabilization of ill-posed inverse problems in various scientific disciplines. Such are the Tikhonov regularization, the ridge regression, the method of Backus (*subjective*), optimal filtering (*stochastic inverse*), the Bayes estimator, and the method of Backus and Gilbert. Most of these methods rely on some kind of statistical *a priori* information on the model. In an earlier study of the 3D inversion of teleseismic traveltimes (Koch, 1989) and of local traveltimes for SSH (Koch, 1993a,b,c) it was demonstrated that a reliable inversion should comprise the concurrent use

## 2.3. SEISMIC INVERSE THEORY

Table 2.1: Effect of regularization (increasing the damping parameter $k$ in the **damped least-squares technique**, or equivalently decreasing the pseudo rank $p$ in the **method of Wiggins**) on seven trade-off characteristics of the model- and the data space as they have been discussed in the previous sections. Group I denotes those parameters which are positively (and the reason why the regularization is done at all) affected by increasing regularization, while group II denotes those (the trade-off) parameters which are adversely (negatively) affected. $||x_s||^2$ is the norm of the solution vector, $S_r$ is the geometrical spread, and the bias$^2$-term. $R_{ii}$ denotes the diagonal elements of the resolution matrix $R$ 2.32 $||r^2||$, the residual sum of squares (2.28), and $R^2$, the coefficient of multiple determination (2.29).

| damping-parameter $k$ ⇑⇔ $pseudorank$ p ⇓ | | |
|---|---|---|
| model space | \multicolumn{2}{c}{data space} | |
| group I (positive) | \multicolumn{2}{c}{group II (negative)} | |
| $||x_s||^2 cov(x_s)$ | $S_r R_{ii}$ bias $^2$ | $||r||^2 R^2$ |
| ⇓⇓ | ⇑⇓⇑ | ⇑⇓ |

of several of these ORT's to obtain stable, optimal models. Because of the different optimum criteria used in each of the methods, they usually do not perform equally well in restricting an 'optimal' solution for some of the models. However, the various ORT's often appear to complete each other and are thus, despite of some mathematical similarities, not redundant. Among the ORT's those of Tikhonov and a new technique called Backus (*subjective*) have been found particularly efficient, whereas the widely used methods of optimal filtering and of Backus and Gilbert have not been found satisfactory, due to the inability of defining an optimal solution appropriately. Unfortunately the method of Backus (*subjective*) relies on a certain amount of *a* priori information on the model which in many applications is not readily available. On the other hand, because of their theoretical inception (see Koch (1989), for details) the ORT's of Tikhonov and of Backus and Gilbert are the only ones which can be used without *a* priori information and they will be applied in the later chapters.

### 2.3.4 Limitations and inherent sources of errors in the SSH traveltime inversion

The approximation of wave theory by ray theory as discussed in Section 2.2 contains some inherent restrictions (Pratt et al., 2002; Spetzler and Snieder, 2004), which rely on the simplification, neglecting some aspects of wave effects.

#### 2.3.4.1 Restrictions of ray theory: Band-limited waves, Fresnel zones and theoretical resolution

Basically it is assumed that the waves are of infinite high frequency which means that the waves propagate along infinitely narrow lines, the rays, through the earth volume, sampling the seismic structure delta function like along the ray paths. However, as discussed above, the waves recorded actually at a seismic station have only a finite frequency content, i.e. they are band-limited. This is due, (1) to the low frequency oscillations of the seismic source and, (2) the instrumental filtering at the seismometer itself. This band limitation implies - as a consequence of Heisenberg's principle - that the propagation of waves extends now to a finite volume around the geometrical ray path, the Fresnel zone. Within this small zone, wave characteristics still play a role, namely diffraction and interference of elementary waves due to Huygen's principle. One of these waves corresponds to the wave that travels along the central geometrical ray path and it will interfere with waves emanating from neighboring points. The envelopes where these interferences are constructive define the first Fresnel zone. Among the latter only the first Fresnel zone, the inner one, also called the Fresnel volume, is the most important one. The first Fresnel zone is then the zone where the diffracted wave arrives approximately in phase with the direct wave. Theory shows that across a cross section perpendicular to the ray most of the wave energy is in fact contained within the first Fresnel zone, falling off then like a Gaussian.

The occurrence of the Fresnel zone means practically that medium structures smaller than this Fresnel volume are then not "seen" by the ray, i.e. cannot be resolved by ray theory since they will not affect traveltimes. Within the Fresnel zone there is a self healing effect at the wave front and all information about the structure becomes invisible. On the other hand, ray theory works well in media with structures that have a length scale larger than the first Fresnel zone of the recorded wave field. The structures should not have smaller spatial extensions then the wavelength recorded. Otherwise there is a self healing effect at the wave front and all information about the structure become invisible.

In a 3D seismic medium the Fresnel volume has the shape of an ellipsoid (for a curved ray the Fresnel volume looks more like a banana) (see for example Husen and Kissling (2001)) whose center line is along the ray and whose foci are at the source and the receiver. The width $r$ of the ellipsoid, i.e. of the first Fresnel zone, at a particular point along the ray is then a function of the distances $d_s$ and $d_r$ to the source and station which are apart by a total distance $L$ and it is given by (Spetzler and Snieder, 2004)

$$r = \sqrt{\lambda d_s d_r / L} \tag{2.35}$$

where $\lambda$ is the wavelength of the wave. Obviously, the width of the Fresnel zone is maximal halfway between the source and the receiver, corresponding to the worst case scenario with respect to the resolving power of the ray theoretical approximation. Using $d_s = d_r = L/2$, one obtains

$$r_{max} = 0.5\sqrt{\lambda L} \tag{2.36}$$

This equation shows that $r_{max}$ increases with increasing source receiver distance $L$ and increasing $\lambda$, which means that the possible resolution decreases. For example, seismogram traces used in ray theory based tomography are usually low pass filtered with a maximal cut off frequency of $\nu = 5\,Hz$. Applying an average seismic velocity of about $v_p = 6\,km/s$ and a distance $L$ between 50 and $200\,km$, as is generally the case for the regional seismic rays used in this investigation, $\lambda = v_p/\nu = 95\,m$ and $r_{max}$ will range between 10 and $20\,km$. This means that structural inhomogeneities smaller than this size cannot be resolved and there is no use of setting up model grids smaller than this value.

#### 2.3.4.2 Seismometer registration characteristics and arrivaltime uncertainties

In the early years of the history of seismology seismometers still had a major influence on the form of the arriving seismic signal so that arrivaltimes could not be picked that precisely. The main limitations in the early days were bandwidth (mainly narrow band LP recording), and limited dynamic range.

Nowadays, fortunately, the operational frequency range of the individual seismometer has become more and more broad-band. Also the recordings of the incoming seismic waves are done digitally with a high sampling frequency of approximately $\nu_s = 40\,Hz$. From the sampling theorem it follows that the maximal resolved wave frequency is given by the Nyquist-frequency $\nu_{Ny} = \nu_s/2$ i.e. $\nu_{Ny} = 20\,Hz$. However, additional band-pass filtering is required to be able to compare different seismometer characteristics of stations of a seismic network - in the routine analysis of the GRSN data used in the present study the band-pass lies in the range $1-8\,Hz$ - the frequency content of an recorded wave is reduced further - down to about $\nu = 1\,Hz$ in the worst case, i.e. a wave-period of $T = 1\,s$. With such a large value of $T$ it is, in most cases, not possible to achieve a picking precision of more than $\approx 0.1\,s$ for the arrivaltime of a P-phase and, somewhat more, for a S-phase. With regard to the tomographic inversion this means that there is no use to fit the traveltime residuals below this threshold-time, otherwise one would start to fit only the inherent noise in the data.

Assuming such an arrivaltime uncertainty of $0.1\,s$ this results, with an average seismic propagation

## 2.3. SEISMIC INVERSE THEORY

velocity of $6\,km/s$, in a spatial error of about $600\,m$ in the crust which is definitively below the block sizes used in tomographic studies so far the latter being more of the order of $10\,km$. One may then ask what is the maximum lateral velocity perturbation one may be able to determine in such a block with the given arrivaltime uncertainty? Assuming, for example, a maximal velocity perturbation of $\delta v = +5\%$ - which is a reasonable value to be expected in the crust- the traveltime through a block of $10\,km$ size is reduced by

$$\Delta t = \frac{10\,km}{v_0(1+\delta v)} - \frac{10\,km}{v_0} = 0.08\,s$$

i.e. a value that comes close to the picking-time uncertainty of $0.1\,s$.

### 2.3.5 Practical considerations of the SSH traveltime inversion

From the theoretical aspects of the inversion theory, some practical or elementary considerations can be deduced. This mainly is how to set up the model space (discretization) and to interpret the results (convergence).

#### 2.3.5.1 Choice of block-discretization of the tomographic model

From the discussion in the previous section above it becomes clear that the tomographic model grid should not be reduced below a block size of $\approx 10\,km$, which is also the lower limit, prescribed by the Fresnel zone condition above. It will be seen in the application of the SSH-inversion, this theoretical lower limit cannot be sustained in practice as, especially, limited ray-coverage in parts of the model[7], additional observational errors (misidentification of the correct phases, etc.) lead to extra instabilities and statistical errors in the inverse solution for the block velocities. Eventually this will require an increase of the block size, at least, in parts of the model. Unfortunately, the present SSH-technique does not allow for variable block sizes in one layer, so a compromise for a reasonable block size has to be found by trial and error which takes into consideration the various conditions discussed in the previous sections. One approach to that will be the use of the checkerboard test discussed earlier. Overall, a good ray distribution which, in turn, depends on the available seismic network and the earthquakes recorded, is still the best guarantee for a successful SSH tomographic inversion.

#### 2.3.5.2 Convergence

The ultimate goal of the nonlinear SSH inversion is to find a local minimum - which, hopefully, would also be a global minimum - of the traveltime residual function $F$, equation 2.17, starting from an initial solution $x_{start}$, the so-called starting model. In 3D seismic tomography, the starting model is usually a 1D, vertically inhomogeneous velocity model. Before considering a 1D Model as a good starting model for the subsequent 3D model inversion, it is necessary to consider the convergence of the original 1D starting models. Steady convergence to a specific, final solution during the nonlinear iteration process by means of the Levenberg-Marquardt or the Gauss-Newton method provide confidence in the starting model, unlike when strong variations of the final solutions occur which hints at a poorly defined minimum of the objective function. It follows that the aim should be to have a series of 1D inversions that converge to a clear minimum, irrespective of the damping sequence used in the LM-technique. However, as with most other seismic inverse problems with non-perfect data, this is not always the case here neither, as will be shown in Chapter 5.

---

[7]A large number of, for example, $n = 1000$ rays traveling all along the same path obviously will give no extra information about the velocity structure than a single ray, other than reducing its statistical error - following the $1/\sqrt{(n)}$ - law, i.e. reducing the effects of statistical noise. So, the possible structural resolution of the specific earth volume of interest strongly depends on the distribution of the rays that sample the individual cells of grid.

### 2.3.5.3 Comparability of the velocity models

The comparison of the various 3D-velocity models through the graphical representation by means of the GMT-graphical software is often problematic. As it turns out, simple changes of single parameters may lead to strong changes in the representation and the colors assigned to the various velocity values. So, small changes in velocity become visible, when choosing an appropriate assignment, by strong contrasts, which resemble a structure that does not exist in this strength symbolized. The velocity values are going from iteration to iteration, thus pretending a structure. This should be avoided.

# Chapter 3

# Anisotropy

An anisotropic behavior of matter basically means that the reaction to an external influence (this could be a force, temperature, magnetic force, light, or a seismic wave traversing the matter) depends on the direction of the influence. In a fluid or solid matter the effects of anisotropy can be caused by different mechanisms. As soon as the matter consists of different chemical elements which often implies an inhomogeneous or crystal structure, a certain volume, which may look homogeneous from the outside, can have an anisotropic behavior.

In seismology, anisotropy relates to the propagation of seismic waves which, as they are traveling through the Earth, may feel the influences of an anisotropic rock material.

## 3.1 Origins of anisotropy

In principle, one has to distinguish two kinds of occurrences of anisotropy. One is inherent to the matter and its molecular and chemical structure.

The other one is of macroscopic origin so, for example, cracks stop seismic waves traveling perpendicular to the major crack directions but not those traveling parallel to the latter.

### 3.1.1 Mineralogical sources of anisotropy

Especially crystalline mineralogical structures often show this effect which is a consequence of special, not unidirectionally acting binding-forces between the different atoms of the material under question (Bystricky et al., 2000; Bina, 1998; Nelson, 2006; Su and Park, 1994). Olivine ($[Mg, Fe]_2 SiO_4$) and olivine-like minerals, for example, which is supposed to be a major component of the Earth's upper mantle and is a common constituent of many basalts and gabbros, exhibits strong seismic and electrical anisotropy (Gatzemeier, 2001). (See Table 3.1)
(http://www.mineralienatlas.de/lexikon/index.php/Olivine)
Some nice aspects of the crystal properties of olivine and olivine-like structures can be found in Bass et al. (1984); Bass (1995). But even a chemical material composed of the same element may show an anisotropic behavior, due to its particular crystalline structure. Graphite, for example, has a specific, not uniform lattice with a $sp2$-Hybrid electron configuration, where the binding forces are strong over the extension of a plane, but weak between the planes. As a consequence, graphite is weak compared to diamond, which has a tetrahedral configuration due to the $sp3$-hybrid electron configuration.

Concerning electrical conductivity, diamond has none, whereas graphite has one, but only along the planes, not rectangular to them, that shows that graphite has strong initial anisotropic behavior, and even its acoustic (seismic) behavior is anisotropic. But how can anisotropy be caused in a heterogeneous structure where the inherent anisotropic volumes probably are mixed up in that way that their anisotropic behavior is equalized and cancel each other in a statistical way? As described in, for

 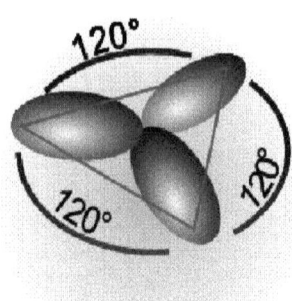

Figure 3.1: An example for anisotropy related to inhomogeneous bindings: The lattice configuration of diamond (left) and graphite (right). Although, these two materials consist of the same element, Carbon, they have completely different physical properties due to their different electron configuration. Whereas the atoms in diamond have a $sp3$-configuration with forces in tetrahedral directions, in graphite they have $sp2$ with strong forces in planar direction. Between the plains of graphite, only weak forces act.

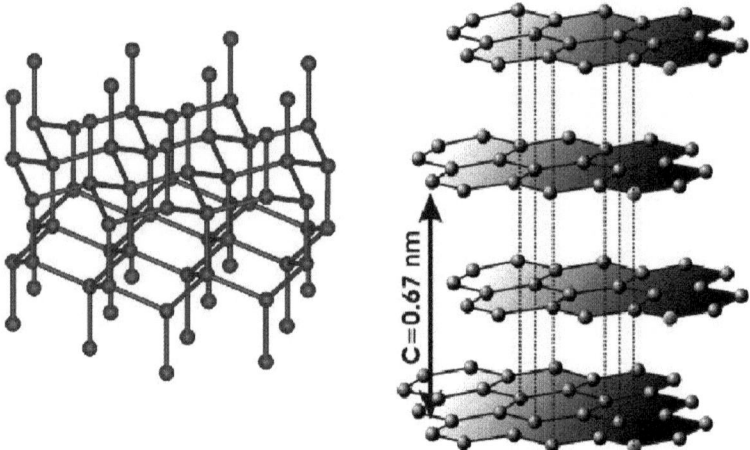

Figure 3.2: The two different lattices of diamond (left) and graphite (right). Whereas diamond has a cubic configuration which gives it a rather isotropic structure, graphite has a planar configuration.

example (Little et al., 2002; Fouch and Rondenay, 2006), there is a relationship between strain and anisotropy.

When strain occurs, minerals and their crystals are oriented in that direction which influences the propagation of seismic waves in various ways. Strain, or deformation that has occurred as a consequence of large-scales tectonic stress over geological times along the crust-mantle (MOHO) boundary, may result in seismic anisotropy by a process which is called lattice-preferred orientation (LPO) of mantle minerals (chiefly olivine and orthopyroxene) (Nicolas and Christensen, 1987; Ribe, 1989a,b; Ribe and Yu, 1991; Kaminski and Ribe, 2002). During this LPO process the crystals in an initially isotropic volume, deformed uniformly, take a preferred orientation[1] (Kaminski and Ribe, 2002), that follows the finite strain ellipsoid (FSE), with the fast a-axis of olivine aligned with the longest axis of the finite strain ellipsoid (Wenk et al., 1991; Ribe, 1992).

In fact, LPO in olivine minerals is most likely the dominant origin of anisotropy observed at the crust mantle boundary and in the first kilometers of the upper mantle in many regions in the world, including central Europe (Fuchs, 1975, 1983; Song et al., 2001a,b). For example, in south western Germany the stress field deduced from fault-plane solutions of regional earthquakes and the ensuing strains lead to LPO in olivine with the preferred gliding plane consistent with the observed direction of anisotropy. Other investigations like that of (Zhang and Karato, 1995) show that at large strain, the mean a-axis orientation no longer follows the finite strain ellipsoid but rotates more rapidly toward the shear direction. This evolution is accompanied by intensive dynamic recrystallization, by sub grain rotation (SGR) and grain boundary migration (GBM) (Zhang et al., 2000). To interpret which kind of anisotropy is the most prominent, one has to look at the time scales, that control the mantle convection (Kaminski and Ribe, 2002). For Germany, the time scale is rather long, so the anisotropic axis should follow the long strain axis of the finite strain ellipsoid (FSE).

### 3.1.2 Anisotropy caused by structural complexity

In addition to the previously discussed chemo-physical-property origin, rock anisotropy can also be caused by macroscopic inhomogeneities in the geological material, such as, for example, cracks and cavities, when these are oriented in a macroscopic way owing to a specifically directed tectonic stress field (Eisbacher, 1991) [2]. These cracks or voids may also be filled with water or magma, causing additional changes in the propagation properties of seismic waves. Normally, the seismic velocity in water is slower than that in a rock, so along the cracks, the waves travel with lower speed than perpendicular to them. In detail, this effect is much more complex than described here, since it also depends on the kind of waves considered, i.e. compressional (P) or shear (S) waves, where in the latter case, anisotropy gives rise to a phenomenon called "shear-wave splitting". In fact, the analysis of S-wave splitting has become nowadays one of the most powerful tools of modern seismology to study anisotropy in the Earth's interior, namely in the upper mantle (c.f. Savage (1999), for a review).

## 3.2 Mathematical description of anisotropy

Linear elasticity is theoretically described by Hooke's law which gives the relation between stress $\sigma$ and strain $\varepsilon$,

$$\sigma_{ij} = C_{ijkl}\varepsilon_{kl} \tag{3.1}$$

where $C_{ijkl}$ is the elasticity tensor, which theoretically has 81 elements but which, because of the symmetry of the stress- and strain tensors and of the deformation energy (Malvern, 1969; Weber et al., 2007) reduce to only 21 elements for a general elastic and anisotropic medium. However, depending

---

[1]A polycrystal responds to an imposed deformation rate tensor by simultaneous intra crystalline slip and dynamic recrystallization.

[2]Disturbances in the homogeneity of the structure of matter ( $\mu$ m up to *meters* and several kilometers), which are much bigger than the length of a chemical bond (Å).

Table 3.1: Anisotropy in Olivine. The velocity variations for the three different body wave types (each of them with their compressional, qP and the two shearing, qSH and qSV parts) along the main crystal plains (A-B, B-C, A-C), where q stands for quasi, just relating the appropriate velocity to the isotropic counterpart. These three body waves arise in an anisotropic medium according to Hooke's Law which has 21 independent elastic moduli in its most general form. (after Kawasaki (1989), page 65)

A-B-plain for axis a
qP:   $V_{max} = 10\ km/s$ at $90°$, $V_{min} = 7.9\ km/s$ at $0°$ and $180°$
qSH:  $V_{max} = 5.3\ km/s$ at $30°$ and $150°$; $V_{min} = 4.9\ km/s$ at $0°$, $90°$ and $180°$
qSV:  $V_{max} = 4.9\ km/s$ at $90°$, $4.45\ km/s$ at $0°$ and $180°$

B-C-plain for axis b
qP:   $V_{max} = 8.5\ km/s$ at $90°$, $V_{min} = 7.9\ km/s$ at $0°$ and $180°$
qSH:  $V = 4.9\ km/s$ nearly constant.
qSV:  $V_{max} = 4.9\ km/s$ at $0°$ and $180°$, $4.45\ km/s$ at $90°$

A-C-plain for axis c
qP:   $V_{max} = 9.9\ km/s$ at $0°$ and $180°$, $V_{min} = 8.6\ km/s$ at $90°$
qSH:  $V_{max} = 5.5\ km/s$ at $45°$ and $135°$; $V_{min} = 4.95\ km/s$ at $0°$, $90°$ and $180°$
qSV:  $V_{max} = 4.9\ km/s$ at $0°$ and $180°$, $4.9\ km/s$ at $0°$ and $180°$

on the kind of symmetry axes that may exist in particular anisotropic media because of its inherent crystalline structure, or as a consequence of LPO through external strain, there is an additional reduction of the number of independent coefficients of the elasticity tensor, namely:

- General anisotropic medium:
  21 independent components

- Symmetry by three orthogonal plains, the so-called orthotropic medium:
  9 independent components

- Axial symmetry like in a hexagonal medium:
  This is also known as transverse isotropy (Bamford and Crampin, 1977). Hexagonal symmetry has an axis of cylindrical symmetry, which implies that the properties of the medium in a given direction depend only on the angle between that direction and the symmetry axis direction. For example, if the direction of symmetry results from horizontal regional stresses (of interest in the present thesis) or vertically parallel cracks, the symmetry axis lies horizontally and the medium has transverse isotropy with a horizontal axis of symmetry. Overall this results in 5 independent components.

- Azimuthal anisotropy:
  Only angular variation of the propagation velocity of a seismic wave between a maximal and minimal value in a plain. This is also called elliptical variation of anisotropy, described later.

- Isotropic medium:
  2 components (the Lamé constants, as discussed in Chapter Theory 2).

The particular case of azimuthal anisotropy where seismic waves are traveling along a horizontal plane, results for its so-called "weak anisotropy", as mentioned by Song et al. (2001a); Song and Every (2000); Silver (1996), in the following named as elliptical correction for the traveltimes $t_{ijk}$ from the

## 3.3. OCCURRENCE OF SEISMIC ANISOTROPY IN THE EARTH'S INTERIOR

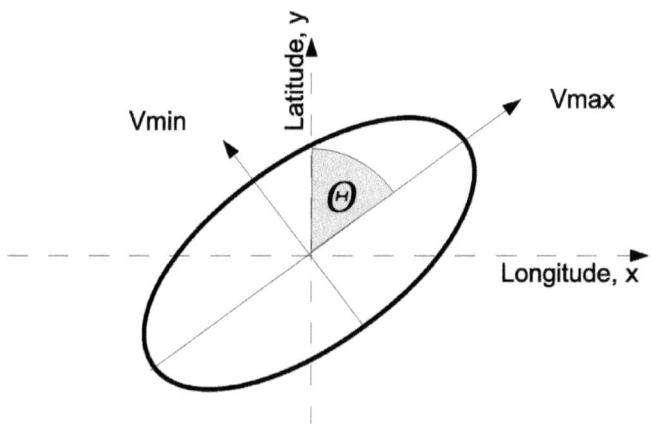

Figure 3.3: The elliptical correction for anisotropy with fast axis $Vmax$ and azimuth $\alpha$. The velocity disturbance is given in percent to the average of $Vmin$ and $Vmax$.

$i_t h$ event to the $j_t h$ receiver, traveling through $k$ cells:

$$\Delta t_{ijk} = a_i + b_j + \sum_k l_{ijk} \Delta s_k \quad (3.2)$$

where $a_i$, $b_j$ are the event and station time term delays, $l_{ijk}$ is the ray length in cell $k$ and $\Delta s = \frac{1}{s_0}(A + Bcos2(\phi - \theta) + Ccos4(\phi - \theta))$ is the slowness perturbation. Here, $\phi$ is the azimuthal angle of a ray direction, $\theta$ the azimuth of the fast axis, $A$ the isotropic constant and $B$ and $C$ are the coefficients for anisotropy. To correct directly for the velocities in the ray-tracing part of the SSH-program, the formula 3.3 is used.

$$v(\phi, v_{min}, v_{max}) = \sqrt{\frac{v_{min}^2 \cdot v_{max}^2}{v_{min}^2 \cdot sin^2\phi + v_{max}^2 \cdot cos^2\phi}} \quad (3.3)$$

## 3.3 Occurrence of seismic anisotropy in the Earth's interior

The occurrence of anisotropy in the Earth is not only an accidental effect but is most often related to special dynamic processes such, for example, the movement of the lithospheric plates on the surface of the Earth. That tectonic movement leads to strain which, eventually gives rise to the anisotropic orientation of the rock fabric and the crystalline structure, especially LPO, in the crust and upper mantle.

### 3.3.1 Anisotropy in the mantle

In the mantle, anisotropy is normally associated with crystals aligned with the mantle flow direction. Due to their elongate crystalline structure, olivine crystals tend to align with the flow due to mantle convection. The mineral composition of the mantle is, as known until now, different from the crust, which is deduced from the different, increasing seismic velocity. The main components of the mantle are Olivine and Pyroxenes which are minerals with strong anisotropic behavior. This different mineral composition is responsible for anisotropic effects in the mantle and therefore affects the head ($P_n$) waves, traveling along the crust-mantle-boundary, the MOHO. In fact, azimuthal anisotropy was initially observed for $P_n$ velocities measured in marine seismic refraction studies (Hess, 1964; Raitt et al., 1969; Keen and Barrett, 1971). As one penetrates deeper into the mantle, Olivine is still the major mineral. At a depth of approximately $410\,km$ the pressure is high enough that Olivine changes from the low pressure configuration to the high-pressure configuration, the $\beta$ Spinel phase. While this entails a dramatic change of the propagation speed of seismic waves, the Spinel polymorph shows also much less anisotropic behavior with the consequence that the middle mantle appears to be pretty much devoid of seismic anisotropy. On the other hand, anisotropy seems again to be prevalent in areas of the D"-layer at the core-mantle boundary (Savage, 1999).

As mentioned previously, experimental and theoretical work suggest that the fast velocity axis of olivine should be oriented in the direction of maximum extensional strain, i.e. orthogonal to the compressional direction of the tectonic stress-field. For Germany and much of central Europe, starting with the work of Bamford (1973) who used a modified time-term analysis of $P_n$ first arrivals, he and the subsequent studies of Fuchs (1983); Enderle et al. (1996a); Babuška and Plomerová (2000); Song et al. (2001a) have revealed an anisotropic behavior of $P_n$ phases with the fast axis of the LPO directed towards $N\,22.5°$. Other investigations like the teleseismic TRANS-ALP experiment in the Eastern Alps also indicate the SW-NE orientation of the fast anisotropy axis in the upper mantle there (Kummerow, 2004).

Seismic anisotropy in the mantle has also been inferred from long-period surface waves (Forsyth, 1975; Nataf et al., 1984; Tanimoto and Anderson, 1885; Montanger and Tanimoto, 1990). These studies suggest that the fast axis of anisotropy is consistent with the present-day flow direction of the major tectonic plates. Moreover, the orientation of anisotropy found in the ocean crust and ocean sub-Moho mantle strongly indicates, that it is due to fossil fabric formed at the mid-oceanic spreading ridges, with the fast axis parallel to the spreading direction.

### 3.3.2 Anisotropy in the crust

In contrast to the mantle the Earth's crust is more brittle and therefore crustal rocks may reveal anisotropic behavior, though its origin here is somewhat different from that observed in the mantle. Anisotropy is caused by mechanical tension or compression by which the rock and its crystal structure accepts a certain organization. Because the Earth's crust is somewhat brittle and has rather heterogeneous consistence, anisotropy may be caused by aligned micro cracks, shear fabric, layered bedding in sedimentary formations, or highly foliated rocks. Crustal anisotropy resulting from aligned cracks can be used to determine the state of stress in the crust. In most cases, cracks preferentially align with the direction of maximum compressive stress, which is also then the orientation of the fast velocity axis if the cracks are filled with fluid (Crampin, 1986). In active tectonic areas, such as near faults and volcanoes, anisotropy can be used to look for changes in preferred orientation of cracks that may indicate a rotation of the stress field. Crustal anisotropy is very important in the detection of oil reservoirs, as fast directions can be synonymous with certain fluid flow directions. The same holds for magma flow in volcanic intrusion, which might influence the propagation of seismic waves in a correlated direction. An example for anisotropy in the crust is the one found under the southern alpine region of New Zealand Okaya et al. (1995).

## 3.4 Shear wave splitting due to seismic anisotropy

Besides the azimuthal impact on $P_n$ waves traveling along the crust-mantle-boundary an anisotropic upper mantle has another important effect on the propagation of seismic waves, namely, shear (S) wave splitting. This phenomenon is comparable to the polarization of light in an optical, birefringent medium where an optical ray inclined to the direction of the optical axis is split into two rays, the

## 3.4. SHEAR WAVE SPLITTING DUE TO SEISMIC ANISOTROPY

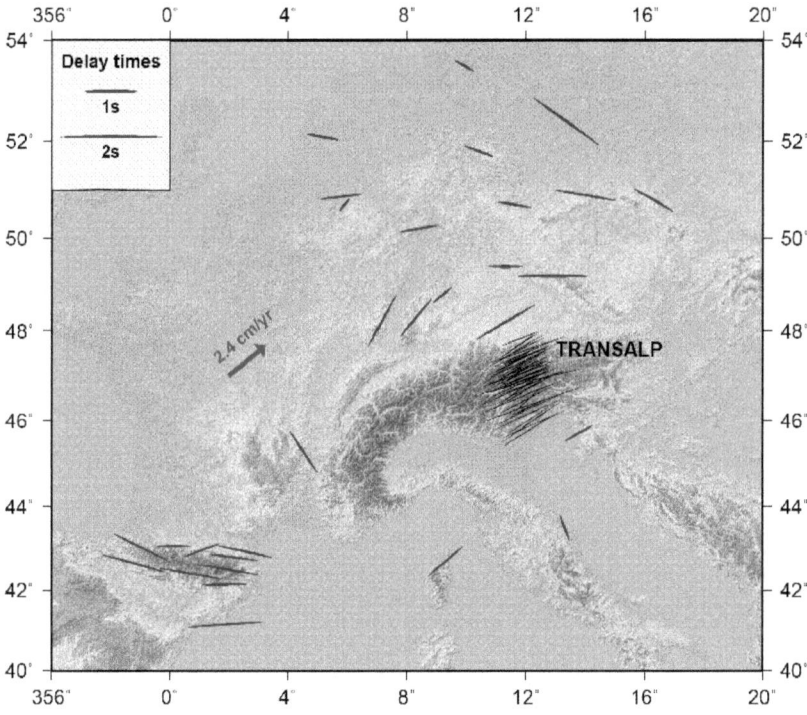

Figure 3.4: Orientation of the fast axis at the TRANS-ALP stations and some seismic stations in Middle Europe (after Kummerow (2004), page 93).

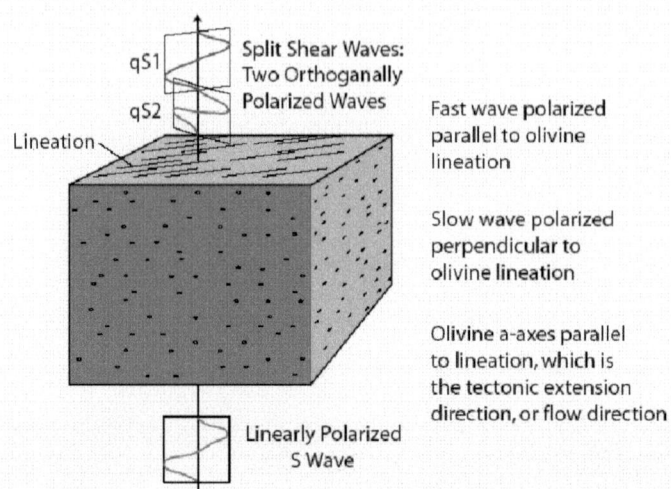

Figure 3.5: The splitting of a P-wave by a heterogeneous medium into two perpendicular components.

ordinary and the extraordinary one, with polarizations that are perpendicular and parallel to the optical axis and have different phase speeds $c_0$ and $c_e$, respectively (Orfanidis, 2008). For an S-wave propagating in an anisotropic medium the incident shear wave also separates into two orthogonally polarized waves, called the quasi-vertical (qSV) and quasi-horizontal (qSH) shear wave, that are traveling with different velocities. From the measurement of the arrival delay time $\delta t$ of these two quasi S-phases, the direction and the size of the anisotropy can then be determined. In fact, most of the knowledge about the Earth's seismic anisotropy, at least in the mantle, where significant large splitting times $\delta t$ are generated[3], has come from analysis of shear wave splitting. These have become numerous over the last two decades (cf. Savage (1999); Audoine et al. (2000); Pulford et al. (2003). For example Ando and Ishikawa (1982) investigated the **anisotropy in the upper mantle** beneath Honshu by measuring the splitting of $ScS$ phases from nearby events in the descending slab. The $SKS$ phase has been widely used to measure the anisotropy beneath seismic stations because of its simple radial polarization and its easy identification if the epicentral distance is larger than 85°. The SKS splitting method, first introduced by Vinnik et al. (1984), has been systematically applied by (Kind et al., 1985; Silver and Chan, 1988, 1991; Vinnik et al., 1989; Savage, 1999). These studies have also led to the conclusion that anisotropy between the D" layer and the $420\,km$ olivine—spinel transition is negligible (Kaneshima and Silver, 1992; Silver. and Bina, 1993).

---

[3]Crustal anisotropy is a poor candidate to explain relative splitting delays $\delta t \approx 1s$, because, for reasonable mineral assemblages, it is too thin, especially for oceanic crust

# Chapter 4
# Data Set

The basic input dataset of a seismic tomography study using regional earthquakes, i.e. SSH, consists of arrivaltimes of seismic signals recorded at various seismometers at the earth's surface in Germany. From this, one can deduce the hypocentral locations and the seismic velocity variations along the ray paths between source and the receivers. However, to get reasonable results with respect to these parameters, one has to consider several aspects with regard to the preprocessing of the original data, as described in the following sections.

## 4.1 General aspects of data requirements in seismic inversion

The initial important question in all tomographic investigations is "how powerful is my database"? A supreme database[1] is essential for any high-quality seismic 3D tomographic investigations. (Koch, 1985a,b; Kissling, 1988; Husen et al., 1999; Tarantola and Valette, 1982b) In fact, there are three main aspects what make a database supreme:

- Distribution
  A good inversion computation using the SSH-method can only be done when the seismic sources in the study volume and the seismic recording stations at the earth's surface are well distributed in space and surface area. Thus, earthquakes concentrated in a small volume of the crust, though registered across a wide surface area, will provide useful information only for a small part of the model-space. More so, the rays within a small tube of the model may not carry independent, but only redundant information. Mathematically such a situation reverts to an "underdeterminacy", i.e. effectively, there are less observations than unknowns in the discretized inverse problem.

- Amount
  The amount of arrivaltime data is important because of the statistical character of the inversion. Because there are errors in the registrations they have to be compensated by numerous though, sometimes redundant, observations. But not only numerous seismic sources are necessary, also the number of registrations per event should be higher than at least four, in order to increase the number of the "degrees of freedom". Mathematically, this situation amounts to the requirement of an effective "overdeterminacy" of the discretized inverse problem.

- Quality
  The measurement and identification of the registered waveforms is the third aspect that limits the accuracy of the tomographic investigation. For example, as discussed in Chapter 2, if the

---

[1]Unfortunately, because of time restrictions, a thorough analysis of the original waveforms to check the reliability of the arrivaltime data set could not been carried out. Even more, because the dataset is a collection of registrations of different seismic surveys, using different standard velocity models, the data will have different interpretation. Doing the waveform and data-analysis by more than one person, there will occur problems with the comparability of the different data (Arlitt, 2001).

Table 4.1: Number of recorded events and the according phases for the 1975-1999 and the 2000-2003 datasets.

**P-phases:**

| period | events | $P_g$ | $P_n$ | $PmP$ | total |
|---|---|---|---|---|---|
| 1975 - 1999 | 6623 | 27919 | 8905 | 432 | 37256 |
| 2000 - 2003 | 3405 | 18631 | 3899 | 463 | 22993 |

**S-phases:**

| period | events | $S_g$ | $S_n$ | $SmS$ | total |
|---|---|---|---|---|---|
| 1975 - 1999 | 6623 | 28284 | 2501 | 0 | 30785 |
| 2000 - 2003 | 3405 | 22025 | 1402 | 0 | 23427 |

arrivaltime signal can only be identified with a time-uncertainty of 0.5 s, the residuals after the inversion should also lie within this range, of course, results in a corresponding model uncertainty. Especially worrisome, and leading to even higher time errors may be the correct identification of the appropriate seismic phase ($P_g$, $P_n$, $S_g$, $S_n$ and so on.) (Arlitt, 2001).

To check the adequacy of the available dataset to fulfill these requirements for a reliable SSH inversion some preliminary studies are done as described in the next sections.

## 4.2 Data used in the present study

For the present tomographic study a huge traveltime data set from regional earthquakes, compiled at BGR ("Bundesanstalt für Geowissenschaften und Rohstoffe"), is available. This data-set provides a rather good coverage of the investigation area, which consists mainly of the "old" BRD part of Germany (Western Germany), however, with a concentration of seismic sources (red points in Figure 4.1) in the south-western triangle of Germany. In contrast to traveltime data available from numerous studies of controlled source seismology (Gajewski and Prodehl, 1985a; working Group, 1989; Gajewski et al., 1990; Enderle et al., 1996a; Enderle, 1998; Enderle et al., 1998; Mechie, 2005) carried out over the last decades, the ray paths of the recorded regional seismic arrivaltime data used here distinguish themselves by a good azimuthal coverage over Western Germany. Nevertheless, because of general low seismicity, only few seismic data are available in the area north of 52° latitude.

More specifically, the dataset used in this thesis contains 6623 seismic events with 37256 P-phases and 22993 S-phases recorded between 1975 and 1999, already analyzed in previous studies of $P_n$-anisotropy by Song et al. (2001a, 2004). Updated by about the same number of regional earthquakes that occurred between 2000 and 2003, this nearly doubles the number of available registrations to 68041 P-phases and 46420 S-phases with 10028 events and a corresponding increase of ray coverage across the study area. This data are published yearly in the "Jahreskatalog" (Henger and Leydecker, 1975); (Hartman et al., 1990)
http://www.seismologie.bgr.de/www/sdac/erdbeben/catalogue_ger.htm, hosted at BGR.

Compared with the numerous active experiments in refraction- and reflexion seismics carried out in recent years across Germany (Enderle, 1998; Edel et al., 1975; Gajewski and Prodehl, 1985a,b; Gajewski et al., 1990; Enderle et al., 1998), the advantage of the present huge regional dataset is its immense density and redundancy in azimuthal ray coverage (Figures 4.1 and 4.2). By virtue of this fact, starting with the first analysis of Schlittenhardt (1999), and the subsequent studies of the $P_n$-anisotropy by Song et al. (2001a), Song et al. (2004), the present dataset has already provided relevant information on the seismic structure of the crust and upper mantle beneath Germany. It

## 4.3 Initial data preprocessing and selection criteria

Because of the immense amount of regional seismic events that occurred in the study region between years 1975 and 2003 with a corresponding high number of seismic phases recorded, the ray coverage of the study area should also be very good. Figure 4.1 shows that, although most of the hypocenters are concentrated in the Rhinegraben region, leading to a dense ray distribution in the form of stripes that extend from the Rhinegraben to the northern and southern parts of Germany, due to the long registration period, a quite good number of rays are also illuminating other parts of the former BRD.

On the other hand, owing to technical innovations and increased seismological experience over the last 30 years with regard to the seismic data recording process and the correct interpretation of the seismic phases in the seismograms ($P_g$ or $P_n$), the quality of the present arrivaltime data has also been improved during the long recording period 1975 - 2003. Thus, event registrations have been relatively poor in the early years (1975 - 1985), namely due to the smaller amount of recording seismic stations. With experience of the seismologist on duty of the routine arrivaltime data processing improving, uncertainties in the picking times of the various phases have also been decreasing significantly in more recent decades.

Thus, whereas for data of the first years of the recording period the scatter around the expected average traveltime distance function - derived from a standard vertically inhomogeneous 1D velocity model for the crust and lithosphere underneath Germany (Schlittenhardt, 1999) - is still large, this becomes less in the case for arrivaltime data recorded in more recent times.

During the course of the analysis of the input arrivaltime data it was recognized by the author that several criteria, as itemized in the subsequent paragraphs, are essential for the selection of a good traveltime set.

- Maximal azimuthal gap (GAP)

  The GAP criterion (see 4.4.1) is probably the most important constraint for a good initial hypocentral relocation of the earthquake by standard relocation routines, such as HYPO71 (used for most of the events in the present study) or HYPOELLIPSE. Applying this criterion, earthquakes with an unequal distribution of rays between source and station are discarded from the SSH inversion as their hypocenters have most likely already been determined badly in the original relocation process. The parameter GAP is defined as the maximally allowed angle between the rays from the epicenter to two adjacent stations. Experience shows that if the GAP is greater than 180°, the hypocentral determination must be taken with a grain of salt.

  Earthquakes recorded by stations where the smallest angle between the stations is greater than 180° are rejected, because the "view" to the hypocenter is only from one "side". This means essentially that a possible arrivaltime error at one station cannot be laterally compensated by the arrivaltime at a diametrically opposite station, leading eventually to a certain trade off between epicentral location perpendicularly to the GAP and the origin time of the earthquake.

  On the other hand, for the SSH inversion, whose reliability depends on the use of a large and well distributed number of seismic events, reducing the GAP too much, would eliminate a large number of marginal earthquakes, i.e. data. Thus, one must strive to strike a proper balance between quality and quantity of the data used in the analysis. This can be seen from the two ray-path plots (Figures 4.1 and 4.2). Whereas Figure 4.1 includes all events with a maximal GAP of 180°, in Figure 4.2 only events with a maximal GAP of 120° are included, resulting in a less well distributed ray coverage in some areas of the study region. To select an adequate GAP for the earthquakes used in the SSH inversion, some further tests will be done in Section 4.4.1.

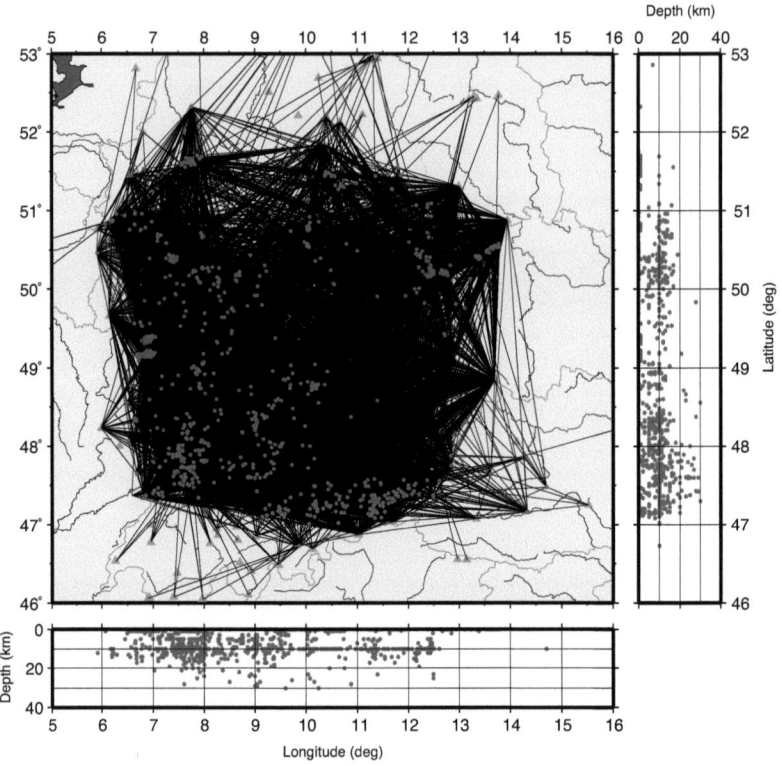

Figure 4.1: Seismic ray coverage of the study region (projection of the ray paths on the earth's surface). Rays are plotted for all observed phases ($P_g, P_n, S_g, S_n$), used in the inversion process if they fulfill two criteria: (1) azimuthal GAP for each earthquake $\leq 180°$ and, (2) minimum number of observations per event $NOBS \geq 8$. Triangles represent the positions of the recording stations. The panels at the right and bottom of the figure show NS- and EW- cross sections of the area.

## 4.3. INITIAL DATA PREPROCESSING AND SELECTION CRITERIA 49

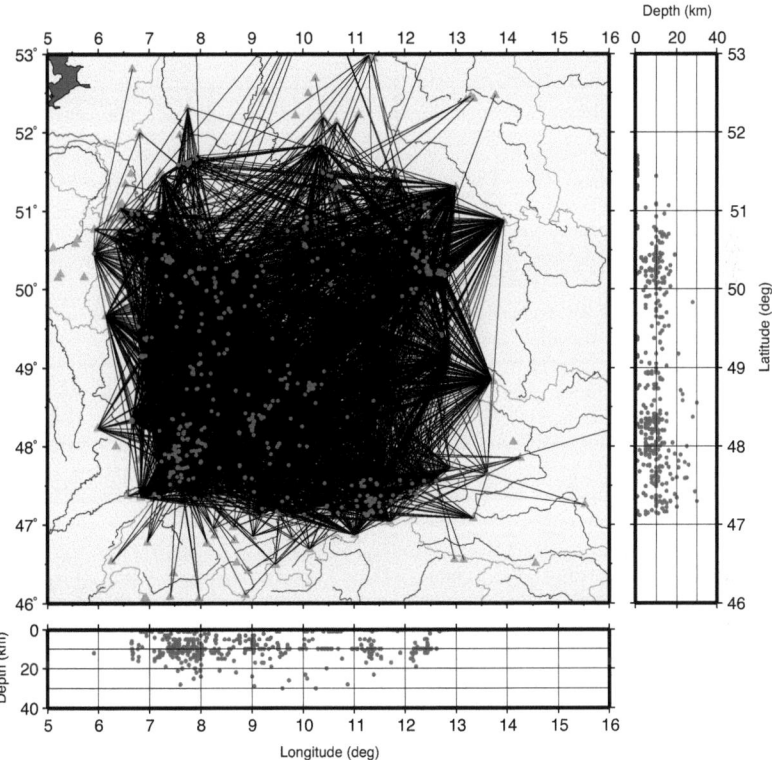

Figure 4.2: Similar to 4.1, but now with a stronger GAP criterion of $\leq 120°$. The ray coverage is now mainly due to seismic rays from regional earthquakes in the south-west, whereas the coverage in the south east and east is thinned out, due to a lack of events there.

- **Number of Observations (NOBS)**

  The number of observations per seismic event, in the following abbreviated as NOBS, is essential to avoid non uniqueness in the inversion of the Fréchet-matrices of the discretized SSH inverse problem. At least NOBS = 4 is necessary to calculate the $p = 4$ hypocentral parameters of an earthquake. But such a low number of NOBS would leave no additional information, also defined by the degrees of freedom NOBS$-p$ to resolve the velocity structure or to statistically reduce the variance of the calculated hypocentral parameters. Therefore it is common to set the NOBS to at least NOBS = 8, which is quite a good choice to have, (1) enough observations to obtain an overdetermined system of equations and, (2) not to reduce the available amount of data too much and, so, to discard informative data. Setting NOBS to higher values, the available data, i.e. number of seismic events will be reduced and the ray coverage may be poor[2]. The investigations for this effect are done in section 4.4.2.

- **Residual limit (RES)**

  Another effective criterion is to limit the maximally allowed residual, in order to reject arrivaltime outliers. The maximally allowed residual used in the tomographic inversion calculation has to be checked quite carefully. It should neither be too big nor too small. If it is too big, erroneous data is projected into the inverse solution; if it is too small, small scale seismic structure in the 3D model would be rejected, i.e. indispensable information about the model space is excluded[3], and even more the information about anisotropy.

  For choosing the residual criterion appropriately, one may have recourse to the time-range within the expected arrivaltimes must theoretically lie. For example, knowing the exact seismic velocity structure of the lithosphere and the exact hypocentral location, i.e. the hypocentral parameters x, y, z and t, one can calculate the exact arrivaltimes for a set of given seismic stations. For a most simple verification of the exactness of the analyzed arrivaltimes, one would use just a vertically inhomogeneous model and create synthetic traveltime- and arrivaltime plots. From these plots crude picking errors can be detected and eventually corrected. All measurements outside that range have to be rejected or sensitively analyzed, also they might still be a hint of strong seismic anomalies.

  A further aspect of this selection criterion consists in the proper detection and separation of different seismic phases, i.e. direct, reflected or refracted ones. For example, when having measurements at distances greater than the crossover distance, i.e. when the $P_n$-phase overtakes the direct $P$-phase as the first arrivals, it is important to clearly separate these two phases. This is sometimes difficult when the epicentral distance is located closely to the critical one.

---

[2]The NOBS-criterion is somewhat competing with the GAP criterion. Setting the GAP to small values, only seismic events with numerous recorded phases will most likely pass, though this is not guaranteed

[3]The residual limit varies with hypocentral distance between source and receiver, because the longer the ray path, the more 3D crustal structure is sampled. Otherwise, after a certain distance, the inhomogeneities may equilibrate themselves so that there will be no further residual variation. Studies about this should be undertaken to improve seismic imaging and to reduce the dependency of the inversion results on the quality of the arrivaltime pickings.

## 4.4 Tests of the impacts of the various selection criteria

In the present section the impacts of the various selection criteria (GAP, residual range and time period of events chosen) on the basic statistics of the dataset to be used in the final tomographic inversion will be tested and illustrated in various plots.

### 4.4.1 Effect of maximal GAP

The GAP criterion reduces the number of low quality data effectively. Low quality means, that the available data are not suitable for a good hypocenter relocation. The major effect of the GAP criterion is to reject such events that have unstable solutions due to an inhomogeneous ray distribution. The worst case would be to 'see' the epicenter or hypocenter only from one side. In this case no ray traveling in an opponent direction would be available for the stabilization of the inverse solution, neither for the hypocentral nor for the velocity part of the model.

Care has to be taken, because the GAP criterion does not reflect the spatial distribution of the observables. Even stations lying on top of the hypocenter could result in a GAP of 180 degree or less but, as the corresponding traveltimes contain nearly the same information, there is no extra contribution (redundant data) to the solution. This would result in a tremendous trade-off between the hypocentral and the velocity part of the model solution, as the rays are traveling to the surface nearly perpendicular within a small cone, leading to the well-known earthquake relocation problem of the trade-off between hypocentral depth and origin time. In the present case of SSH[4], the trade-off situation is even more intricate, as the seismic velocity is an additional solution parameter[5].

The results of the various tests for different GAP are summarized in Tables 4.2, 4.3, 4.4, 4.5 and Figure 4.4. From Tables 4.2, 4.3 one can see that the biggest reduction of the $TSS_{SSH}$ is obtained with a GAP of around 120°, but the amount of available records is nearly half than that with GAP = 180°. Reducing the GAP to only 180° or 150° leads to a higher RMS, which indicates that no improvement of data quality is achieved, but reduces the amount of remaining data not to much. The smaller value of the data-fit for 360° in Table 4.2 probably is a consequence of hypocenter foci not well determined because of missing opponent records and where the measured arrivaltimes with GAP > 180° can be fit better to the computed ones whereas in Table 4.3 the lower data-quality in earlier years is dominant.

Another aspect of the GAP selection is the influence of the hypocenter locations onto the data-fit as the number of hypocenters (each with 4 unknown hypocentral parameters) is reduced together with its number of recorded phases, when decreasing the GAP. Each hypocenter contributes with its recorded arrivaltimes to the total RMS. If the RMS increases with decreasing GAP below 180°, this means that some good data are rejected, with the rest of the data showing a worse model fit. Therefore, a GAP with 120° seems to be suitable to select good data, but concerning the amount of available data, the GAP = 180° is more suitable.

So, for subsequent tomographic SSH-calculations the standard choice of a GAP of 180° is chosen, and in some cases compared with results using a GAP of 120°, because this seems to be a good compromise between the available amount of data and the ray coverage.

In Figure 4.4 the results of two different GAP criteria (180° and 120°, each applied with a minimum of observations, NOBS = 8 per event) are shown to demonstrate its influence onto the data (earthquake) selection for use in the subsequent tomographic inversion. As one can see in the figures, as the GAP is decreased from 180° to 120°, the number of outliers and the scatter of the traveltimes is reduced. Otherwise, below a GAP of 180°, the amount of data, i.e. number of seismic events, is significantly reduced, whereas the scatter in the traveltimes does not diminish significantly. This proves that the

---
[4]SSH = Simultaneous inversion for Structure and Hypocenters
[5]To eliminate the trade off between depth and origin time, later investigations are done with Wadati diagrams to externally fix the origin time.

Table 4.2: Statistics of data variation as a function of the GAP for the earthquakes recorded between 2000 and 2003. $TSS_{SSH}$ is the squared sum of the calculated residuals, $TSS_{hypo}$ the squared sum of the residuals from the original data set. $RMS_{aver}$ is $(TSS_{SSH}/Nphase)^{1/2}$, $RMS_{unbias}$ is $(TSS_{SSH}/Nphase - Nmod)^{1/2}$, $RMS_{event_i}$ is $\sum \left(Res^2_{SSH}(i)/Nphase - 4\right)^{1/2}$. Nmod is the number of velocity parameters, Nphase the number of recorded phases and Nevent the number of used seismic sources. Ifix = 0, hypocenters free.

|              | GAP (degree) | | | | | | | |
|---|---|---|---|---|---|---|---|---|
|              | 360 | 180 | 150 | 130 | 120 | 115 | 90 | 60 |
| I            | 1 | 1 | 1 | 1 | 1 | 1 | 1 | 1 |
| K            | 0 | 0 | 0 | 0 | 0 | 0 | 0 | 0 |
| Nevent       | 666 | 504 | 428 | 348 | 254 | 236 | 91 | 22 |
| Nphase       | 12284 | 9670 | 8308 | 6744 | 4851 | 4398 | 2246 | 557 |
| $TSS_{hypo}$ | 29412 | 24817 | 22491 | 14625 | 6023 | 5565 | 2897 | 658 |
| $TSS_{SSH}$  | 35701 | 30193 | 26525 | 17961 | 9478 | 9037 | 5212 | 555 |
| $RMS_{unbias}$ | 1.9245 | 1.9793 | 2.0059 | 1.8326 | 1.5729 | 1.6185 | 1.6659 | 1.0926 |
| $RMS_{aver}$ | 1.7029 | 1.7618 | 1.7868 | 1.6319 | 1.3978 | 1.4334 | 1.5233 | 0.9983 |
| $RMS_{event}$ | 1.3338 | 1.2663 | 1.2359 | 1.1392 | 1.1515 | 1.1626 | 1.1987 | 0.9241 |

Table 4.3: Similar to Table 4.2, but for the complete dataset between the years 1975 - 2003.

|              | GAP (degree) | | | | | | | |
|---|---|---|---|---|---|---|---|---|
|              | 360 | 180 | 150 | 130 | 120 | 115 | 90 | 60 |
| I            | 1 | 1 | 1 | 1 | 1 | 1 | 1 | 1 |
| K            | 0 | 0 | 0 | 0 | 0 | 0 | 0 | 0 |
| Nevent       | 1824 | 1223 | 992 | 791 | 632 | 592 | 275 | 78 |
| Nphase       | 30447 | 22136 | 18184 | 14800 | 12049 | 11340 | 6582 | 2208 |
| $TSS_{hypo}$ | 95348 | 55621 | 46159 | 32882 | 22251 | 21225 | 12077 | 3770 |
| $TSS_{SSH}$  | 97219 | 58436 | 48143 | 34525 | 24216 | 23243 | 13530 | 3228 |
| $RMS_{unbias}$ | 2.0559 | 1.8399 | 1.8405 | 1.7228 | 1.5951 | 1.6099 | 1.5716 | 1.3063 |
| $RMS_{aver}$ | 1.7925 | 1.6238 | 1.6271 | 1.5273 | 1.4177 | 1.4317 | 1.4337 | 1.2092 |
| $RMS_{event}$ | 1.6707 | 1.3779 | 1.3480 | 1.2960 | 1.3045 | 1.3108 | 1.3188 | 1.1350 |

GAP criterion is essential, but not the only one, to select high quality data for the inversion process and to fulfill the quality conditions mentioned above. The most important is to set a maximal limit to at least GAP = 180°, or better to 120° to stabilize the hypocenter localization. This will assure to have at least two opposite stations (in case of a gap of 180°) or a triangular geometrical setup (in case of a gap of 120° ) of the station-event configuration.

### 4.4.2 Influence of the number of observations per event (NOBS)

The GAP criterion alone is not sufficient as it only has to be used together with the number of observations per event (NOBS). Theoretically, at least NOBS = $n$ = 4 observations per seismic event are necessary to solve the system of equations just for the $m$ = 4 hypocenter parameters, however, because of errors in the data and poor ray distribution which induce colinearities in the corresponding system matrix, the effective number of degrees of freedom $p = n - m = n - 4$ should in fact be larger than zero, even more so, when performing a full SSH inversion, as only $p$ extra observations can contribute to the velocity structure. A good choice of the number NOBS is about 8 and more, so some redundancy is available for the full inversion. On the other hand, setting NOBS too high, a lot of useful events will be rejected, as one can see in Table 4.7, where the number of events available for the inversion steadily decreases, as NOBS is increased. An increase of NOBS from 6 to 8 reduces

## 4.4. TESTS OF THE IMPACTS OF THE VARIOUS SELECTION CRITERIA

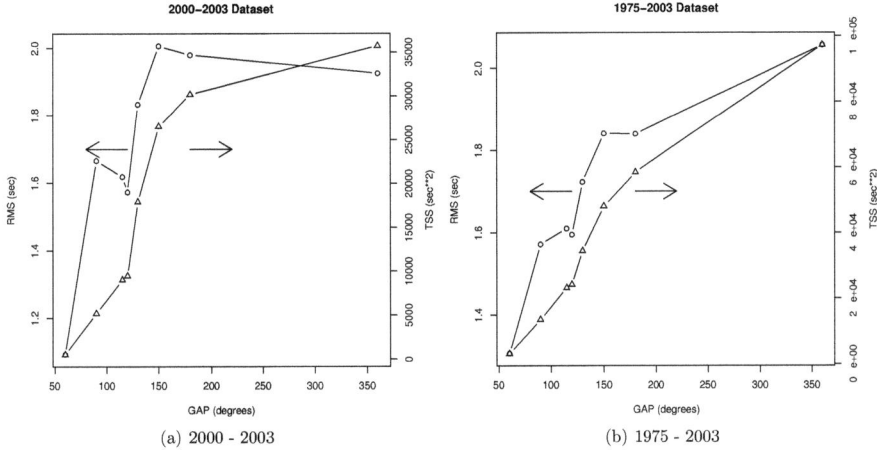

(a) 2000 - 2003

(b) 1975 - 2003

Figure 4.3: The reduction of RMS (circles, arrow to the scale on the left) and TSS (triangles, arrow to the right scale), respectively with decreasing maximal azimuthal gap between two adjacent stations (GAP). The non linearity of the line with the saddle point like region may be interpreted as the best choice for GAP, which is indeed between 120° and 180°.

Table 4.4: Number of events and various phases as a function of the parameter NOBS for 1975 - 1999 dataset. Only Pg, Pn and PmP-phases.

|  | 1) NOBS$\geq$ 1 GAP $\leq$ 360° | 2) NOBS$\geq$ 8 GAP $\leq$ 360° | 3) NOBS$\geq$ 8, GAP $\leq$ 180° | 4) NOBS$\geq$ 8, GAP $\leq$ 120° |
|---|---|---|---|---|
| Number of events | 6623 | 1146 | 719 | 378 |
| Pg-Phases | 27919 | 11703 | 8488 | 5105 |
| Pn-Phases | 8905 | 5777 | 3544 | 1892 |
| PmP-Phases | 432 | 420 | 367 | 178 |
| all P-phases | 37256 | 17900 | 12399 | 7175 |
| Sg-Phases | 28284 | 11491 | 8213 | 5349 |
| Sn-Phases | 2501 | 1765 | 1121 | 60 |
| SmS-Phases | 0 | 0 | 0 | 0 |
| all S-phases | 30785 | 13256 | 9334 | 5409 |
| all phases | 68041 | 31156 | 21733 | 12584 |

Table 4.5: Number of events and various phases as a function of the parameter NOBS for 2000 - 2003 dataset. Only Pg, Pn and PmP-phases.

|  | 1) NOBS>= 1<br>GAP <= 360° | 2) NOBS>= 8<br>GAP <= 360° | 3) NOBS>= 8,<br>GAP <= 180° | 4) NOBS>= 8,<br>GAP <= 120° |
|---|---|---|---|---|
| Number of Events | 3405 | 666 | 504 | 254 |
| Pg-Phases | 18631 | 8576 | 6950 | 3601 |
| Pn-Phases | 3899 | 3224 | 2207 | 1042 |
| PmP-Phases | 463 | 453 | 384 | 172 |
| Sum P-phases | 22993 | 12253 | 9541 | 4815 |
| Sg-Phases | 22025 | 8899 | 6969 | 3991 |
| Sn-Phases | 1402 | 1188 | 865 | 45 |
| SmS-Phases | 0 | 0 | 0 | 0 |
| Sum S-phases | 23427 | 10097 | 7834 | 4036 |
| Sum all Phases | 46420 | 22350 | 17375 | 8851 |

Table 4.6: Number of events and various phases as a function of the parameter NOBS for 1975 - 2003 dataset. Only Pg, Pn and PmP-phases.

|  | 1) NOBS>= 1<br>GAP <= 360° | 2) NOBS>= 8<br>GAP <= 360° | 3) NOBS>= 8,<br>GAP <= 180° | 4) NOBS>= 8,<br>GAP <= 120° |
|---|---|---|---|---|
| Number of Events | 10028 | 1712 | 1223 | 632 |
| Pg-Phases | 46550 | 20279 | 15438 | 8706 |
| Pn-Phases | 12804 | 9001 | 5751 | 2934 |
| PmP-Phases | 895 | 873 | 751 | 350 |
| Sum P-phases | 60249 | 30153 | 21940 | 11990 |
| Sg-Phases | 50273 | 20390 | 15182 | 9340 |
| Sn-Phases | 3903 | 2953 | 1986 | 105 |
| SmS-Phases | 0 | 0 | 0 | 0 |
| Sum S-phases | 54212 | 23343 | 17168 | 9445 |
| Sum all phases | 114461 | 53496 | 39108 | 21435 |

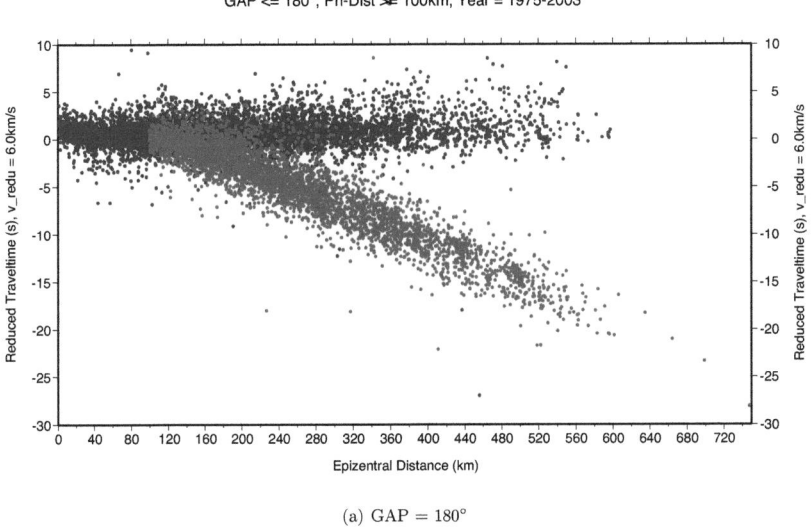

(a) GAP = 180°

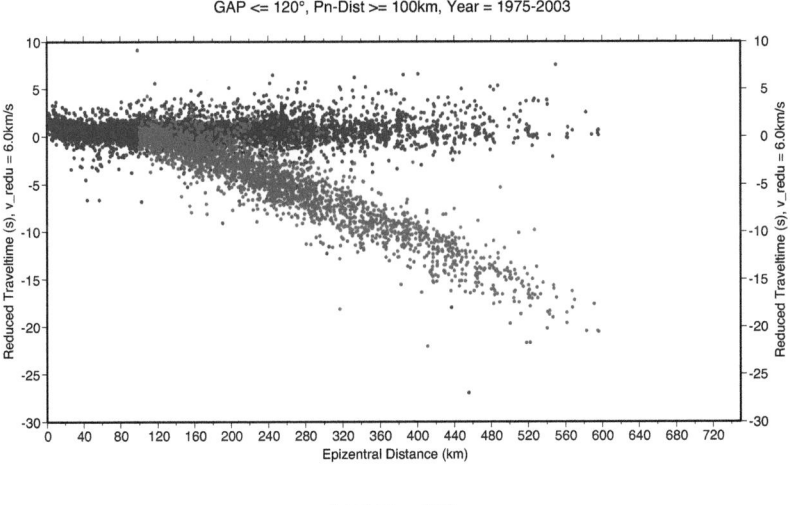

(b) GAP = 120°

Figure 4.4: Traveltime plots with reduction velocity at $6\,km/s$. Upper plot: Data with maximally allowed GAP of 180°. $P_g$ is plotted in blue, $P_n$ in red. Lower plot: Data with maximally allowed GAP of 120°.

Table 4.7: Several different NOBS and their influence on the data-fit. All computations are done with a GAP of 180°. The increase from 6 to 8 observations per event reduces the amount to nearly half, but only reduces the amount of phases by one sixth. $TSS_{SSH}$ is the square sum of the calculated residuals, $TSS_{hypo}$ the sum of the residuals from the data set. $RMS_{aver}$ is $(TSS_{SSH}/Nphase)^{1/2}$, $RMS_{unbias}$ is $(TSS_{SSH}/Nphase - Nmod)^{1/2}$, $RMS_{event(i)}$ is $\sum\left(\left(Res_{SSH}^2(i)/Nphase - 4\right)\right)^{1/2}$. $Nmod$ is the number of velocity parameters, $Nphase$ the number of recorded phases and $N_{event}$ the number of used seismic sources.

| NOBS | Number of event | Number of phases | $TSS_{SSH}$ $s^2$ | $RMS_{unbias}$ $s$ | $RMS_{aver}$ $s$ | $RMS_{event}$ $s$ | $TSS_{hypo}$ $s^2$ |
|---|---|---|---|---|---|---|---|
| 6 | 2014 | 27165 | 63049 | 1.8166 | 1.5235 | 1.2730 | 60187 |
| 8 | 1223 | 22136 | 58408 | 1.8406 | 1.6244 | 1.3782 | 55592 |
| 10 | 910 | 19519 | 52218 | 1.8137 | 1.6356 | 1.3855 | 49233 |
| 12 | 750 | 17845 | 48782 | 1.8130 | 1.6534 | 1.3839 | 46372 |
| 14 | 635 | 16424 | 45469 | 1.8099 | 1.6639 | 1.3827 | 43384 |
| 16 | 546 | 15133 | 42039 | 1.8021 | 1.6667 | 1.3722 | 40838 |
| 18 | 451 | 13569 | 38095 | 1.7998 | 1.6756 | 1.3730 | 36620 |

the amount to nearly half, but reduces the number of phases by only one sixth. For NOBS = 8, 1223 events with 22136 phases are left over, whereas these values decrease to 910 events with 19519 phases for NOBS = 10. As for the RMS values, they are about constant and $\approx 1.8$ for all NOBS tested.

Another point to consider is the ray coverage of the earth volume investigated. It strongly depends on the amount of events used, so less events lead to a poorer ray coverage. Considering both, the need for a certain overdeterminacy of the least-squares inversion problem and a reasonable ray coverage, a judicial choice for NOBS appears to be NOBS = 8, which is also the value to be used in the later SSH inversions. In Figure 4.5, the decrease of available data is visualized and the detailed results for different limits of NOBS are in Table 4.7.

### 4.4.3 Influence of the maximally allowed residuals

The traveltime residuals of the original regional earthquake dataset carry the information on the seismic structure of the crust and lithosphere, beyond the seismic velocity model used in the original HYPO event relocation process. On the other hand they represent errors, due to bad registrations, wrongly picked phases, incorrect analyzed waveforms and so on. As discussed earlier, it is *a priori* not easy to separate these two components in the residuals from each other. Thus, reducing the maximal allowed residual - annotated by the parameter RES in the following - below a certain limit may, in fact, reject useful information about the seismic structure, setting it too high will allow for errors.

A detailed list for several residual limits RES is provided in Tables 4.8 and 4.9. Further details about the effects of different RES can be seen from Figure 4.6, where the traveltimes are plotted for two different limits RES (5 s and 1 s). Compared to the impacts of two other restriction parameters discussed so far, i.e. NOBS and GAP, the residual limit appears to be a very efficient tool to eliminate wrongly assigned phases. The reason to chose a residual limit of 5 s is the beginning of an effective reduction of the data-fit, but also outliers are rejected and don't contribute to the inversion.

In Figure 4.7, the RMS values for the original dataset (red bars) and after the ray tracing ($TSS_{SSH}$, blue bars) are shown and the individual results are in Tables 4.8 and 4.9. The RMS for the original data becomes smaller together with the reduction of the maximally allowed residual, as expected. By ray tracing, using the standard input velocity model MKS 2004, a new set of residuals for each limit is computed. Because of the use of the average velocity model instead of local velocity models for the distinct data analysis, the RMS does not become less than the original one, the RMS even increases for small residual limits. This is because a lot of phases are rejected having a residual limit

## 4.4. TESTS OF THE IMPACTS OF THE VARIOUS SELECTION CRITERIA

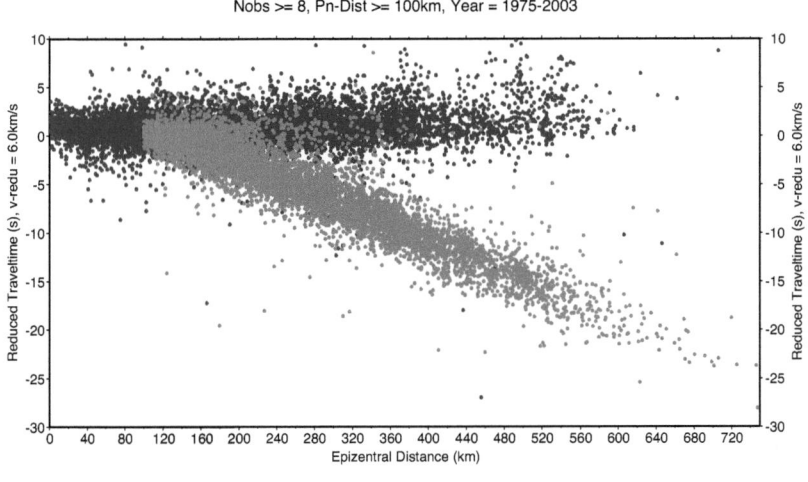

(a) NOBS = 8

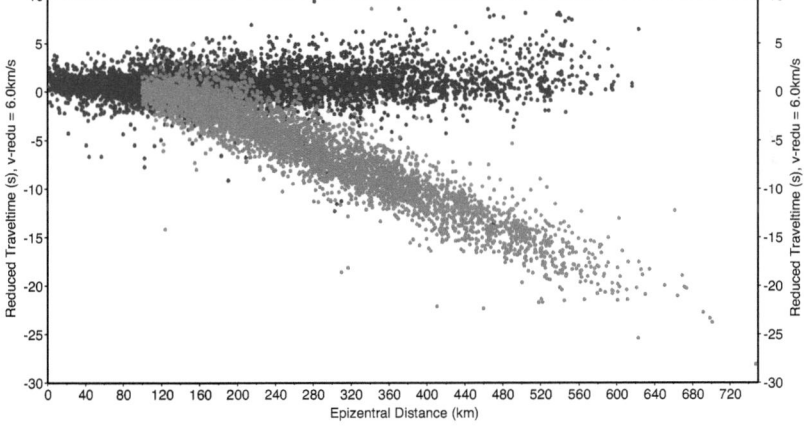

(b) NOBS = 16

Figure 4.5: Traveltime distance plots with different limits for the number of records per event. Mainly outliers are eliminated, but the influence is rare.
Figure 4.5(a): more than 8 observations per event,
Figure 4.5(b): more than 16 observations per event. $P_g$ is plotted in blue, $P_n$ in green.

Table 4.8: Data-statistics as a function of the upper residual limit RES, using NOBS $\leq 8$ and GAP $\leq 120°$. Remarkable is the improvement of the data-fit, as quantified by RMS, for residuals less than $10\,s$, which is caused by rejecting about 200, but probably erroneous observations.

| Residual $s$ | Number of events | Number of phases | $TSS_{SSH}$ $s^2$ | $RMS_{unbias}$ $s$ | $RMS_{aver}$ $s$ | $RMS_{event}$ $s$ | $TSS_{hypo}$ $s^2$ |
|---|---|---|---|---|---|---|---|
| 1  | 419 | 6984  | 19390.6 | 1.9120 | 1.6663 | 1.2421 | 1606.8 |
| 2  | 551 | 10282 | 16551.3 | 1.4318 | 1.2688 | 1.0235 | 6666.6 |
| 3  | 594 | 11352 | 16724.9 | 1.3653 | 1.2138 | 1.0651 | 10404.0 |
| 4  | 613 | 11686 | 17748.7 | 1.3867 | 1.2324 | 1.1160 | 12756.2 |
| 5  | 618 | 11804 | 17089.7 | 1.3535 | 1.2032 | 1.1458 | 14264.2 |
| 10 | 630 | 12023 | 20949.6 | 1.4851 | 1.3200 | 1.2513 | 18967.1 |
| 15 | 631 | 12036 | 21611.6 | 1.5076 | 1.3400 | 1.2622 | 19673.5 |
| 20 | 632 | 12046 | 22557.3 | 1.5398 | 1.3684 | 1.2842 | 20600.9 |
| 25 | 632 | 12048 | 23399.6 | 1.5681 | 1.3936 | 1.2956 | 21450.1 |
| 30 | 632 | 12049 | 24215.5 | 1.5951 | 1.4177 | 1.3045 | 22251.0 |

Table 4.9: Similar to Table 4.8, but for a GAP $\leq 180°$. The use of this value for GAP nearly doubles the amount of available events and recorded phases. For description of the parameters see Table 4.8

| Residual $s$ | Number of events | Number of phase | $TSS_{SSH}$ $s^2$ | $RMS_{unbias}$ $s$ | $RMS_{aver}$ $s$ | $RMS_{event}$ $s$ | $TSS_{hypo}$ $s^2$ |
|---|---|---|---|---|---|---|---|
| 1  | 827  | 12997 | 33862.4 | 1.8699 | 1.6141 | 1.1784 | 2984.2 |
| 2  | 1057 | 18639 | 28436.2 | 1.4049 | 1.2352 | 1.0406 | 12069.7 |
| 3  | 1148 | 20657 | 28752.8 | 1.3380 | 1.1798 | 1.0950 | 19725.3 |
| 4  | 1191 | 21450 | 32073.8 | 1.3866 | 1.2228 | 1.1683 | 25418.7 |
| 5  | 1204 | 21730 | 33835.6 | 1.4145 | 1.2478 | 1.2136 | 28545.5 |
| 10 | 1222 | 22121 | 40821.9 | 1.5393 | 1.3585 | 1.3095 | 37537.9 |
| 15 | 1222 | 22131 | 42223.6 | 1.5650 | 1.3813 | 1.3224 | 38970.1 |
| 20 | 1223 | 22143 | 43754.1 | 1.5928 | 1.4057 | 1.3367 | 40455.5 |
| 25 | 1224 | 22159 | 45746.5 | 1.6281 | 1.4368 | 1.3522 | 42456.8 |
| 30 | 1224 | 22160 | 46562.4 | 1.6425 | 1.4495 | 1.3568 | 43257.7 |

below $5\,s$, but the remaining ones are started to be anisotropically corrected. Below the residual limit of $3\,s$, almost all anisotropic phases are eliminated, and the correction changes the original residuals in a negative way. This might also be the influence of wrongly picked phases, where the original interpreter "corrected" the phase registrations to an isotropic one by introducing a small shift to the picked phase.

But over all we can see a minimum at about $3\,s$ which is representative for the best compromise between residual limit and number of records, eliminating the most of erroneous data. This rather nicely coincides with the results of Section 4.6, "Testing for Anisotropy", where the result is about the same limit, but now considering the available range for the residuals to include anisotropic effects[6].

---

[6]In fact, the variation of the traveltime between an isotropic and anisotropic medium with velocity variation of $\pm 2.5\%$ is about $\pm 1.75\,s$ at a distance of $600\,km$, giving the upper limit for the study area.

## 4.4. TESTS OF THE IMPACTS OF THE VARIOUS SELECTION CRITERIA

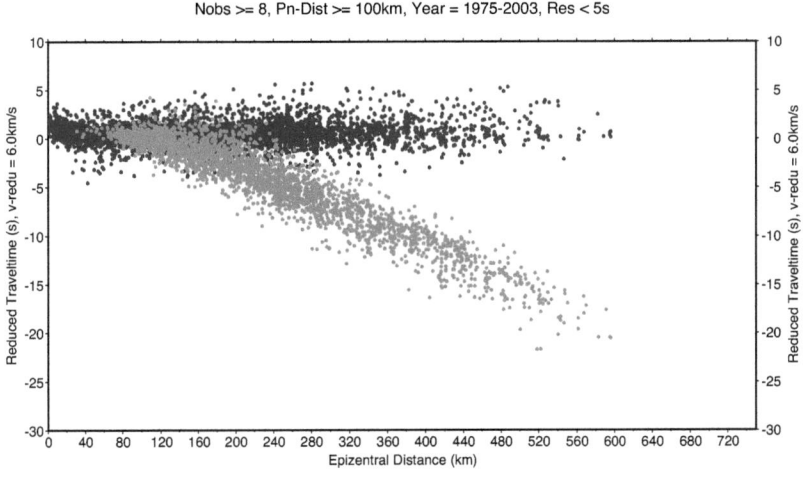

(a) Res = 5 s

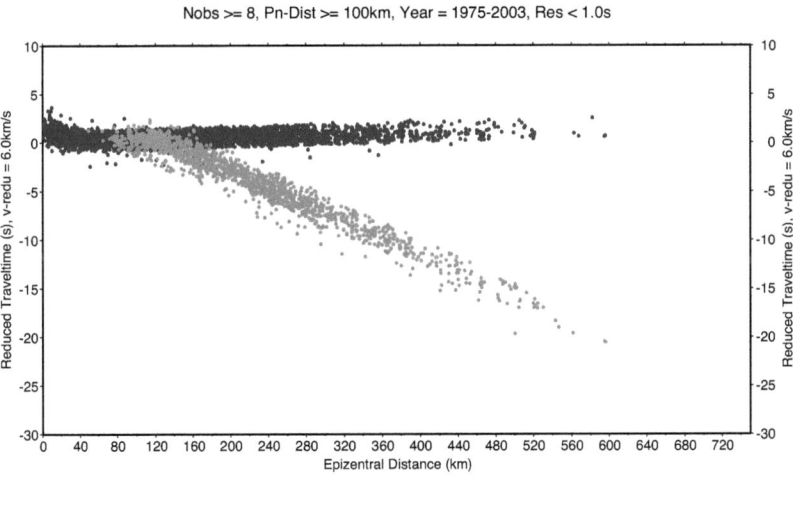

(b) Res = 1 s

Figure 4.6: Travel time distance plots for two different limits for the maximally allowed residual of $5\,s$ (a) and $1\,s$ (b). Limiting the range to $5\,s$ reduces the amount of misinterpreted phases. Going beyond $5\,s$ might reduce the available information for heterogeneity and anisotropy, as will be shown later. $P_g$ is plotted in blue, $P_n$ in yellow.

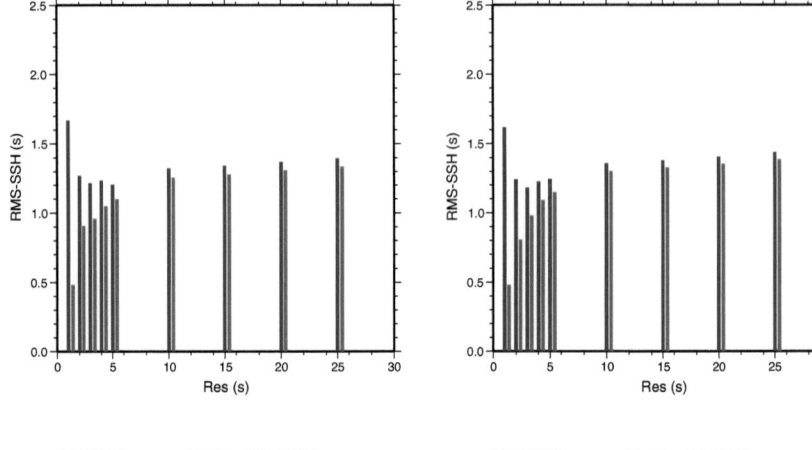

(a) RMS over residuals with G120  (b) RMS over residuals with G180

Figure 4.7: The RMS values (Root Mean Square of the residuals) for the original dataset (red bars) and after the ray-tracing (RMS$_{SSH}$, blue bars), sorted by different residual limits. The RMS for the original data becomes smaller together with the reduction of the maximally allowed residual, as expected. After the ray-tracing, using our standard input velocity model MKS 2004, the RMS does not become smaller, that is an effect of the anisotropic correction for $P_n$-phases. The minimum for the RMS is below $5\,s$ which is just the limit for the occurrence of anisotropy.

## 4.5 Testing for ambiguous $P_n$-phases

A special important parameter, $\Delta_{cut-off}$, is introduced to limit the epicentral distance where $P_n$-phases are included in the inversion computations. For an average depth of about $30\,km$ for the Moho, depending on the depth of the hypocenter, the cross-over distance for a $P_n$-phase is about 100 to $150\,km$. Thus, renaming all phases that are signed as $P_n$, but whose epicentral distances are below $\Delta_{cut-off}$ to $P_g$, should avoid the use of such possibly wrongly signed phases which could bias the inversion results. In the following the influence of the cross-over distance $\Delta_{cut-off}$ on the statistics of the data is tested for values 100, 120 and $150\,km$ (Table 4.10). One notices that, depending on the value of GAP, a slight minimum of the RMS can be identified in the epicentral range of 100 to $120\,km$. Obviously, setting $\Delta_{cut-off}$ to values higher than $150\,km$ would reject useful $P_n$-phases or wrongly rename them to $P_g$.

As it turns out, a more detailed analysis of the parameter $\Delta_{cut-off}$ has to await the full SSH inversion presented in the later chapters, when the hypocenters will be relocated simultaneously. There it will be shown that phases with epicentral distances smaller than $100\,km$ and signed as $P_n$ should not to be trusted and are renamed as $P_g$. This appears to have a very positive effect on the results of the tomographic inversion.

## 4.6 Testing for anisotropy

One simple way to check the input event data for signs of anisotropy is to plot the residuals of the $P_n$-phases in the raw data set against the azimuth before and after the ray tracing step with an anisotropic correction included, i.e. before any inversion is effected. This is done here for two different time ranges of the data-set, one between 1975 until 2003, and the range of the new data set between

## 4.6. TESTING FOR ANISOTROPY

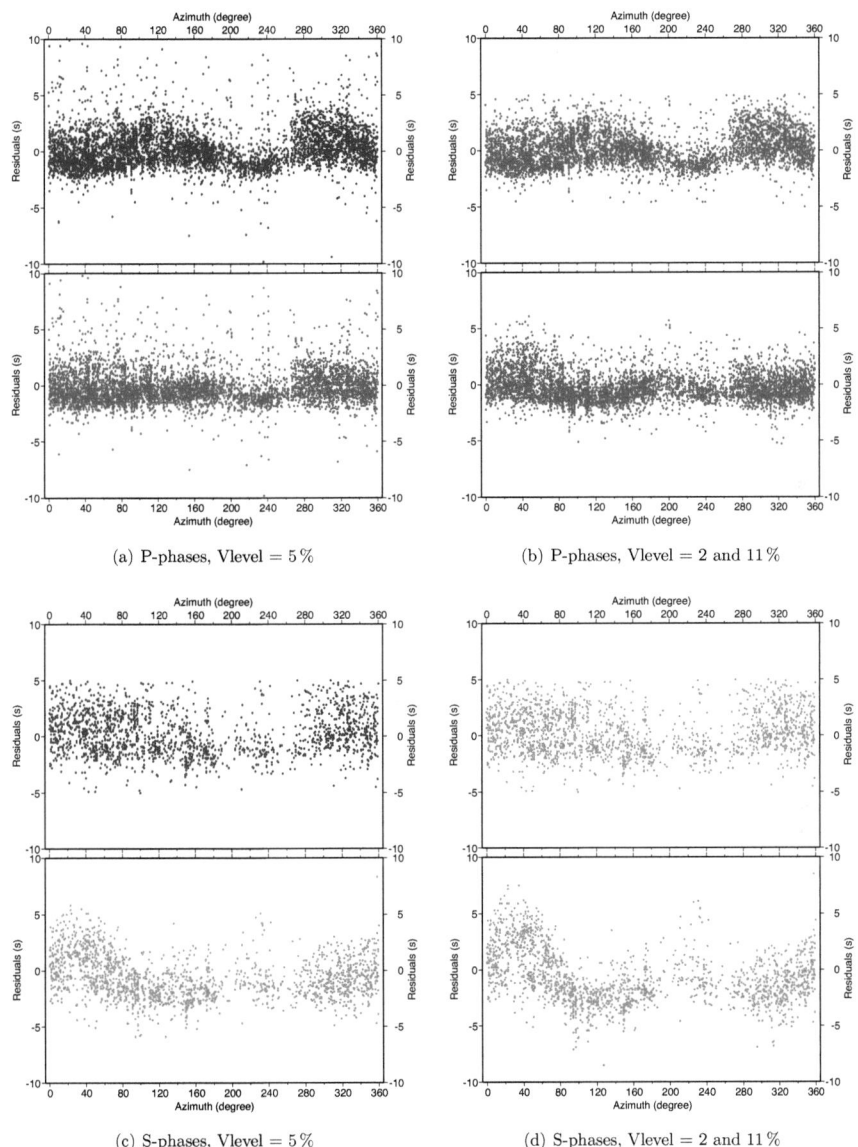

Figure 4.8: Original data (blue) with three different velocity levels of the anisotropic correction (2, 6 and 11%) for $P_n$ (top, in red) and $S_n$-phases (bottom, in green). The fast axis for the elliptical correction is at 35°. The effect of anisotropy for $S$-phases is less clear. GAP is at 180° and NOBS is less or equal 8.

Table 4.10: The increase of the data-fit, measured by TSS and RMS after the initial ray tracing in SSH with different values for $\Delta_{cut-off}$ and the two standard GAP limits at 120° and 180°. The strongest reduction of the TSS is between 100 and 120 $km$. Phases below 100 $km$, but signed as $P_n$ are not to be trusted. In this case, the phases are renamed as $P_g$. So, the minimum represents the best elimination for wrongly analyzed $P_n$-phases.

| GAP = 120° | \multicolumn{8}{c}{$\Delta_{cut-off}$ (km)} | | | | | | | |
|---|---|---|---|---|---|---|---|---|
| | 0 | 100 | 110 | 115 | 120 | 130 | 140 | 150 |
| $TSS_{SSH}$ ($s^2$) | 24257 | 24215 | 24178 | 24160 | 24159 | 24178 | 24332 | 24590 |
| $RMS_{unbi}$ ($s$) | 1.5965 | 1.5951 | 1.5939 | 1.5933 | 1.5933 | 1.5939 | 1.5989 | 1.6074 |
| $RMS_{aver}$ ($s$) | 1.4189 | 1.4177 | 1.4166 | 1.4160 | 1.4160 | 1.4166 | 1.4210 | 1.4286 |
| GAP = 180° | | | | | | | | |
| $TSS_{SSH}$ ($s^2$) | 58548 | 58408 | 58348 | 58323 | 58345 | 58356 | 58535 | 58886 |
| $RMS_{unbi}$ ($s$) | 1.8428 | 1.8406 | 1.8397 | 1.8393 | 1.8396 | 1.8398 | 1.8426 | 1.8482 |
| $RMS_{aver}$ ($s$) | 1.6263 | 1.6244 | 1.6235 | 1.6232 | 1.6235 | 1.6237 | 1.6261 | 1.6310 |

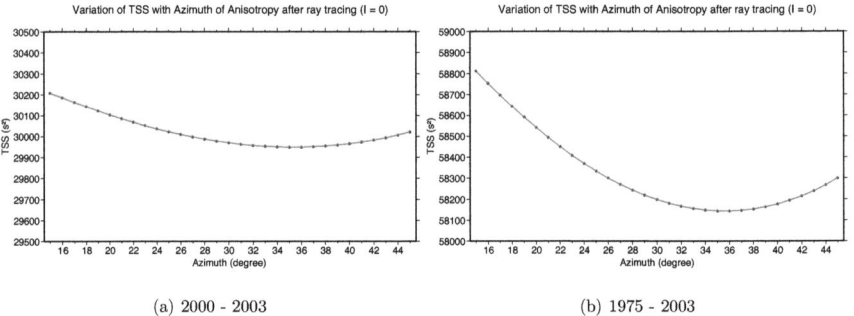

(a) 2000 - 2003          (b) 1975 - 2003

Figure 4.9: The reduction of TSS by choosing different values for the azimuth of the fast axis for anisotropic correction. The residuals are taken from the original dataset between the years 1975 to 2003 and tested for a second sub dataset with data between the years 2000 and 2003, to check the influence of changes in the recording techniques, which is negligible. Both have the minimum at about 35°.

## 4.6. TESTING FOR ANISOTROPY

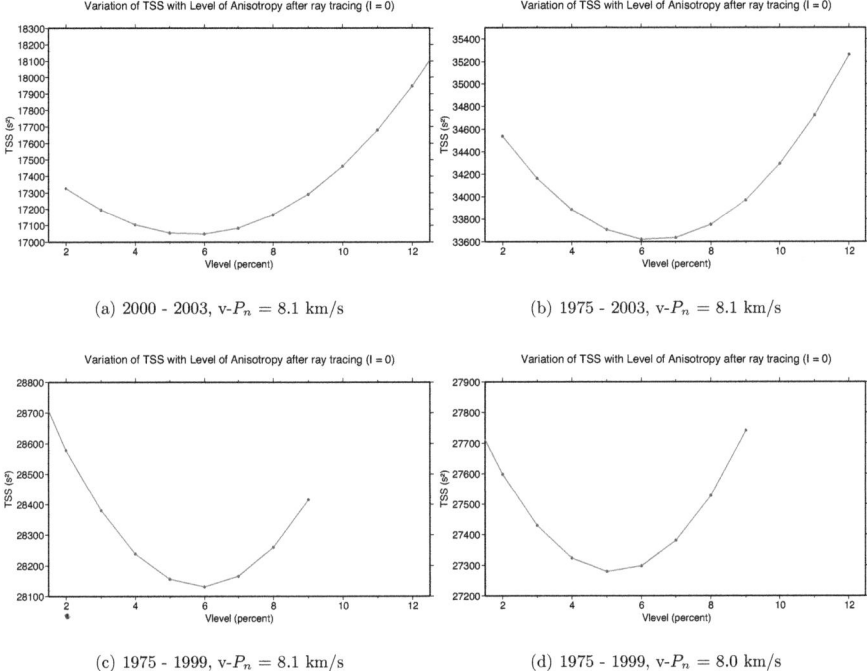

Figure 4.10: The TSS over the amplitudes of the anisotropic correction. The minimum and therefore the best solution for the original data is around 6%. The level of the velocity variation is tested for two different datasets, the complete one between 1975 to 2003, and the newer one between 2000 - 2003. The lower two are for data between 1975 and 1999, the first available data set. The difference there is the level of the $P_n$-velocity, which is at $8.1\,km/s$ at the left and the upper plots, and at $8.0\,km/s$ at the lower right plot. This results in a shift of the optimum for the percentage of anisotropic correction to 5%.

2000 and 2003, and two different residual limits, ($5\,s$ and $10\,s$) in Figure 4.8.

To completely include the elliptical correction for anisotropy, the azimuth and the amplitude of the velocity variation is necessary. This is included by the parameters ANI for the azimuth and VLEVEL, which gives the amplitude in $\pm 0.5$ VLEVEL % from the initial $P_n$-velocity. To test this parameter different levels for the anisotropic velocity variation between 2 and 11 % are tested (see Figures 4.10). There, the minimum for the velocity variation (depending on the initial $P_n$-velocity, see Figure 4.10) can be identified between 5 and 6 %, which results in the best compensation in the residual plots, which is the optimal result for this parameter to be used in most of the further studies.

Out of this follows, that first, the azimuthal anisotropy of the $P_n$-phases in the upper mantle underneath Germany has its fast axis at 35° (see Figure 4.9) in the raw data (which is rather different to the later results after a complete simultaneous inversion) and, secondly, the range of variation of the traveltime residuals is less than $\pm 10\,s$. Limiting the residuals below $5\,s$ would start to reject necessary information being important for the next steps in this study. Further reducing the residual limit, the anisotropic correction will have an inverse effect on the data and the result will become worse again. This variation is of major importance, because if one were to neglect the anisotropic effect, one might be taken to consider these residuals as pure arrivaltime errors, instead of carriers of useful and important information on the earth's structure. The raw residual scatter plot shows an undulation with a rough period of 180°, which is consistent with a first order anisotropy ellipse, namely, minima in the azimuthal range of 20° to 60° and 200° to 250° and maxima 90° apart. On the other hand, the amplitude of these periodic residual variations is reduced clearly, once the anisotropic correction has been applied.

As will be shown later, applying a full SSH inversion for structure and hypocenters together, the optimal value for the fast axis of the elliptical correction for anisotropy will be at azimuth 27° which is near the value of 25° found by Song et al. (2001a) and others (Fuchs, 1975, 1983; Enderle et al., 1996b,a).

In the plots, a slight shift upwards can be seen, which is caused by the used initial velocity model MKS2004 with $P_n$-velocity at $8.10\,km/s$, that is too fast to explain the observed arrivaltimes and results in average in positive residuals[7].

The same procedure can be applied to $S_n$-phases, but due to the smaller amount of these phases (1868 instead of 5671), the result is not as clear as for $P_n$-phases, so the effect of anisotropy can not be identified. Further, there are two components of the S-wave, ($S_V$ and $S_H$) which are influenced differently and have to be separated. The TSS shows minimal differences between the isotropic (TSS = 9138) and the anisotropic case (TSS = 9154). This small difference probably is the result of two different anisotropic influences with a phase shift.

The results clearly show again the importance of the anisotropic correction, which will result at most in different 1D velocity models and, in the case of the 3D tomography, in different structures. This has some influence onto the relocation of the hypocenters, that will differ from that of the isotropic relocation in some cases.

---

[7]The residual $\Delta$ t is calculated as $t_{obs} - t_{calc}$, the difference between the observed and the computed arrivaltime.

## 4.7 Summary of the data preprocessing analysis

As one can infer from the discussions of the previous sections, the most powerful tool to reduce the data variation or to remove possibly erroneous data is to limit the range of the traveltime residuals. In fact, most of the residuals are lying in the narrow range of $\pm 5\,s$ which, as Figure 4.8 shows, is also the range of the anisotropy traveltime effect. This figure also illustrates the interval for the fast axis of the $P_n$ anisotropy to be used later in the tomographic modeling.

The NOBS-criterion for the number of observations per event shows that NOBS $= 8$ is a good value as it is big enough for eliminating most of the earthquakes with poor initial relocations[8], while still providing at least 4 degrees of freedom (beyond the four hypocentral parameters) to be used for the seismic velocity determination. Setting NOBS to a higher value will not significantly reduce the spread of the data, but will cut down the number of seismic events too much, resulting in poorer resolution of the tomographic model.

The GAP criterion has some importance to stabilize the event relocation and to reduce the number of events with unstable determination by the arrivaltime recordings. For example, the worst case would be to have more than eight observations at stations that lie on a straight line along one direction from the seismic source. This might fulfill the NOBS criterion and the RES criterion, but not the GAP criterion. In fact, such a configuration will most likely result in an erroneous relocation, as there is a high degree of parallelism of the seismic rays in that direction, leading to collinearity and instability in the hypocentral system matrix. Also, the anisotropy traveltime effect, which is about $\pm 5\,s$ (see Figure 4.8), becomes more significant in such a case of a very small azimuthal ray coverage as such a large residual anomaly would project partly into a corresponding epicentral dislocation and partly into the origin-time of the event.

The results for the influences of the choice of the parameters GAP, RES and NOBS are shown in detail in Tables 4.7, 4.8, 4.9, 4.11 and 4.12. As the TSS directly depends on the number of the remaining data, after the individual selection criterion has been applied, it decreases steadily when the former become more and more restrictive, i.e. when increasing NOBS or decreasing GAP and RES.

Another measure than the TSS is the RMS (see Table 4.11), computed either as the RMS-average or as the RMS-unbiased, defined both in the caption of Table 4.2. Compared to TSS, the RMS is normalized to the amount of phases (records), so the different solutions can be judged independent of the amount of data. This supports some comparability, for example when using different subsets of the whole data set. For example, when increasing NOBS, the value of RMS-average increases (see Table 4.7), whereas TSS decreases. This is an unexpected and astonishing result, because a larger NOBS should improve the data-quality. Also the value of RMS-unbiased diminishes as expected but augments again in the gap-range of $120 - 150°$ when NOBS is chosen too high.

It is interesting to see that the RMS reacts rather strong by a decrease when the value of GAP is reduced. This indicates that the criterion GAP has much more influence on the appropriate selection of accurate data than the NOBS criterion. Obviously, a smaller GAP criterion effectively weeds out poorly relocated earthquakes with large original traveltime residuals.

As for the RES criterion, the table shows a clear response of the RMS. Reducing the latter rejects bigger residuals, i.e. the RMS gets lower until it reaches a minimum at about RES $= 5\,s$. As discussed earlier, this is also the threshold value of the anisotropic traveltime effect in Figure 4.8, thus, using a lower RES would in fact start to cut away this data information.

So, among the three criteria, the limits on RES, GAP and NOBS, the first one appears to be the most effective one to reduce the number of outliers. Notwithstanding, the three criteria are not independent of each other. A GAP value less or equal $180°$ comes together with a decreasing number of observations and a reduction of the residuals. Also, a small residual limit RES might correlate with a good interpretation of the recorded signals. But limiting it too much (lower than about $5\,s$) would

---
[8]Done for example with HYPO71 or HYPOINVERSE

Table 4.11: Left table: The data variation (RMS-average) depending on NOBS (number of observations per event) and GAP. The residual limit is at $90\,s$ for all to include almost all phases.
Right table: The data variation depending on RES (the maximally allowed residual per phase) and GAP, NOBS is eight for all.

| GAP | 360° | 180° | 150° | 120° |
|---|---|---|---|---|
| NOBS | RMS | RMS | RMS | RMS |
|  | s | s | s | s |
| 6  | -      | 1.5235 | 1.5157 | 1.3606 |
| 8  | 1.7986 | 1.6238 | 1.6271 | 1.4177 |
| 10 | 1.7706 | 1.6356 | 1.6305 | 1.4204 |
| 12 | 1.7671 | 1.6534 | 1.6615 | 1.4372 |
| 14 | 1.7338 | 1.6639 | 1.6885 | 1.4474 |
| 16 | 1.6959 | 1.6667 | 1.7016 | 1.4335 |

| GAP | 360° | 180° | 150° | 120° |
|---|---|---|---|---|
| RES | RMS | RMS | RMS | RMS |
| s | s | s | s | s |
| 1  | 1.9508 | 1.6158 | 1.5407 | 1.6663 |
| 2  | 1.3540 | 1.2376 | 1.2309 | 1.2688 |
| 3  | 1.2956 | 1.1792 | 1.1805 | 1.2138 |
| 4  | 1.3385 | 1.2212 | 1.2116 | 1.2324 |
| 5  | 1.4039 | 1.2407 | 1.2193 | 1.2032 |
| 10 | 1.4969 | 1.3561 | 1.3188 | 1.3200 |
| 15 | 1.5808 | 1.3789 | 1.3419 | 1.3400 |
| 20 | 1.6203 | 1.4034 | 1.3656 | 1.3684 |
| 25 | 1.6459 | 1.4346 | 1.3937 | 1.3936 |
| 30 | 1.6688 | 1.4473 | 1.4096 | 1.4177 |
| 90 | 1.7986 | 1.6238 | 1.6271 | 1.4177 |

Table 4.12: Left table: Data variation (RMS-unbiased) depending on NOBS (number of observations per event) and GAP.
Right table: Data variation depending on RES (the maximally allowed residual per phase) and GAP, where NOBS is chosen as eight for all cases.

| GAP | 360° | 180° | 150° | 120° |
|---|---|---|---|---|
| NOBS | RMS | RMS | RMS | RMS |
|  | s | s | s | s |
| 6  | -      | 1.8166 | 1.8044 | 1.5883 |
| 8  | 2.0628 | 1.8399 | 1.8405 | 1.5951 |
| 10 | 1.9800 | 1.8137 | 1.8049 | 1.5655 |
| 12 | 1.9468 | 1.8130 | 1.8205 | 1.5661 |
| 14 | 1.7339 | 1.8099 | 1.8366 | 1.5667 |
| 16 | 1.8378 | 1.8021 | 1.8397 | 1.5425 |

| GAP | 360° | 180° | 150° | 120° |
|---|---|---|---|---|
| RES | RMS | RMS | RMS | RMS |
| s | s | s | s | s |
| 1  | 1.9510 | 1.8718 | 1.5410 | 1.6667 |
| 2  | 1.3540 | 1.4077 | 1.2311 | 1.2690 |
| 3  | 1.2956 | 1.3373 | 1.1806 | 1.2140 |
| 4  | 1.3386 | 1.3848 | 1.2117 | 1.2326 |
| 5  | 1.4039 | 1.4065 | 1.2194 | 1.2034 |
| 10 | 1.4969 | 1.5366 | 1.3190 | 1.3202 |
| 15 | 1.5808 | 1.5624 | 1.3421 | 1.3402 |
| 20 | 1.6203 | 1.5902 | 1.3657 | 1.3687 |
| 25 | 1.6459 | 1.6255 | 1.3938 | 1.3939 |
| 30 | 1.6688 | 1.4475 ? | 1.4098 | 1.4179 |
| 90 | 2.0628 | 1.8399 | 1.8405 | 1.4179 |

## 4.7. SUMMARY OF THE DATA PREPROCESSING ANALYSIS

Table 4.13: Left table: The data variation (RMS-hypo) depending on NOBS (number of observations per event) and GAP.
Right table: The data variation depending on RES (the maximally allowed residual per phase) and GAP, NOBS is eight for all.

| GAP  | 360°   | 180°   | 150°   | 120°   |
|------|--------|--------|--------|--------|
| NOBS | RMS    | RMS    | RMS    | RMS    |
| 6    | -      | 1.4543 | 1.4860 | 1.3090 |
| 8    | 1.7838 | 1.5847 | 1.5932 | 1.3589 |
| 10   | 1.7479 | 1.5882 | 1.5888 | 1.3534 |
| 12   | 1.7432 | 1.6120 | 1.6181 | 1.3670 |
| 14   | 1.4994 | 1.6253 | 1.6384 | 1.3701 |
| 16   | 1.6925 | 1.6427 | 1.6568 | 1.3677 |

| GAP | 360°   | 180°   | 150°   | 120°   |
|-----|--------|--------|--------|--------|
| RES | RMS    | RMS    | RMS    | RMS    |
| 1   | 1.4002 | 0.4792 | 0.4751 | 0.4797 |
| 2   | 1.1778 | 0.8047 | 0.8040 | 0.8052 |
| 3   | 1.2312 | 0.9772 | 0.9682 | 0.9573 |
| 4   | 1.3157 | 1.0886 | 1.0731 | 1.0448 |
| 5   | 1.3877 | 1.1461 | 1.1276 | 1.0992 |
| 10  | 1.5228 | 1.3027 | 1.2690 | 1.2560 |
| 15  | 1.5943 | 1.3270 | 1.2750 | 1.2785 |
| 20  | 1.6252 | 1.3517 | 1.3180 | 1.3077 |
| 25  | 1.6383 | 1.3842 | 1.3472 | 1.3343 |
| 30  | 1.6472 | 1.3972 | 1.3634 | 1.3589 |
| 90  | 1.7838 | 1.5847 | 1.3480 | 1.3589 |

reject worthful information (effects of anisotropy) and may lead to instabilities in the inversion (see the computations with residuals about $1\,s$), with a subsequent increase of the RMS again.

Taking all things together, RES $\leq 5\,s$, GAP $\leq 180°$ and NOBS $\geq 8$ are chosen as basic values for the selection of an optimal data set in the tomographic SSH inversion computations in the following chapters. Also, despite the thorough analysis presented so far, these base values will still be adjusted slightly, in order to study their final effects on the statistical quality of the inversion models obtained.

# Chapter 5

# SHH-inversions for 1D vertically inhomogeneous velocity models

## 5.1 General approach of SSH

At the beginning of the simultaneous inversion for 3D seismic structure of the crust and upper mantle and hypocenter locations (SSH), one should search for a so-called optimal starting 1D velocity model. This starting model consists of a simple, vertically stratified, crustal model which fits the measured arrival data best. The procedure consists in finding a minimum of the RMS, the root mean square of the total sum of squared differences between measured and computed arrivaltimes (TSS), divided by the number of degrees of freedom $Nphase - p$ ($Nphase$ = number of recordings; $p$ = total number of model parameters), i.e. RMS = $\sqrt{TSS/(Nphase - p)}$. Doing so, one can assure that the later optimal 3D solution is a stable and optimal one which is a rather important issue, because we deal with a linearized inverse least-squares problem, with the possibility to end up with complete different solutions. This may happen, in particular, when the starting 1D model is chosen too far away from the optimal 1D velocity solution so that slightly different global minima of the final solutions might be obtained. This occurs, as will be shown later, when using the old SKKS 2001 and the MKS 2004 1D solutions as starting models. Whereas the former one always results in low $P_n$ velocities below $8.00\,km/s$, the latter one has the tendency to converge to higher $P_n$ velocities, while still providing lower, more reasonable velocities in the crustal layers.

The theoretical basis of the SSH inversion program of Koch (1993a), essentially allows three kinds of inversion computations, which will be effected in the present thesis:

- *Inversion for seismic velocities only (IFIX=1)*
  In this case - annotated henceforth by the parameter IFIX = 1 in the SSH-program - the inversion is done only for the 1D seismic velocity structure, whereas the hypocenters are fixed to their original locations as listed in the original event list of the BGR-catalogue. With this approach the computational burden is reduced tremendously, as the number of inversion parameters is lowered by $4 * N_{hyp}$ ($N_{hyp}$ = number of hypocenters). One then gets an average velocity distribution which represents the average of all velocity models used in the original determination of the hypocenters in question. This velocity model should give reasonable values for future regional hypocenter relocations, although it may be slightly different from local velocity models[1] used in part of the study regions.

- *Full inversion for seismic velocities and hypocenters (IFIX = 0)*
  The full inversion consists in the true simultaneous inversion for both velocity structure and hypocenters - annotated henceforth by the parameter IFIX = 0 in the SSH-program. Naturally,

---
[1]Because they result e.g. from solutions done by local institutions using local velocity models fitting the local structure better than a simple regional velocity model ever could

this - computationally very demanding - approach should lead to a better fit of the model to the data, as measured by the minimal RMS or TSS, because of the additional degrees of freedom provided by the numerous ($4 * N$, with $N$, the number of hypocenters) hypocentral parameters. The best solution for the 1D velocity model may then be used for simple earthquake location procedures, where the velocity model is fixed. In fact, compared to a full 3D seismic model, the use of a 1D seismic velocity model is often sufficiently accurate for hypocenter relocation, up to an error range of a few kilometers.

- Relocation of hypocenters with fixed velocity structure (IFIX = 2)
  For the earthquake location procedure, the velocity model is fixed, as previously mentioned. With sufficient accuracy, an optimal 1D vertical stratified velocity model, as it will be found in the present Chapter, can be used. This procedure is fast compared to the complete inversion process for structure and hypocenters, especially relocating few hypocenters. For a more precious relocation, a 3D structure model can be used, too.

Commonly used seismic velocity distributions in the crust and first sections of the upper mantle are based on PREM[2] and the more recent AK135 earth model, the latter listed in Table 5.1. The PREM model was computed in 1981 by Dziewonski and Anderson (1981) with a huge dataset of about 1000 normal modes, 500 registered traveltimes, 100 normal mode Q values and mass and moment of inertia. The AK135 model is the updated version of PREM, with small changes for some velocities to better fit the traveltime curves for the seismic rays through the earth.

Table 5.1: Seismic velocities for the upper kilometers in the earth from the AK135-model for traveltimes (continental structure). P for P-velocities, S for S-velocities respectively. AK135 is an updated version from IASP91 (Kennett, 2005; Kennett and Engdahl, 1991; Kennett, 1991).

|  | Depth $km$ | P $km/s$ | S $km/s$ | density $kg/dm^3$ | older density $kg/dm^3$ |
|---|---|---|---|---|---|
| Crust | 0.000 | 5.8000 | 3.4600 | 2.4490 | 2.7200 |
|  | 20.000 | 5.8000 | 3.4600 | 2.4490 | 2.7200 |
|  | 20.000 | 6.5000 | 3.8500 | 2.7142 | 2.9200 |
|  | 35.000 | 6.5000 | 3.8500 | 2.7142 | 2.9200 |
| Upper Mantle | 35.000 | 8.0400 | 4.4800 | 3.2976 | 3.3198 |
|  | 77.500 | 8.0450 | 4.4900 | 3.2994 | 3.3455 |
|  | 120.000 | 8.0500 | 4.5000 | 3.3013 | 3.3713 |

## 5.2 Control of important input parameters in the SSH-program

In the SSH-program of Koch (1993a) various input control parameters need to be set *a priori* which may have a strong effect in the results of the inversion. For example, in addition to the parameters GAP, NOBS, RES, and $\Delta_{cut-off}$, examined in the previous chapter, one has to find out the sequence of optimal damping factors $\lambda$ and the number of nonlinear iterations in the inversion. Further parameters relate to the $V_P/V_S$ ratio and anisotropy (Angle of the axis and amount of perturbation of the fast velocity) and will be discussed extra in later sections of this chapter. The same holds for the average depth of the Moho discontinuity which is varying over several km across the model region of Germany.

---

[2]Preliminary Reference Earth Model

## 5.2. CONTROL OF IMPORTANT INPUT PARAMETERS IN THE SSH-PROGRAM 71

- $V_P/V_S$ *ratio*
  This is an important parameter to be fixed when including $S$-phases in the inversion. The SSH-code allows the use of different values for two depth-sections of the velocity model to mimic the possibility of different Poisson-ratios $\sigma$ across the crust and the upper mantle.

- $\Delta_{cut-off}$
  As discussed, this parameter is to exclude wrongly associated $P_n$ phases beneath a certain critical distance $\Delta_{cut-off}$ which are then automatically renamed to a direct $P_g$ phase. The analysis of the previous chapter indicated that a value $\Delta_{cut-off} \approx 100\,km$ is a reasonable choice and has been used hence in the inversions.

- *GAP, NOBS and RES*
  As a consequence of the investigations in Chapter 4 the parameter GAP, the widest angle between two neighboring recording stations, is set to 180°; NOBS, the number of observations per event to 8, and RES, the upper limit for the allowed residuals to $10\,s$. As these values still turn out not to be restrictive enough, they are changed in some of the computations to more stringent values, GAP = 120°, NOBS = 10 and RES = $5\,s$ or even RES = $2\,s$, in the case of which significant (unrealistic) changes in the inverted velocity model, "low velocity layer" in the lower crust arise (see Section 4.4.3, for details).

- *Constraints on hypocentral depth HYPIMP and velocity variations VELIMP*
  To constrain the maximal variations in the SSH-solution, i.e. the resulting hypocentral and velocity perturbations, the parameters HYPIMP and VELIMP are available. HYPIMP sets the limit for the depth shift in each iteration to avoid hypocentral depth instabilities which often plague hypocenter determination. VELIMP limits the velocity perturbations to avoid unreasonable high values. Both HYPIMP and VELIMP act especially at the beginning of the damping (regularization) sequence, when the solutions are still more or less undamped, whereas with increasing damping the solution perturbations are usually constrained naturally below these limiters.

- *Anisotropy parameters ANI and VLEVEL*
  The azimuthal anisotropy ellipse, used for the anisotropic correction of the seismic traveltimes is specified by its azimuthal angle ANI and its relative difference in the major and minor velocity axis, VLEVEL = $(v_{max} - v_{min})/(v_{max} + v_{min})/4 * 100$ (in percent).

- *Number of external iterations ITM and interior damping sequences NEPILO*
  As discussed in Chapter 2, Theory, within one linearized (external) inversion step ($IT \leq ITM$) of the nonlinear least-squares (Levenberg-Marquardt) procedure the system (Fréchet) matrix is inverted several NEPILO times with different values of the damping factor to stabilize the solution perturbations, thus preventing overshooting beyond the minimum of the objective function (calculated by nonlinear ray tracing). After finishing the interior damping sequence and extracting the solution with the local minimum, a new external inversion sequence IT = IT + 1 is carried out using an updated Fréchet matrix **G**, by tracing rays through the optimal model of the previous external iteration. More specifically, the interior damping sequence is controlled by two parameters, EPANF and EPFAK. EPANF denotes the starting damping factor $\lambda$ and EPILON the multiplier to calculate the damping factor for the next interior iteration step. Thus $\lambda$ increases in form of a geometric series throughout the NEPILO steps. For most of the computations, it has been sufficient to set the number of interior damping iterations to NEPILO = 9, EPANF = 0.1 or 0.2, and EPFAK = 2, and the number of external iterations to ITM = 4, beyond which no further reduction of the misfit function could be achieved. This minimum should theoretically lie above the range of the statistical noise of the traveltime data.

## 5.3 Analysis of the azimuthal $P_n$-anisotropy

The two parameters (Azimuth and velocity perturbation between slow and fast axis) of the azimuthal ellipse, describing the $P_n$-anisotropy of the upper mantle, are not the explicitly computed in the SSH program. They have to be known *a priori* in the anisotropic SSH inversions. Thus, before proceeding to the SSH inversion for a 1D seismic velocity model, these anisotropy parameters, the angle of the major (fast) velocity axis ANI and the anisotropy contrast VLEVEL have to be controlled.

### 5.3.1 Previous evidence of $P_n$-anisotropy across Germany

As discussed in Chapter 3, for Germany and much of central Europe, starting with the work of (Bamford, 1973) who used a modified time-term analysis of $P_n$ first arrivals from refraction experiments, he and, later on, other seismologists (Fuchs, 1983; Enderle et al., 1996b,a; Babuška and Plomerová, 2000; Song et al., 2001a, 2004; Kummerow, 2004) revealed azimuthal anisotropy of the seismic $P_n$ velocity with a contrast of 5 - 6 % and the fast velocity axis oriented in $N23°E - N27°E$ direction. While most of these findings have been obtained from numerous refraction experiments criss-crossing the county (see below) the more recent $P_n$-time term inversions of Song et al. (2001a, 2004) are based on a subset of the regional earthquake traveltimes also used in the present thesis. Moreover, the more refined 2D laterally inhomogeneous $P_n$-velocity analysis of Song et al. (2004) indicated that the anisotropy ellipses are remarkably invariant across the upper mantle underneath Germany and can be replaced by one large anisotropy ellipse with the parameters as stated, namely, angle ANI of the fast axis at $N25°E \pm 3°$ and anisotropy contrast VLEVEL = $\pm 2.5\%$.

### 5.3.2 Control of the $P_n$-anisotropy ellipse

To test the anisotropic correction and the positive influence onto the data-fit, the fast axis is changed in steps of 5° degree in the whole azimuthal range between 0° and 360°. The anisotropic correction in the SSH program, at the moment, is only included for phases, traveling along the boundary between the Crust and the Upper Mantle, the Mohorovicic discontinuity, or shortly known as Moho ($P_n$-phases or Head waves). The correction for anisotropy is done block-wise, depending on the actual angle at which the ray is traveling through the individual block. Doing so, the optimal value, which is suitable for the data set, used in this inversion procedure, can be resolved.

For the computations, the former computed 1D velocity model SKKS2001 (Song et al., 2001b) is used. At this stage of the search for the fast axis of anisotropy, the influence of the velocity model is of minor importance, so this is valid. Using different initial 1D models give the same results. Refining the axis could be done later with the optimal model, but will be unnecessary because the accuracy of estimating the angle of the fast axis within some degrees is sufficient enough.

The result in Figure 5.1(b) is a sinusoidal distribution of the RMS-values, where the minima are at about 30° and 210° azimuth. This makes the range for the fast axis better visible than the analysis of the residual plot 5.1(a). In the following, some more computations are done for a narrow area, zooming into the range of the optimal RMS values in Figure 5.1. This is done for two cases, the first one with hypocenters fixed (Figures 5.2(a) and 5.2(b)), the second one with hypocenters free (Figures 5.2(c) and 5.2(d)). For each case, two non linear inversions are done to clarify the computations. Especially in the case with free hypocenters, the situation is less clear, because the anisotropic correction might be compensated by a new arrangement of the hypocentral parameters. Therefore, the line is somewhat flatter, the minimum becomes more representative after the second non-linear inversion.

**Results:** The range for the fast axis for V-$P_n$-max is at azimuth between 25° and 35°, where the influence of fixing the hypocenters or performing a simultaneous inversion onto the results can be clearly separated. For the two cases (hypocenters fixed and hypocenters free), a strong difference in the azimuth is the result, the minimum for fixed hypocenters is about 35°, which is much different

## 5.3. ANALYSIS OF THE AZIMUTHAL $P_N$-ANISOTROPY

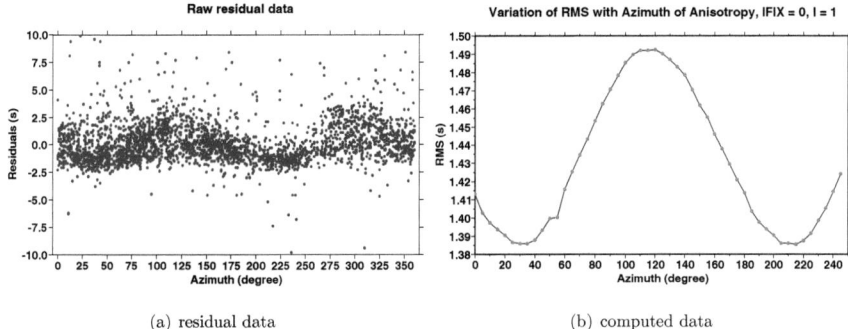

(a) residual data

(b) computed data

Figure 5.1: Distribution of TSS for simultaneous inversions with changes of the fast axis in steps of 5° and the original residual data. The typical sinusoidal curve for elliptical correction for $P_n$-phases appears.

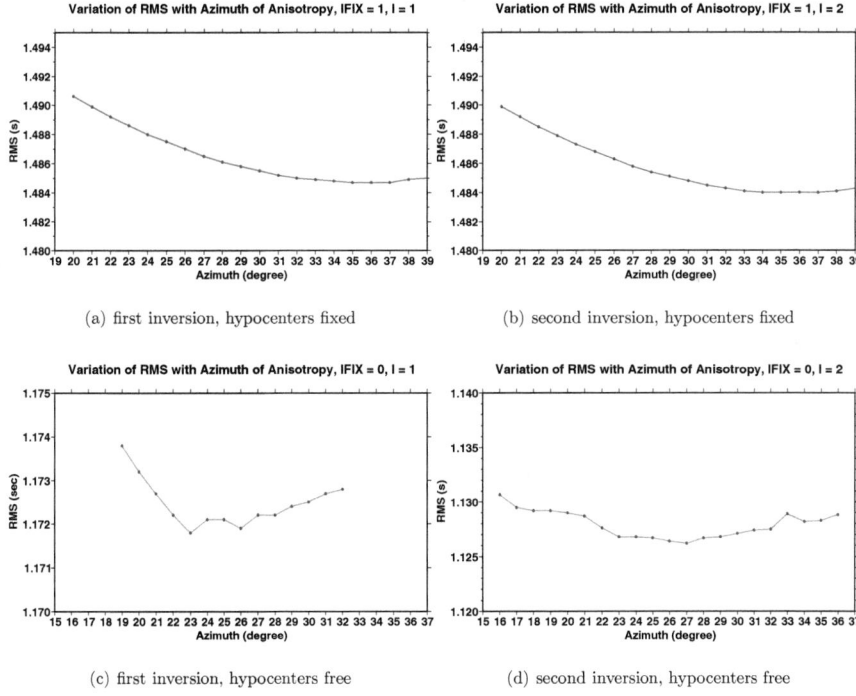

(a) first inversion, hypocenters fixed

(b) second inversion, hypocenters fixed

(c) first inversion, hypocenters free

(d) second inversion, hypocenters free

Figure 5.2: The distribution of RMS values for changes in the azimuth of the fast axis at 5°-steps in a more narrow range to refine the previous investigations, hypocenters fixed (upper plots) and free (lower plots). Figures 5.2(a) and 5.2(c): Optimal result after the first damping sequence. Figures 5.2(b) and 5.2(d): Optimal result after the second damping sequence.

from other results, and at about 27° for free hypocenters (complete inversion), that is confirmed by other results. This might be the effect of some biasing in the hypocenter location in the data set.

The result with free hypocenters, i.e. doing a complete simultaneous inversion, nicely coincides with the results of Enderle et al. (1996a), who had found nearly the same value for the fast axis of anisotropy for $P_n$-phases in Germany, while interpreting data of several refraction profiling experiments done in South-Western Germany and the results of Song et al. (2001a,b).

## 5.4 Finding the optimal average depth of the Moho

One of the constraints of the SSH inversion code of Koch and Kalata (1992) is that the depths and thicknesses of the layers of a 1D velocity model have to be fixed *a priori* and these are kept invariant during the iterative inversion process. Moreover, as the layer interfaces are to be chosen as horizontal, local variations of the depth of a seismic discontinuity in the crust, they cannot be considered appropriately in the SSH-inversion, which might lead to some bias and ambiguity in the inverted velocity model. While this deficiency of the SSH-code might be partly circumvented in the crust by using more and finer layers there, it is more relevant at the crust-mantle boundary (Moho), where refracted $P_n$-phases arise whose traveltimes are strongly affected by the depth of the former. For example a deeper lying Moho in the theoretical velocity model leads to longer geometric ray paths for the modeled $P_n$ traveltimes which the SSH-program attempts to compensate by increasing the $P_n$-velocity. In the case that the Moho boundary is inclined in a certain direction, as it is the situation in sections of the crust underneath Virginia, Koch and Kalata (1992) have incorporated there a traveltime correction into the SSH procedure for the up- and downhill traveling $P_n$-rays.

### 5.4.1 Information on the Moho depths across Germany

To get some preliminary information about the crust mantle boundary underneath Germany, several seismic investigations are presented in the next sections. These are data from surface wave investigations (section 5.4.1.1) and some results from refraction seismic experiments (sections 5.4.1.2 and 5.4.1.3). A third one, the Eifel Plume tomographic experiment (5.4.1.4), and the related receiver function studies gave another important information about the crust mantle boundary.

#### 5.4.1.1 Moho depths based on the global CRUST 2.0 model

By inverting surface wave-, crustal seismic- and sedimentary data collected up to the year 1998, the crustal thickness has determined all over the world (Bassin et al., 2000; Mooney et al., 1998). From this exhaustive analysis the CRUST 5.1 model, defined on a 5° × 5° grid was established and, later on, was refined to the 2° × 2° CRUST 2.0 model which incorporates 1° × 1° model for the global sediment thickness as well as for the crustal thickness (Moho depth) Bassin et al. (2000); Mooney et al. (1998); Tsoulis (2004). It is available at
http://mahi.ucsd.edu/Gabi/rem.dir/crust/crust2.html

In Figure 5.3(a), the Moho depths across Germany as taken from the raw data of the CRUST 2.0 model of Bassin et al. (2000) are plotted on a 1° × 1° grid. For better visibility, an interpolated Moho map has been created which is shown in Figure 5.3(b). This map clearly shows that for most of the study region the Moho depths are located between 26 and 32 $km$, with a slight dip-down across southern Germany toward the Alpine region, where the Moho goes down to about 40 $km$. In the Rhinegraben, the average Moho depths are more shallow, going up to 26 $km$ around the Kaiserstuhl volcano, being a lithospheric finger-print of the asthenospheric upwelling that occurred during the Miocene (see Koch (1993c), for more details).

In any case, an average Moho depth of 30 $km$ for most of Germany appears to be reasonable and this value is being used in the set up of most of the velocity models in the SSH-inversion. As the Moho map shows, major deviations from this value arise only at the southern (German-Austrian) border of the

## 5.4. FINDING THE OPTIMAL AVERAGE DEPTH OF THE MOHO

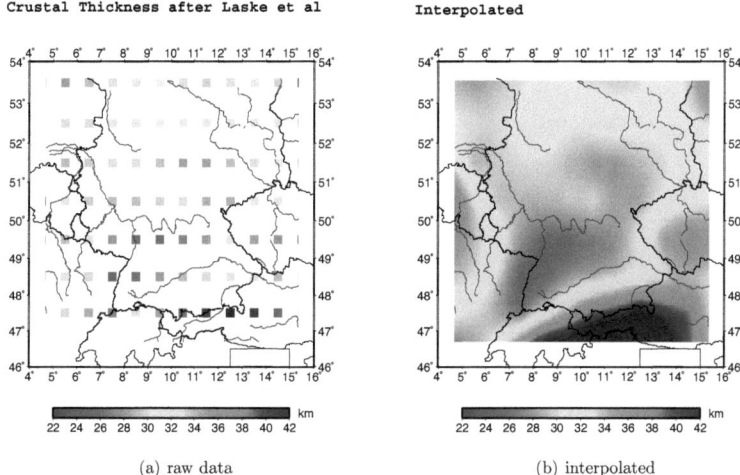

Figure 5.3: Figure (a): The Moho depth underneath Germany and adjacent areas taken from the raw data of the CRUST 2.0 model of Laske et al. (2000). Figure (b): Interpolated Moho depths using the raw data of (a).

model region which, fortunately, has only a minor impact on the reliability of the SSH inversion. In fact, since $P_n$-phases are observed only beyond an epicentral distance of about $100\,km$, the effectively $P_n$-resolved model region is smaller by about $100\,km$ from each border of the Moho-map shown. With this border-trimming the true SSH model region covers $7.0°$ and $13.0°$ Longitude and $47.5°$ and $51.5°$ Latitude and one obtains, indeed, a calculated mean for the Moho depth of $29.7\,km$, with a standard deviation $\sigma = 1.86\,km$. As will be shown later in Section 5.4.2, a depth value between 29 and $30\,km$ results also in the best data fit for the inverted velocity models.

### 5.4.1.2 Moho depths from the GRANU95 refraction study

In 1995, Enderle et al. (Enderle, 1998) performed a 2D refraction experiment, called GRANU95[3] whose profile extended along a line between the cities Dresden and Bamberg, crossing the Saxonia.

Figures 5.5 and 5.6 show a few velocity models some derived of the various profiles drawn in Figure 5.4. Again, within a range of $\pm 1\,km$, the average depth of the Moho discontinuity deduced from these studies is again about $30\,km$, with the upper mantle $P_n$-velocities ranging between 7.9 and $8.0\,km$ and showing the anisotropy property as discussed in the previous sections.

Other results of the reinterpretation of these seismic experiments by Enderle (1998) are, (1) a crystalline bedrock in the $2.5 - 3.5\,km$ depth range, (2) a zone of low seismic velocity (LVZ) in the $8-15\,km$ depth range and, (3) an intra-crustal seismic reflector at about $20\,km$ depth. These results will be important for the later interpretation of the results of the 3D computations.

While these findings (for more details see also Enderle et al. (1996a)) will be compared later in Section 5.4.2 with those obtained from the 1D- and 3D tomographic inversions, at this stage they already provide some preliminary information about the initial set-up of the velocity model in the SSH inversion code.

---

[3]The name GRANU comes from the name Granulitgebirge

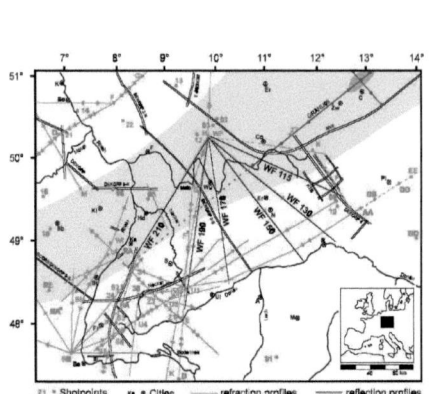
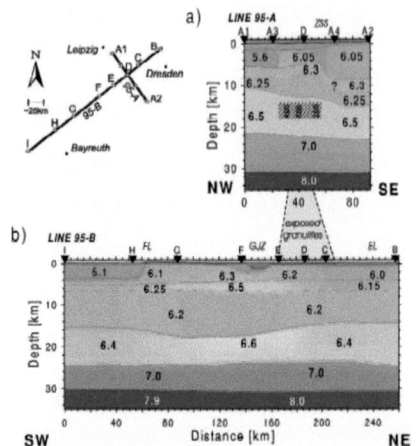

Figure 5.4: Historical refraction profiles used in the reinterpretation analysis of Enderle (1998). The green profiles pertain to the 1982 Wildflecken experiment, and the red line indicates the GRANU95 experiment.

Figure 5.5: 2D-velocity model sections along the two GRANU95 profiles (Enderle, 1998).

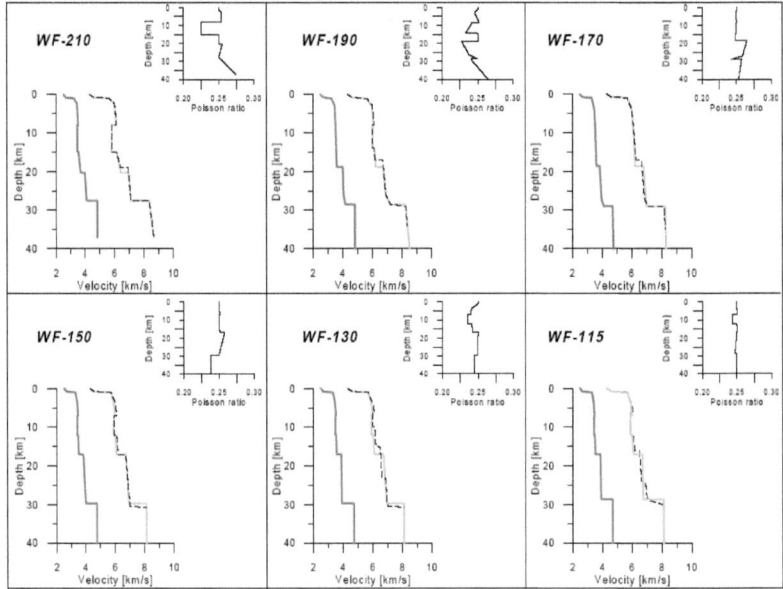

Figure 5.6: 1D velocity models derived from historical refraction seismic investigations, partly reinterpreted by Enderle (1998).

## 5.4. FINDING THE OPTIMAL AVERAGE DEPTH OF THE MOHO

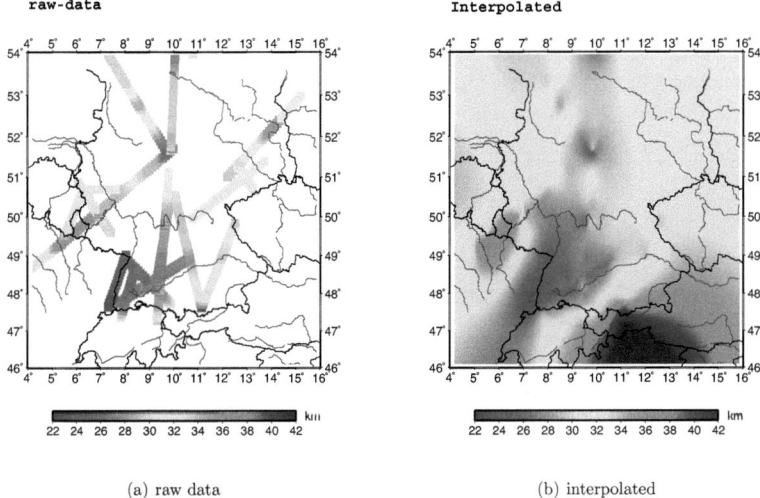

(a) raw data     (b) interpolated

Figure 5.7: Figure (a): Moho depths collected from the raw refraction profile data of J. Mechie (personal communication). Figure (b): Interpolated and smoothed Moho depths of the raw profile data.

### 5.4.1.3 Compilation of Moho depths of J. Mechie

Another informative collection of Moho depths data is that of J. Mechie (personal communication) who compiled a Moho map using results from numerous seismic refraction profiles across Germany, including also the GRANU95 experiment. In Figure 5.7(a), the raw data of Mechie are plotted and, for better visibility, interpolated again in Figure 5.7(b). These maps show similar depths for the Moho as those of the CRUST 2.0 model, namely, a depth of 22 $km$ in the Black Forest region and the upper Rhine Graben area but increases to about 30 $km$ in Northern Germany and to about 40 $km$ in the Southern Alpine region. As expected, compared to the Crust 2.0 Moho model, the Mechie-model is more detailed showing, for example, the sharp rise of the Moho in the Kaiserstuhl region (indicated as a dark red spot in Figure 5.7(b)), where volcanic activity took place during the Miocene epoch. Overall, the differences between the CRUST 2.0 and the Mechie model are about ±1 $km$, especially in the northwestern area of Germany, where the Moho is dipping a little more in the Mechie model than in the CRUST 2.0 model. Due to a lack of information, the Moho in the South is determined by only one blue dot in the southern profile (Figure 5.7(b)). So, the Moho depths South of 47.5° latitude are not well defined. Notwithstanding, the average depth of the Moho discontinuity in the Mechie model turns out to be 29.8 $km$ with a standard deviation $\sigma = 1.4\,km$.

### 5.4.1.4 Moho depths from the Eifel tomographic experiment

The Eifel tomographic experiment was set up during November 1997 to June 1998 to probe for the famous Eifel plume. This experiment delivered a huge and dense arrivaltime dataset from teleseismic earthquakes recorded at mobile stations across the Eifel. From these data, the local average Moho depth was derived. Here it is worth to mention the results of Budweg (2003), who tested P-S conversions using receiver function analysis. Using this technique, the average depth of the Moho underneath the Eifel was shown to be around 30 $km$ depth with, surprisingly, only little vertical variations.

Table 5.2: Variations of the $V_P$-velocity as a result of varying the Moho depths between 27 and 33 km using the datasets between 1975 - 2003 and 1997 - 2003, V-mod-in SKKS2001 and the difference between the correction for $P_n$-phases with $\Delta_{cut-off} = 100\,km$ can be seen. A detailed list of the computations for the different Moho depth is in Appendix A.

| Moho km | V1 km/s | V2 km/s | V3 km/s | V4 km/s | RMS s | $\Delta_{cut-off}$ km | data -set |
|---|---|---|---|---|---|---|---|
| Layer | L1 | L2 | L3 | L4 | | | |
| Depth | 0-10 km | 10-20 km | 20-30 km | ≥30 km | | | |
| 27 | 5.93 | 6.07 | 5.91 | 7.94 | 1.2866 | 0 | 1975 - 2003 |
| 27 | 5.90 | 6.07 | 6.05 | 7.94 | 1.3269 | 0 | 1997 - 2003 |
| 29 | 5.90 | 6.04 | 6.09 | 7.95 | 1.2876 | 100 | 1975 - 2003 |
| 29 | 5.91 | 6.05 | 6.14 | 7.97 | 1.3168 | 100 | 1997 - 2003 |
| Layer | L1 | L2 | L3 | L4 | | | |
| Depth | 0 - 10 km | 10 - 20 km | x - 10 km | x km | | | |
| 29 | 5.91 | 6.03 | 6.10 | 7.96 | 1.2838 | 100 | 1975 - 2003 |
| 29 | 5.93 | 6.05 | 6.05 | 7.98 | 1.2824 | 100 | 1997 - 2003 |

### 5.4.2 SSH-control of the Moho depth

To test the influence of the depth of the Moho discontinuity on the results of the 1D SSH inversion, several computations with various initial Moho depths (ranging between 27 km and 34 km) are done, using the 1D starting model SKKS2001, a gap limit of 180 degrees and two different periods of registered data (Years between 1997 to 2003 and 1975 to 2003, respectively). All the computations, which are listed in detail in Appendix A, are done with anisotropic correction at 27° with ±5% velocity variation. The idea now is to choose the solution with the best data fit (lowest RMS) to represent the optimal average for the depth of the Moho. The result will not be only a test for the best average depth of the Moho to be used in further computations, but also to find a first approximation to the optimal 1D $P_n$ velocity model, which will be refined and tested in following sections.

The results of these tests are listed in Table 5.3, together with the final optimal solutions for the velocity model. Two sets of computations were done: One without setting the limit $\Delta_{cut-off}$ for $P_n$-phases and the other one with renaming all phases below $\Delta_{cut-off} = 100\,km$ to $P_g$ (see Chapter 4, Section 4.5). The later procedure became necessary, because quite a lot of $P_n$-denoted phases are below that distance[4], which is impossible for the 1D velocity model used. The tables show that, whereas the best result for the Moho depth without the $\Delta_{cut-off}$ correction for the $P_n$-phases is 27 km, using the former, this depth becomes 29 km, which is also more in line with the results obtained from the other seismic experiments discussed before. More so, the velocity values are also more realistic[5] (cf. Koch, (Koch and Kalata, 1992; Koch, 1993a; Song et al., 2001a,b; Enderle et al., 1996b,a; Enderle, 1998) than those computed without eliminating the wrongly picked $P_n$-phases.

Also, whereas the velocities in layers L1 and L2 are rather stable, those in layers L3 and L4 are less, where an increase of the $P$ velocity goes hand in hand with a decrease of the $P_n$ velocity, respectively. For the models without the $\Delta_{cut-off}$ correction there occurs a "low-velocity-zone", which is not realistic and disappears when the $P_n$-cut-off is used. The deeper the Moho, the faster these layer velocities become. Obviously this is the influence of a longer travel path in layer L3 which is compensated by faster velocities to explain the arrivaltimes. Thus it is appropriate to test the influence of the thickness of the third layer. To this avail, additional computations are done by changing the

---

[4]In the study of Song et al. (2001b) where reduced traveltime plots were qualitatively analyzed the limit for $P_n$-phases to occur was set at 150 km distance

[5]More realistic means that they are more like those of the other seismic experiments and correlate better with the expected geological and geophysical rock properties

## 5.4. FINDING THE OPTIMAL AVERAGE DEPTH OF THE MOHO

Table 5.3: The optimal results of the different model computations. The importance of the $\Delta_{cut-off}$ parameter to allow for $P_n$-phases only beyond a critical distance $\Delta = \Delta_{cut-off}$ becomes obvious.

| Layer | L1 | L2 | L3 | L4 | | | | | | |
|---|---|---|---|---|---|---|---|---|---|---|
| Depth | 0-10 km | 10-20 km | 20-30 km | ≥30 km | | | | | | |
| Moho km | V1 km/s | V2 km/s | V3 km/s | V4 km/s | RMS s | GAP deg | Year a | Δ km | V-in | L3 |
| *(1) Models without $\Delta_{cut-off}$ for $P_n$-phases.* | | | | | | | | | | |
| 27 | 5.93 | 6.07 | 5.91 | 7.94 | 1.2866 | 180 | 1975-2003 | 0 | SKKS2001 | const |
| 27 | 5.90 | 6.07 | 6.05 | 7.94 | 1.3269 | 180 | 1997-2003 | 0 | SKKS2001 | const |
| 27 | 5.93 | 6.075 | 5.975 | 7.945 | | | < average | | | |
| *(2) Models with $\Delta_{cut-off}$ at 100 km for $P_n$-phases. Below they are renamed to $P_g$* | | | | | | | | | | |
| 29 | 5.91 | 6.05 | 6.14 | 7.97 | 1.3168 | 180 | 1997-2003 | 100 | SKKS2001 | const |
| 29 | 5.90 | 6.04 | 6.09 | 7.95 | 1.2876 | 180 | 1975-2003 | 100 | SKKS2001 | const |
| 29 | 5.92 | 6.05 | 6.11 | 7.96 | | | < average | | | |
| *(3) Models with $\Delta_{cut-off}$ at 100 km for $P_n$-phases. L3 is changed parallel with L4 (Moho)* | | | | | | | | | | |
| 29 | 5.91 | 6.03 | 6.10 | 7.96 | 1.2838 | 180 | 1975-2003 | 100 | SKKS2001 | var |
| 29 | 5.93 | 6.05 | 6.05 | 7.98 | 1.2824 | 180 | 1997-2003 | 100 | SKKS2001 | var |
| 29 | 5.92 | 6.04 | 6.075 | 7.97 | | | < average | | | |

depths of the boundary between layer L2 and L3 and L3 and L4 in parallel, so that the thickness of L3 stays constant.

The variation of the velocity V3 in L3 is somewhat independent of the thickness of that layer, the results for only varying the thickness of L4 (Moho discontinuity) are almost the same compared to the results of keeping that layer thickness constant, i.e. changing the border to L2 parallel with the depth of L4 (Moho).

In summary, one can conclude that the $P_n$-velocity in the upper mantle now is close to $8.0\,km/s$ $\pm 0.5\,km/s$ for an average Moho depth of $29.5\,km$. For convenience, a Moho depth of $30\,km$ is used in the later inversions. Overall remarkable is the good stability of the velocities in the upper two layers at 10 and $20\,km$ depth of the model. Opposite to this, the velocities in the third and fourth layer increase with increasing depth of the Moho [6]. This is the consequence of a low coverage of the lower crust (third layer) by rays, namely only the down- and up going branches of the $P_n$-phases.

The fourth layer - the upper mantle - is again more stable because of the horizontal ray paths along the Moho, despite the fact, that these rays are also affected by the velocities in layer L3. There is no way to reduce this ambiguity further without the use of additional phases which allow for an independent resolution of the lower crust and the upper mantle, such as $PmP$ reflections from the Moho discontinuity, as they have been used with success by Koch (1993a,b,c) in the SSH inversion within the upper Rhinegraben. Since such phases have not been sufficiently recorded in the present dataset, at this stage, the best one could do, based on the previous discussion, is to fix the Moho to its average depth.

We conclude this section by noting the optimal 1D velocity model derived from the tables discussed

---
[6] As will be shown later in Section 5.5.1, fixing the Moho at a certain depth gives rather stable solutions, which shows again the importance of clarifying the initial model parameters

(see appendix A for details). These are, namely, $5.90 \leq V1 \leq 5.93\,km/s$ in L1, $6.03 \leq V2 \leq 6.05$ in L2, $6.05 \leq V3 \leq 6.14\,km/s$ in L3 and $7.95 \leq V4 \leq 7.98\,km/s$. These values will be more refined in the following section.

### 5.4.3 Effects of different initial models on the inversion results

Before doing the SSH-inversions for the optimal 1D model, the input model space will be tested with regard to the inversion stability by using different 1D starting velocity models. The range of results obtained with these models gives an overview about the stability of the inversion process. Stability in that way means, that the outcome of the inversion is independent of the initial model space to a certain degree, and depends only marginally on the measured data. The narrower the range of the final models is, the more stable is the inversion process.

To this avail, inversions are carried out with (1) a low velocity model, (2) a high velocity model, (3) the earlier computed velocity model MKKS 2001 of Song et al. (2001a); Song and Koch (2002) and, (4) the MKS2004 velocity model which is an update of MKKS 2001 by considering the bias in the average residual shift. Table 5.4 summarizes the specifications and the outcome of these four model tests. Moreover, for every case, isotropic and anisotropic inversions (The value of 27°, as estimated in previous studies, is used.) with hypocenters free (IFIX=0, full SSH-inversion) and fixed (IFIX = 1, only velocity inversion) are done and compared to each other.

1. **Low velocity model Vp −**
   In these computations, the model V - in Table 5.4 is used, which has significant lower velocities than normally expected for the continental crust. The sequence of results of $V_P$ over RMS for the different iterations is illustrated in Figures 5.9(a) and 5.9(b). One notes that the solutions for the P-wave velocities cluster together for minimal RMS. This model will test the velocity space from the "bottom up", checking the convergence by starting with lower velocity values than expected, giving in some way the lower limit for the target velocity area (marked as the "target model"[7] in Figure 5.8). In the present case, the final optimal velocity model is nearly reached just after the first iteration, where the velocities in each layer increase towards the expected model increasing less with further ongoing iterations.

   After the inversion for both hypocenters and velocities, two inversions are done for fixed hypocenters. The different result, which has a completely different value for the $V_P$ velocities for the final velocity model clearly can be seen in Figures 5.8(a) and 5.8(b), where the optimal solution for V3 in layer three is at about 6.4 and 6.5 $km/s$, indicated by the blue line.

2. **High velocity model Vp +**
   The inversion process, now starting with a high velocity model, which has strongly increased velocities (about 0.5 $km/s$) for layer one to four, nicely shows the stability of the inversion process and the good quality of the available data. In spite of the fact that the velocity values are now shifted to much higher values than one would normally assume, they return to more normal values (similar to the ones computed earlier as optimal 1D velocity model MKS2005, used in previous investigations for Germany and marked as the "target model" in Figure 5.8), already after the first iteration of the inversion. The optimal values for the individual layer velocities are again represented by the clustering of the solutions at minimal RMS, as can be seen in Figures 5.9(e) and 5.9(f).

3. **SKKS2001 and MKS2004 models**
   The SKKS2001 model was found by Song, Koch, Koch and Schlittenhardt in 2001 as best average in a time-term-analysis, as result of a 2D analysis for $P_n$-anisotropy. (Song et al., 2001b). This model consists again of only four layers with $5.8 km/s$ between $0 - 10 km$, a small increase

---
[7]The term "target model" here stands for an averaging, petrologically reasonable velocity model, computed earlier as the optimal 1D velocity model MKS2005, used in previous investigations.

## 5.5. SENSITIVITY TESTS FOR THE 1D VELOCITY MODELS

Table 5.4: Input and final (optimal) velocity models for the four different inversion tests. "Ini" stands for the starting model, "opt" for the optimal model.

| Model | Layer Depth (km) | | | | Statistical values | |
|---|---|---|---|---|---|---|
|  | L1 | L2 | L3 | L4 | RMS | $\mathrm{RMS_{hypo}}$ |
|  | $0-10\,km$ | $10-20\,km$ | $20-30\,km$ | $30-40\,km$ | s | s |
|  | Velocity (km/s) | | | | | |
| V− (ini) | 5.00 | 5.50 | 6.00 | 7.50 | 5.1322 | 1.4928 |
| V− (opt) | 5.84 | 5.94 | 6.06 | 7.87 | 1.3429 | |
| V+ (ini) | 6.00 | 6.50 | 7.00 | 8.50 | 2.2055 | 1.5842 |
| V+ (opt) | 5.96 | 6.12 | 6.26 | 8.08 | 1.2989 | |
| SKKS2001 (ini) | 5.80 | 6.00 | 6.00 | 8.00 | | |
| SKKS2001 (opt) | 5.89 | 6.04 | 6.21 | 7.97 | | |
| MKS2004 (ini) | 5.90 | 6.10 | 6.10 | 8.10 | | |
| MKS2004 (opt) | 5.92 | 6.05 | 6.05 | 7.99 | | |

to $6.0\,km$ between $10-30\,km$ depth (the two layers between $10-20$ and $20-30\,km$ have the same values, and finally $8.0\,km/s$ lower than $30\,km$, representing the $P_n$-wave velocity.

The MKS2004 model was derived from the former model SKKS 2001, as some delay in the computed residuals $t_{obs} - t_{calc}$ (they become negative) for the $P_n$-phases were recognized which required an increase of the $P_n$ velocity in the upper mantle from 8.0 to $8.1\,km/s$. However, as the results in 5.4 indicate, the SSH inversion reduces this high value again to slightly underneath $8.0\,km/s$. Again, the clustering of the solution sequence at the minimal RMS in Figure 5.10 points to the optimal solution for the $V_P$-velocities. It is interesting to note that for the SKKS2001 and MKS2004 models the SSH-solutions jumps are much less than those of the low- and high input velocity models of Figure 5.9 above, as the initial models SKKS2001 and MKS2004 are much closer to the final optimal SSH-velocity model.

**Conclusive results for the different input models:**
Over all, the results for the different input models show good convergence to a petrologically reasonable velocity model. The values for the different optimal results are in Table 5.4, where the result for the $P_n$-velocity is about $8 \pm 0.8\,km/s$. The values for the other three layers are $V1 = 5.9 \pm 0.03\,km/s$, $V2 = 6.05 \pm 0.05\,km/s$ and $V3 = 6.15 \pm 0.05\,km/s$. So, despite the huge difference of the initial models to earlier computed models, the optimal results converge to reasonable values. Comparing the different results, the reasonable range where to find the optimal result can be tested.

## 5.5 Sensitivity tests for the 1D velocity models

In the following sections, some sensitivity tests are performed to check the influence of several parameters on the SSH inversion results. These are, respectively, (1) the residual limit, (2) the layer discretization and (3) the variation of the input velocities in layers L3 and L4. The last two tests are applied because the previous inversions indicated some degree of relative instability of the velocity in layer L3. Thus, varying the starting velocity for L3 will give the uncertainty range for the velocity of this layer.

### 5.5.1 Test of the convergence of various isotropic and anisotropic input models

As discussed, the estimated values of the velocities in layers three and layer four, in particular, show a certain degree of instability, in contrast to the rather stable solutions for layers one and two. For this reason, several input models with different initial V3 and V4, as well as slightly different values

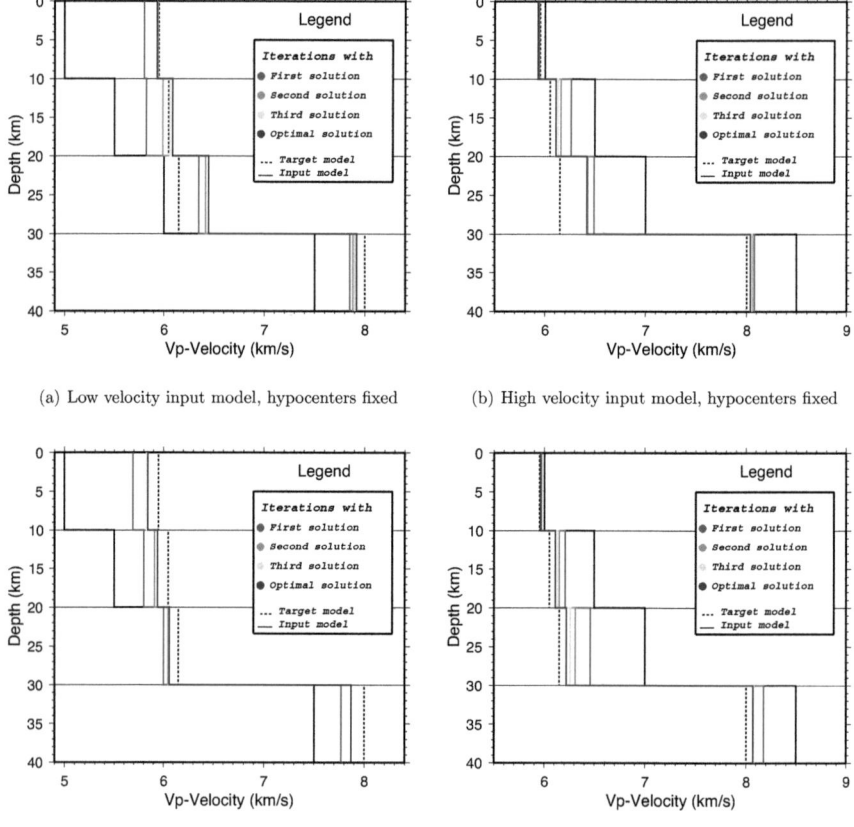

(a) Low velocity input model, hypocenters fixed
(b) High velocity input model, hypocenters fixed
(c) Low velocity input model, hypocenters free
(d) High velocity input model, hypocenters free

Figure 5.8: 1D SSH-inverted velocity models using the V− and V+ input models with hypocenters fixed and hypocenters free, respectively. In each of the diagrams the input velocity model and the different optimal solutions for various iterations are shown. The target model denotes the earlier computed optimal 1D velocity model MKS2005, used in previous investigations.

## 5.5. SENSITIVITY TESTS FOR THE 1D VELOCITY MODELS

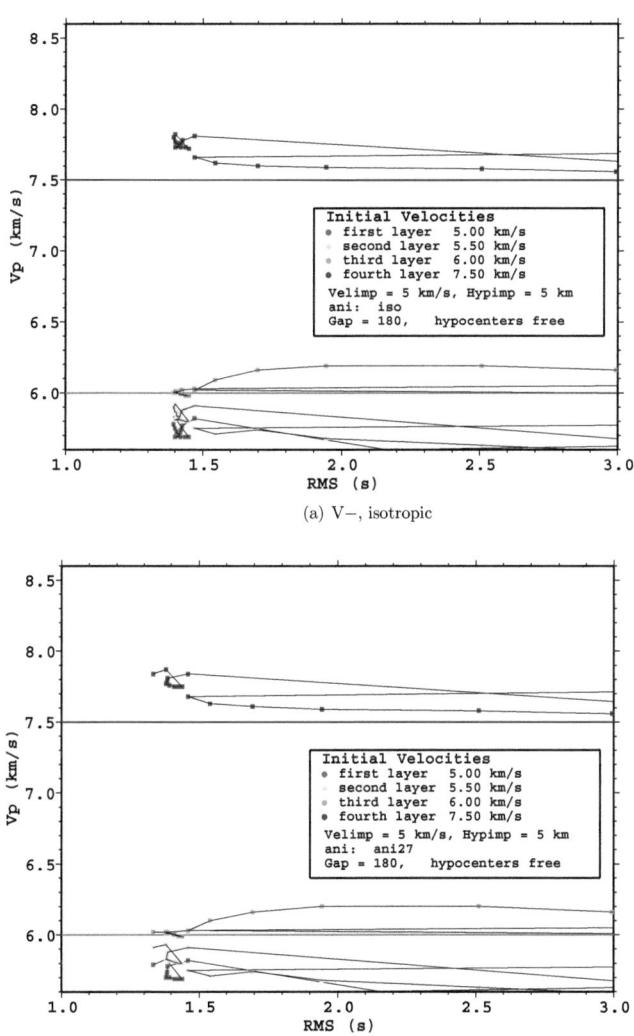

(a) V−, isotropic

(b) V−, anisotropic

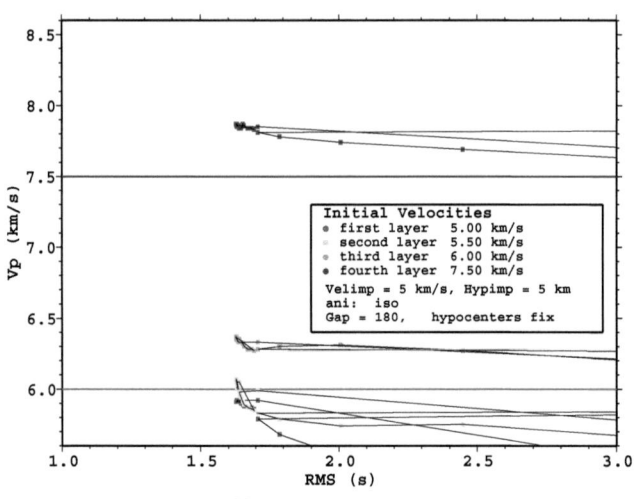

(c) V−, isotropic, hypocenters fixed

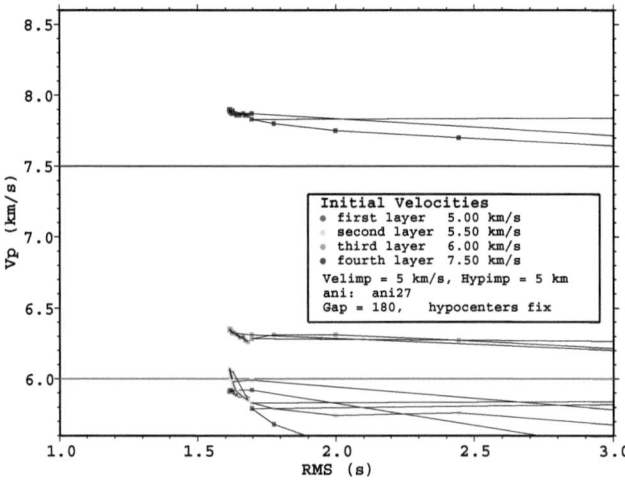

(d) V−, anisotropic hypocenters fixed

## 5.5. SENSITIVITY TESTS FOR THE 1D VELOCITY MODELS

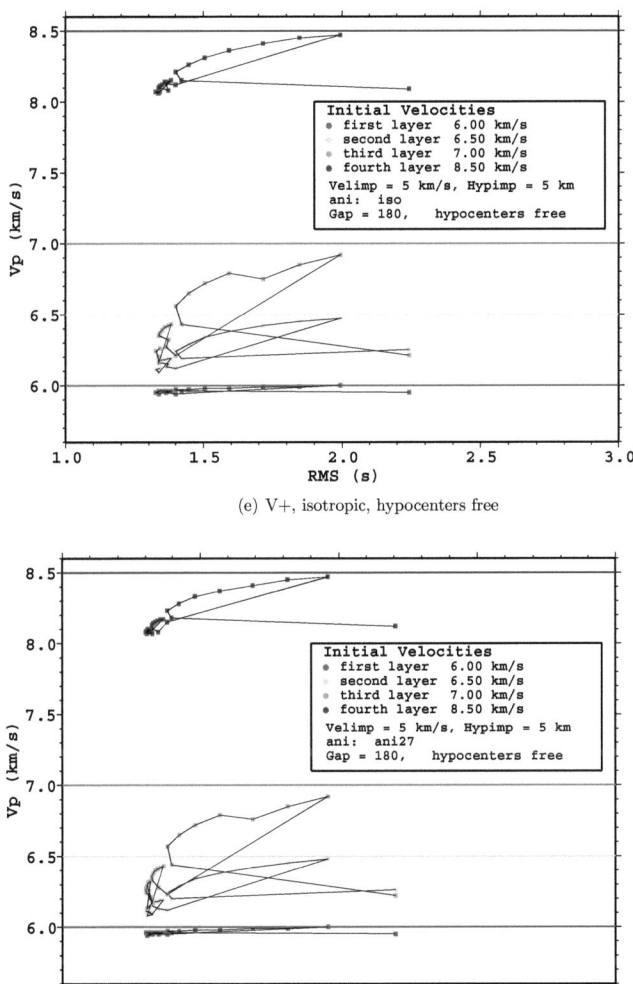

(e) V+, isotropic, hypocenters free

(f) V+, anisotropic hypocenters free

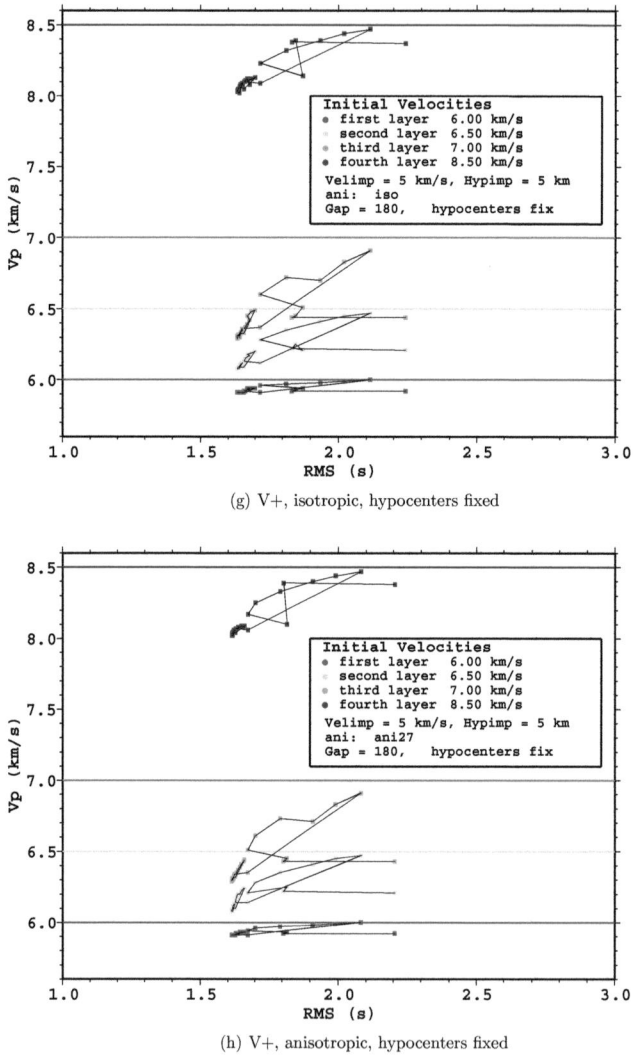

(g) V+, isotropic, hypocenters fixed

(h) V+, anisotropic, hypocenters fixed

Figure 5.9: Variation of the layer velocities over RMS for various SHH inversion types (hypocenters free or fixed, without and with anisotropy included) for the low velocity model V− and for the high velocity model V+. The result for the velocities is a cluster of points about the minimal RMS, which represents the final solution.

## 5.5. SENSITIVITY TESTS FOR THE 1D VELOCITY MODELS

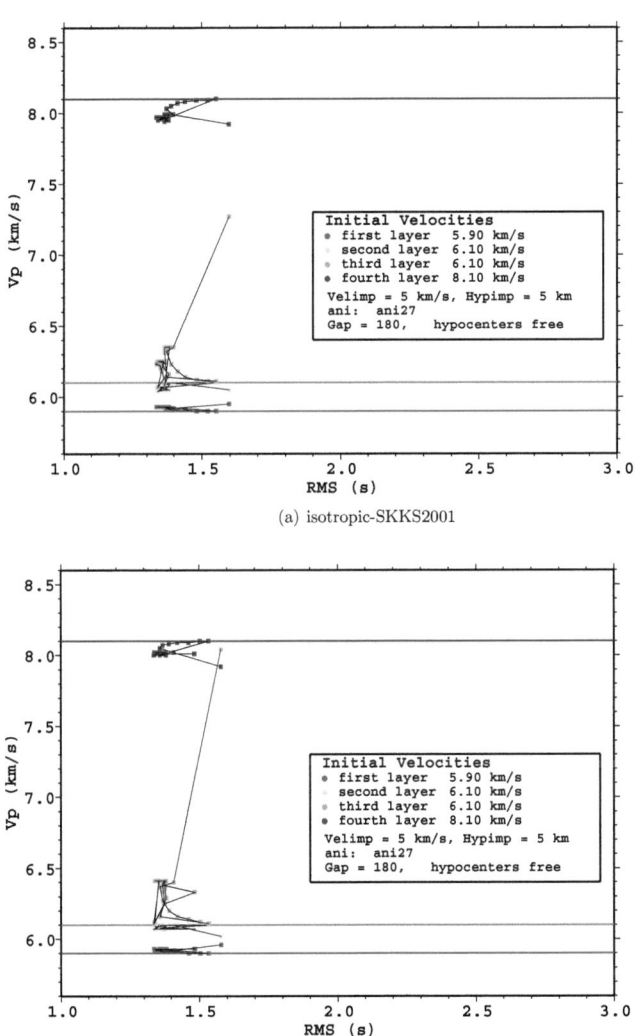

(a) isotropic-SKKS2001

(b) anisotropic-SKKS2001

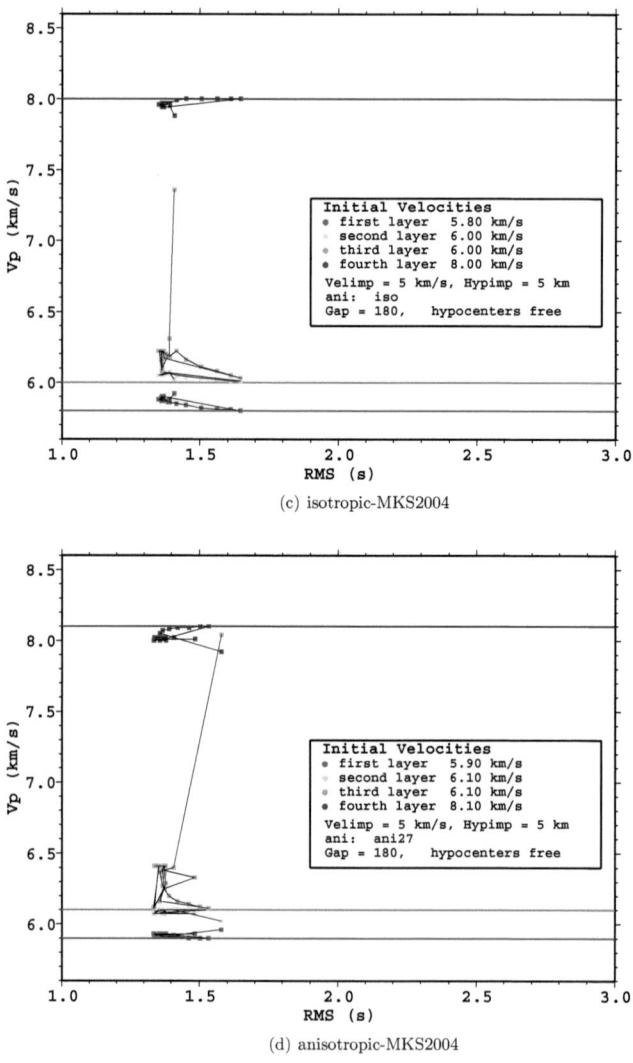

Figure 5.10: Variation of the velocities over the RMS for the SKKS2001 and the MKS2004 models, without and with anisotropy included.

## 5.5. SENSITIVITY TESTS FOR THE 1D VELOCITY MODELS

for V1 and V2, are tested to check the sensitivity of the final result on the starting velocity in these layers.
The results of these simulations are shown in Figure 5.12. The extreme input models V− and V+ are added here too. One notes that, despite the huge range obtained for the velocities in layers L3 and L4, convergence is pretty good. Based on these results, the average velocities and the variances for each of the four layers are calculated, both of which are listed in Table 5.5 for the isotropic case and the anisotropic case. With regard to the differences between the isotropic and the anisotropic models, the following conclusions can be drawn:

- The seismic velocity in layer four has slightly higher values relative to the isotropic case and is closer to the expected 8.00 $km/s$, as found in other aforementioned studies for the upper mantle in Germany.

- The velocity in layer three suffers less variation, i.e. the result is more stable for the anisotropic than for the isotropic models.

- Due to better convergence sequences of the anisotropic than of the isotropic models, the variances of the former are overall smaller than for the latter.

Table 5.5: Statistical summary of the convergence tests using different initial velocity models. The velocities V1 to V4 are the average of all velocities for the different initial models (see Figure 5.11), calculated using all available results, as specified by the number of "Counts" in the distinct iteration I. The corresponding standard deviations are $\sigma 1$ to $\sigma 4$.

| Counts | I | V1 $km/s$ | V2 $km/s$ | V3 $km/s$ | V4 $km/s$ | $\sigma 1$ $km/s$ | $\sigma 2$ $km/s$ | $\sigma 3$ $km/s$ | $\sigma 4$ $km/s$ |
|---|---|---|---|---|---|---|---|---|---|
| | | | | isotropic model | | | | | |
| 20 | 1 | 5.909 | 6.073 | 6.138 | 7.942 | 0.0556 | 0.0666 | 0.1383 | 0.0799 |
| 16 | 2 | 5.906 | 6.055 | 6.151 | 7.939 | 0.1383 | 0.0798 | 0.1170 | 0.0726 |
| 16 | 3 | 5.545 | 5.673 | 5.756 | 7.449 | 1.4320 | 1.4650 | 1.4893 | 1.9241 |
| 12 | 4 | 5.912 | 6.044 | 6.113 | 7.944 | 0.0329 | 0.0328 | 0.0752 | 0.0641 |
| | | | | anisotropic model | | | | | |
| 20 | 1 | 5.905 | 6.075 | 6.174 | 7.965 | 0.0648 | 0.0685 | 0.1190 | 0.0838 |
| 17 | 2 | 5.909 | 6.061 | 6.157 | 7.966 | 0.0519 | 0.0560 | 0.0879 | 0.0747 |
| 17 | 3 | 5.917 | 6.058 | 6.127 | 7.972 | 0.0261 | 0.0425 | 0.0700 | 0.0570 |
| 15 | 4 | 5.923 | 6.055 | 6.124 | 7.981 | 0.0135 | 0.0102 | 0.0661 | 0.0484 |

From these results we can conclude overall that using the $P_n$-anisotropic correction stabilizes the inversion results and leads to more plausible 1D velocity models for both the crust and the upper mantle across Germany.

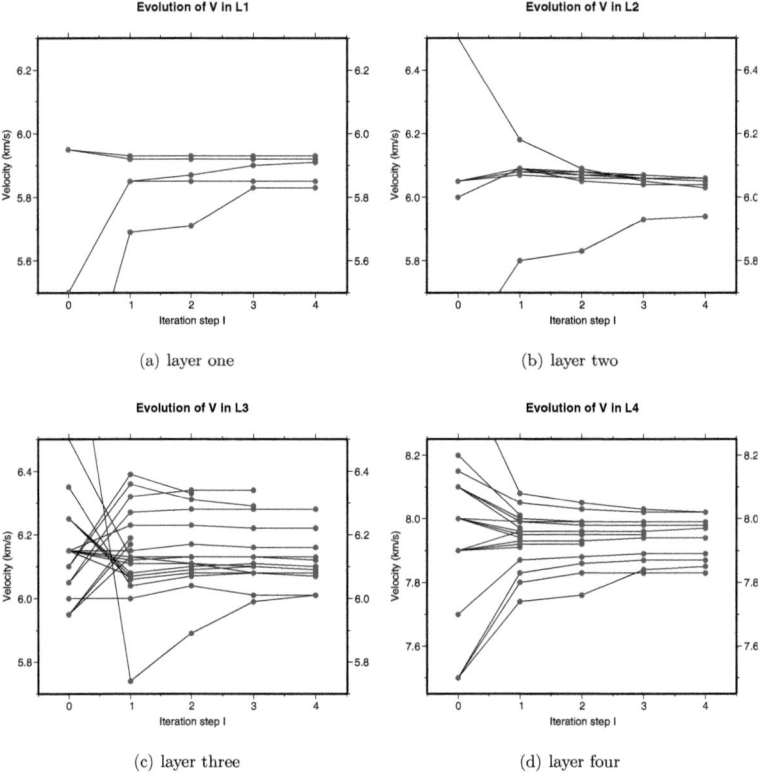

Figure 5.11: The convergence of various velocity input models, where the starting values can be recognized at iteration step 0 (only ray tracing) and the final solutions at iteration step 4, without anisotropic correction.

## 5.5. SENSITIVITY TESTS FOR THE 1D VELOCITY MODELS

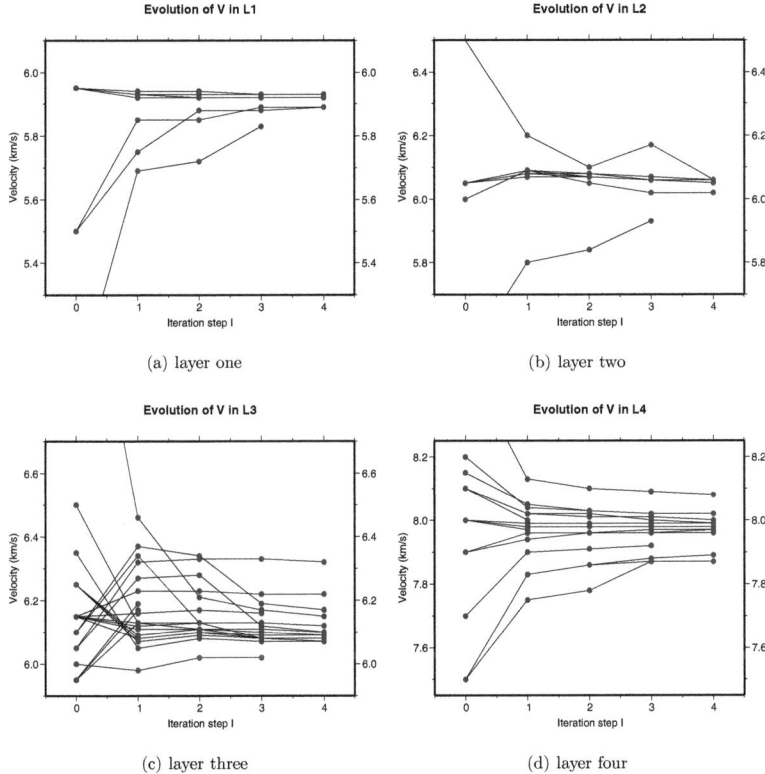

(a) layer one

(b) layer two

(c) layer three

(d) layer four

Figure 5.12: The convergence of various velocity input models, where the starting values can be recognized at iteration step 0, and the final solutions at iteration step 4, with anisotropic correction of ±5% velocity variation and fast axis at 27° azimuth.

Table 5.6: Optimal velocities for the seven-layer model, with (ani=1) and without (ani=0) anisotropic $P_n$-correction included, as well as hypocenters free (ifix=0) and fixed (ifix=1).

| layer depth | | | 0 km | 5 km | 10 km | 15 km | 20 km | 25 km | 30 km |
|---|---|---|---|---|---|---|---|---|---|
| | | | velocity (km/s) | | | | | | |
| input model | | | 5.70 | 5.80 | 6.00 | 6.05 | 6.10 | 6.15 | 8.10 |
| ani | ifix | RMS | | | | | | | |
| no  | 1 | 1.5680 | 5.89 | 6.02 | 6.03 | 6.11 | 6.13 | 6.94 | 7.94 |
| yes | 1 | 1.5550 | 5.89 | 6.02 | 6.04 | 6.12 | 6.11 | 6.88 | 7.99 |
| no  | 0 | 1.3372 | 5.86 | 5.92 | 5.99 | 6.02 | 5.99 | 6.11 | 7.93 |
| yes | 0 | 1.3132 | 5.89 | 5.94 | 5.98 | 6.02 | 6.00 | 6.04 | 7.97 |

### 5.5.2 1D velocity models with finer layer discretization

To fine-tune and to see whether a better resolution for the vertical velocity distribution for the 1D crustal model beneath Germany can be obtained with the present dataset, additional SSH-inversions with seven instead of four layers, resulting in a layer thickness of 5 km, are performed in this section.

The idea now is that allowing for more layers in the model should provide more degrees of freedom for the natural velocity distribution which, in reality, could either be gradually varying or have some "jumps" (Muench, 2000).

Within this context, it should be mentioned, that the term "layer boundary" has to be carefully used. Because of the artificial discretization of the velocity model, these boundaries may not represent natural boundaries. In strongly heterogeneous areas like the crust in Germany, with a lot of fault zones and other crustal anomalies (for example Rheingraben, Eifel, Central German Uplands), one can not expect this idealization to hold. Thus, the results of a regional seismic velocity study will always only be an approximation. To get now further details on possible lateral heterogeneities a complete 3D study, as carried out in Chapter 6 has to be done.

The results of this seven-layer model are shown in Figure 5.13 and listed in Table 5.6. As can be seen, the velocity distribution in the depth-range 5 to 25 km is rather continuous, with only a small gradient-like increase. Thus it is appropriate to pack together these five layers in three, as done in the previous SSH-inversions. The same holds for the upper two layers where the velocity of the first (surface) $0-5$ km layer increases smoothly to about 5.9 km/s in the second, $5-10$ km, layer, similar to the "thicker" $0-10$ km-layer model (see Figure 5.8).

More interesting is the velocity increase in the sixth layer, which is reduced in the SSH-inversion case with free hypocenters. Thus, portions of the traveltime residuals that occur in this layer are partly taken over by corresponding hypocentral shifts. But after all, velocities in layers below 20 km strongly depend on the $P_n$-velocity, due to the poor horizontal ray distribution there with $P_g$ phases (see Figure 4.1 in Chapter 4). Despite the numerous rays in these layers, most of them here only form a section of the overall $P_n$-ray path which starts from a seismic event in the layers above and then extends into the upper mantle. Therefore, as already mentioned earlier, independent information on the seismic velocity in the lower crust, as could be provided by $PmP$-reflection, is not available in the present data set. Moreover the velocities in layers below 20 km strongly depend on each other, so applying the anisotropic correction to the part of $P_n$-phases, traveling in the crust-mantle-boundary, just affects the velocities in the layers above.

For the SSH-inversion case with the anisotropic correction for $P_n$-phases included, a small increase of the $P_n$-velocity in the upper mantle to about 8 km/s is obtained, which represents a more realistic value like that of other studies (Song et al., 2001a, 2004; Enderle et al., 1996a; Enderle, 1998), but as well a $Pn$-velocity somewhat too low for the isotropic inversion, likewise to the SSH models of the previous sections. Also the need to do a complete inversion, i.e. for hypocenters and structure

## 5.5. SENSITIVITY TESTS FOR THE 1D VELOCITY MODELS

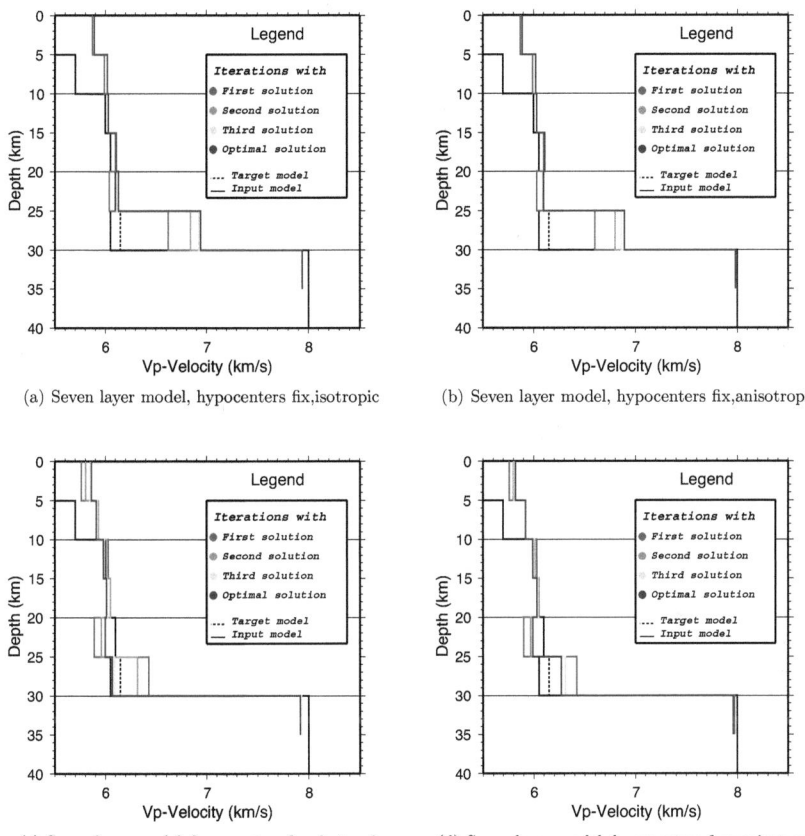

(a) Seven layer model, hypocenters fix, isotropic

(b) Seven layer model, hypocenters fix, anisotropic

(c) Seven layer model, hypocenters free, isotropic

(d) Seven layer model, hypocenters free, anisotropic

Figure 5.13: The velocity distribution for the seven layer model to test the model space for the vertical velocity variation. Figures 5.13(a) and 5.13(b) show the solutions for seven layers with hypocenters fixed. In Figures 5.13(c) and 5.13(d) the same input model is used, but inversion is done together with hypocenters.

# CHAPTER 5. SSH-INVERSIONS FOR 1D VELOCITY MODEL

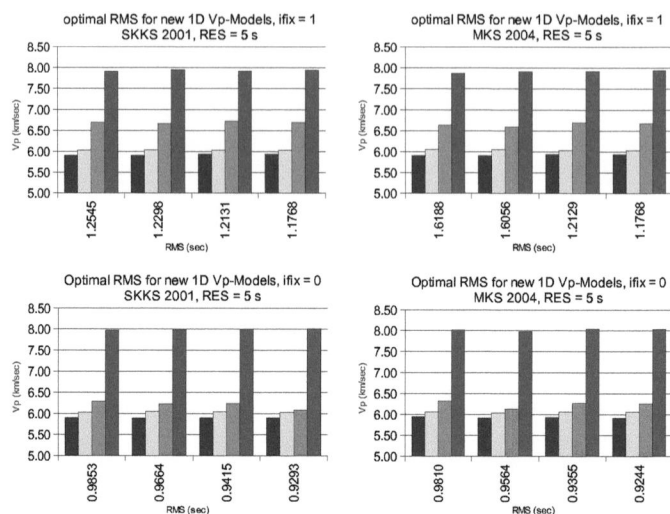

Figure 5.14: The optimal results for the computations for two different input parameters: Gap = 120 and 180 degree, both with and without anisotropic correction. Left figure with SKKS2001, right figure with MKS2004 as starting model, both with RES = 5 s. Hypocenters are fixed. The colored bars represent the layer velocities. From blue for the first layer to red for the last. Lower two plots now with hypocenters free, i.e. complete inversion.

together is obvious by looking at the velocities in layer L7. Whereas in the case with free hypocenters they are in the range of $6.1 - 6.3\,km/s$, they go up to a - petrological unreasonable one for the lower crust as a whole - high value of more than $6.8\,km/s$ when the hypocenters are kept fixed.

## 5.6 Selection of the optimal 1D velocity model

Having performed SSH-inversions with four different input models, (V−, V+, SKKS2001 and MKS2004), with a gap of 180°, here additional SSH-inversions with a gap = 120° for the MKS2004 input model are carried out for both "hypocenters fixed" and "hypocenters free", with and without $P_n$-anisotropy correction included. The optimal results for the various SSH-inversions using all possible combinations of these parameters are shown in Figure 5.14.

One notes that results for the computations with fixed hypocenters have a complete different result for the distinct velocities in layer three and four, compared to those of the inversion with correction for $P_n$-phases. The velocity in layer three (L3) is higher than that for the anisotropic corrected case and the one in layer four (L4) is smaller, below $8.0\,km/s$. Compared with the velocity models of Song et al. (2001a, 2004) or that of Enderle (1998), the velocities for the crust mantle boundary, which is represented by layer four, is a little to small. Moreover, the optimal result is obtained using a GAP = 120° and including the $P_n$ anisotropic correction.

Comparing the RMS, which represents the fit of the inverted model to the observed arrivaltime data and, as such, is the ultimate parameter to measure the "objective" quality of the SSH-model, one recognizes that a gap of 120° is mostly better than that of 180° (except in the case of the full SSH-inversion with the MKS2001 starting model) and that the anisotropic velocity models are always better than the isotropic ones. This is another indication - corroborating the studies of Song et al. (2001a,

## 5.6. SELECTION OF THE OPTIMAL 1D VELOCITY MODEL

Table 5.7: The values for the different velocity models in Figure 5.14

SKKS2001

|  | ifix = 1 | | | | ifix = 0 | | | | unit |
|---|---|---|---|---|---|---|---|---|---|
| Ani | 0 | 0 | 1 | 1 | 0 | 0 | 1 | 1 | |
| Gap | 180 | 180 | 120 | 120 | 120 | 180 | 120 | 180 | $deg$ |
| TSS | 34196 | 32862 | 17371 | 16346 | 11458 | 20296 | 10463 | 18766 | $s^2$ |
| TSS-hypo | 28545 | 28545 | 14264 | 14264 | 14264 | 28545 | 14264 | 28545 | $s^2$ |
| V1 | 5.91 | 5.91 | 5.94 | 5.94 | 5.90 | 5.89 | 5.90 | 5.90 | $km/s$ |
| V2 | 6.03 | 6.04 | 6.03 | 6.03 | 6.03 | 6.05 | 6.04 | 6.03 | $km/s$ |
| V3 | 6.70 | 6.67 | 6.73 | 6.70 | 6.29 | 6.23 | 6.24 | 6.08 | $km/s$ |
| V4 | 7.91 | 7.95 | 7.92 | 7.94 | 7.97 | 7.99 | 7.99 | 8.01 | $km/s$ |
| RMS | 1.2545 | 1.2298 | 1.2131 | 1.1768 | 0.9853 | 0.9664 | 0.9415 | 0.9293 | $s$ |

MKS2004

|  | ifix = 1 | | | | ifix = 0 | | | | unit |
|---|---|---|---|---|---|---|---|---|---|
| Ani | 0 | 0 | 1 | 1 | 0 | 0 | 1 | 1 | |
| Gap | 180 | 180 | 120 | 120 | 180 | 120 | 180 | 120 | $deg$ |
| TSS | 58010 | 57063 | 17365 | 16347 | 19859 | 11359 | 18570 | 10330 | $s^2$ |
| TSS-hypo | 55591 | 55591 | 14264 | 14264 | 28487 | 14264 | 28487 | 14264 | $s^2$ |
| V1 | 5.91 | 5.91 | 5.94 | 5.94 | 5.95 | 5.92 | 5.93 | 5.92 | $km/s$ |
| V2 | 6.05 | 6.05 | 6.03 | 6.03 | 6.06 | 6.04 | 6.06 | 6.06 | $km/s$ |
| V3 | 6.64 | 6.59 | 6.70 | 6.68 | 6.32 | 6.13 | 6.27 | 6.26 | $km/s$ |
| V4 | 7.87 | 7.91 | 7.92 | 7.94 | 8.02 | 7.98 | 8.04 | 8.03 | $km/s$ |
| RMS | 1.6188 | 1.6056 | 1.2129 | 1.1768 | 0.9810 | 0.9564 | 0.9355 | 0.9244 | $s$ |

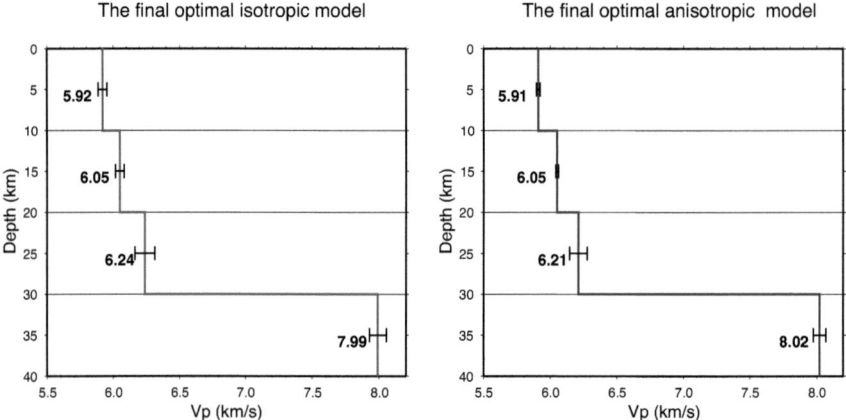

Figure 5.15: The optimal 1D velocity models for the isotropic and anisotropic computations. The first two layers nearly have the same result. The main difference is between the velocities for layer three and four and the uncertainty of the results, indicated by the horizontal error bars. The detailed values are in Table 5.5.

2004) - that $P_n$ anisotropy of the upper mantle beneath most of Germany must be taken for granted.

From the tests of the residual limit RES of the previous section, a further improvement for the velocity models is obtained by reducing the residual limit to $5\,s$. This value is about the limit not to reject the anisotropic information but over all low quality arrivaltime data, that will disturb the inversion process, in the data set, as shown in Chapter 4. The $P_n$-velocity in layer four shows a slight increase to $8.03\,km/s$ by $0.03\,km/s$, whereas the velocity in layer four slightly decreases to $6.08\,km/s$ by $0.08\,km/s$. The velocities for the isotropic case show again lower values at $6.13\,km/s$ in layer three and $7.98\,km/s$ in layer four.

Depending on the various parameters to be selected in the inversion process (GAP, RES, NOBS, and so on), the choice of the correct optimal 1D velocity model is not an easy task. The results of the previous computations allow one to nail the velocity model to a closer range of possible models, out of which the optimal 1D model is selected. The results listed in Table 5.7 together with those of the tests for convergence (Table 5.5), where the possible range for the distinct velocities in each layer is fixed by $\sigma$, explicitly show the problematics. Nevertheless, a judicious choice, based on the SSH-models with the lowest RMS and allowing for free hypocenters (i.e. full SSH inversion) as well as the incorporation of the $P_n$ anisotropic correction results in the following average final 4-layer velocity model listed in Table 5.8 and shown in Figure 5.15. Also indicated are standard deviations for the layer velocities, as computed from four anisotropic solutions with RES $= 5\,s$.

In Figure 5.16, there are the original and with a full inversion newly computed hypocenters. For simplicity and better visibility, only the anisotropic case is shown, where in detail the different hypocenter localizations (blue dots for the newly computed hypocenters and red dots for the original hypocenters) are visible. More details about the hypocenter relocation are in Chapter 7 together with some synthetic test to show the accuracy of the relocation and the improvement by the application of anisotropic correction of $P_n$-phases.

## 5.6. SELECTION OF THE OPTIMAL 1D VELOCITY MODEL

Table 5.8: The final optimal 1D velocity model MKS2007, including anisotropy in the inversion with hypocenters free. For comparison, the optimal isotropic computed 1D model is included in the first line. The errors are from Table 5.5

| ani | V1 km/s | V2 km/s | V3 km/s | V4 km/s | RMS s | TSS $s^2$ | $\sigma_{V1}$ km/s | $\sigma_{V2}$ km/s | $\sigma_{V3}$ km/s | $\sigma_{V4}$ km/s |
|---|---|---|---|---|---|---|---|---|---|---|
| no  | 5.92 | 6.05 | 6.24 | 7.99 | 0.9743 | 20627 | 0.033 | 0.033 | 0.075 | 0.064 |
| yes | 5.91 | 6.05 | 6.21 | 8.02 | 0.9385 | 19140 | 0.014 | 0.010 | 0.066 | 0.048 |

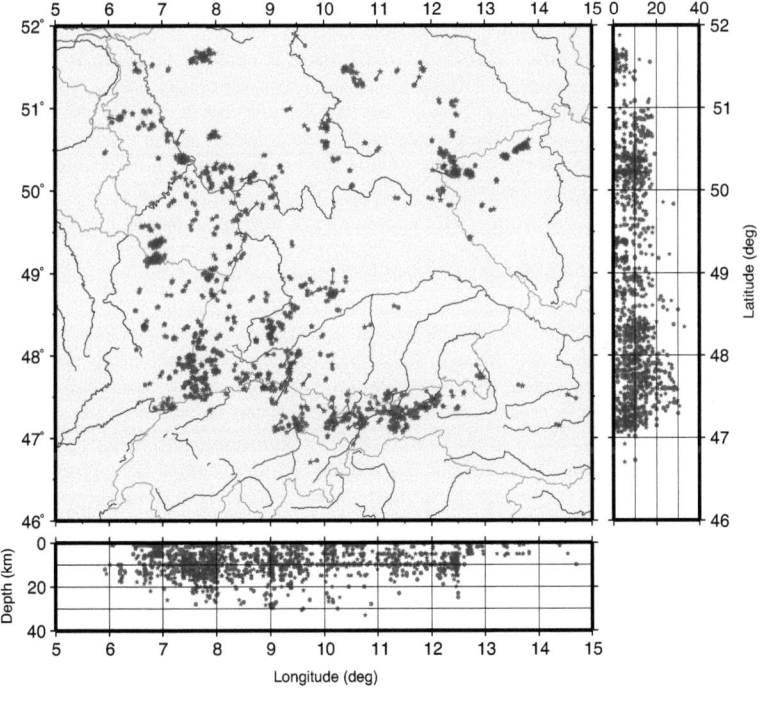

(a) anisotropic relocation

Figure 5.16: The original (red) and new computed (blue) hypocenters and vertical cuts for the anisotropic case. More details are in Chapter 7.

## 5.7 Inversions with $S$-phases and tests of the $V_P/V_S$-ratio

After the SSH-estimation of the optimal 1D P-wave velocity model, some computations are done including $S$-phases. Because of the overall minor quality of the $S$-arrivaltime pick-ups than those of the $P$-arrivals, it is not advisable to perform an independent $S$-wave inversion alone, but the $S$-phases need to be coupled with the $P$-phases over the $V_P/V_S$-ratio in the SSH-inversion.

Using the definitions $V_P = \sqrt{(\lambda+2\mu)/\rho}$ and $V_S = \sqrt{\mu/\rho}$ (Weber et al. (2007), page 39) one notes that the $V_P/V_S$-ratio can be directly related to the Poisson ratio $\nu = \lambda/2(\lambda+\mu)$ as

$$\nu = 1/2 \cdot [(V_P/V_S)^2 - 2]/[(V_P/V_S)^2 - 1] \tag{5.1}$$

For the classic elastic solid $\nu$ is equal 0.25, which is also the average value for the Earth's crust and upper mantle, resulting in $V_P/V_S$-ratio of 1.73, whereas for a pure fluid $\nu$ is equal 0.5. Thus $\nu$ and the $V_P/V_S$-ratio are highly dependent on the presence of fluids in a rock and this property has been used in many studies of the earth's interior to infer water or partial melts there (cf. Koch (1992), Monna et al. (2003) or Eberhart-Phillips and Reyners (1997)). Temporal changes of the $V_P/V_S$-ratio have also been employed in earthquake prediction, again due to the fact, that opening micro cracks may fill up with water leading to subsequent lubrication of the shear zones and an increased propensity for earthquake rupture. In any case, Poisson's ratio is a useful indicator of the lithology and the pore fluid pressure in the Earth's interior. As indicated in the various tables in Appendix B, solid rocks themselves show a large variation of the $V_P/V_S$-ratio with, for example, a $V_P/V_S$ ratio as high as 2.17 and lower one for granitic rocks. In any case, the Poisson ratio is a useful indicator of the lithology and the pore fluid pressure in the Earth's interior. As indicated in the various tables in Appendix B, solid rocks themselves show a large variation of the $V_P/V_S$-ratio with, for example, a $V_P/V_S$ ratio as high as 2.17 and lower one for granitic rocks.

The thorough analysis of Koch (1992) indicates large vertical as well as horizontal variations of the $V_P/V_S$-ratio in the crust underneath the Rhinegraben in south western Germany. For the average $V_P/V_S$-ratio the author obtained a value of 1.73 across the region.

Traditional methods for computing $V_P/V_S$ and thus Poisson's ratio, are either direct or indirect. Direct inversion methods estimate $V_P$ and $V_S$ separately, although, as discussed above, because of the often poor quality of the observed S-wave arrivals, these direct methods are mostly not reliable in practice. Indirect methods, on the other hand, invert for $V_P/V_S$ using known or assumed P-wave velocity models and are more commonly used (cf. Walck (1998). This is also the approach used here.

The main task at this stage of the SSH analysis is to compute an optimal $V_P/V_S$ value to be used as an initial parameter in the 3D SSH inversion in Chapter 6 where lateral variations of the $V_P/V_S$ ratio across Germany will be estimated. To that avail several 1D SSH-inversions with different $V_P/V_S$ ratios ranging between 1.67 and 1.74 for both the anisotropic and isotropic case are carried out and the effect on the final RMS is investigated. The results are shown in Figure 5.17 and, in more detail, in Table 5.9. One notes that the optimal $V_P/V_S$ ratios with the minimal RMS is $V_P/V_S = 1.70$ for the isotropic case, and slightly higher ($\approx 1.71$) for the anisotropic case. The isotropic case shows a smaller minimum for the RMS than the anisotropic case which is not significant to decide whether the isotropic or anisotropic case is the best.

As will be shown in Chapter 7 where the relocation of the hypocenters is discussed and reduced traveltime plots for $P_g$- and $P_n$-phases over $S_g$-, respective $S_n$-phases provide similar results in the same range for the $V_P/V_S$-ratio as above. In Chapter 6 the in blocks lateral distribution of the $V_P/V_S$ ratio will be determined.

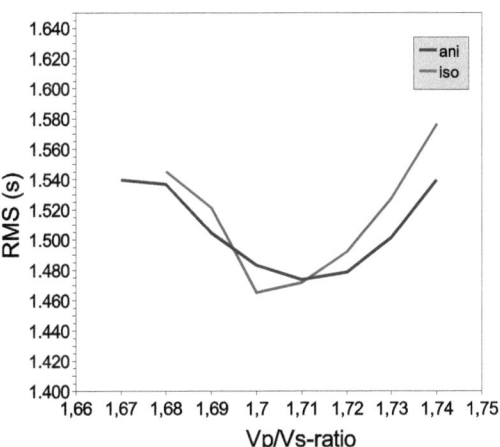

Figure 5.17: Plot of the RMS for different $V_P/V_S$ ratios, red for the isotropic and blue for the anisotropic case. The minimum for the isotropic case is about $V_P/V_S = 1.70$ and $V_P/V_S = 1.71$ for the anisotropic one.

Table 5.9: Different values for $V_P/V_S$. A + after the $\text{RMS}_{\text{av}}$ denotes the best solution.

| $V_P/V_S$ | TSS | $\text{RMS}_{\text{un}}$ | $\text{RMS}_{\text{av}}$ | $\text{RMS}_{\text{hyp}}$ | $\text{TSS}_{\text{hypo}}$ |
|---|---|---|---|---|---|
| | $s^2$ | $s$ | $s$ | $s$ | $s$ |
| | Results for isotropic $P_n$-phases | | | | |
| 1.67 | 54588.0 | 1.7007 | 1.5843 | 1.4156 | 75699 |
| 1.68 | 51963.2 | 1.6594 | 1.5457 | 1.3638 | 75699 |
| 1.69 | 50365.2 | 1.6336 | 1.5218 | 1.3307 | 75699 |
| 1.70 | 46686.8 | 1.5729 | 1.4652 + | 1.2655 | 75699 |
| 1.71 | 47111.8 | 1.5800 | 1.4718 | 1.2729 | 75699 |
| 1.72 | 48419.3 | 1.6018 | 1.4921 | 1.2983 | 75699 |
| 1.73 | 50767.6 | 1.6402 | 1.5279 | 1.3421 | 75699 |
| 1.74 | 54063.0 | 1.6925 | 1.5767 | 1.3999 | 75699 |
| | Results for anisotropic corrected $P_n$-phases | | | | |
| 1.67 | 51553.6 | 1.6528 | 1.5396 | 1.3556 | 75699 |
| 1.68 | 51379.8 | 1.6500 | 1.5370 | 1.3300 | 75699 |
| 1.69 | 49249.4 | 1.6154 | 1.5048 | 1.2922 | 75699 |
| 1.70 | 47858.4 | 1.5925 | 1.4834 | 1.2666 | 75699 |
| 1.71 | 47229.1 | 1.5820 | 1.4737 + | 1.2593 | 75699 |
| 1.72 | 47555.0 | 1.5874 | 1.4787 | 1.2724 | 75699 |
| 1.73 | 49031.8 | 1.6119 | 1.5015 | 1.3054 | 75699 |
| 1.74 | 51553.6 | 1.6528 | 1.5396 | 1.3556 | 75699 |

## 5.8 Final considerations about the optimal 1D velocity model

At the end of the search for the optimal 1D velocity model, usable for precise hypocenter localization and 3D earth structure modeling, there are still some points to mention how the goal was reached.

The most important criterion to find the optimal solution in the SHH inversion is the RMS, but this parameter is not the only one. The fact that the models along the linear and non linear iteration sequence show good convergence is another criterion to judge about the quality of the inversion solution. Having different solutions with nearly similar RMS might be possible and will make it difficult to find the optimal solution.

Another criterion is the trade-off between the data- and the model fit, that will be considered in more details in Chapter 6. The optimal 1D velocity model is calculated as the average of the optimal results of the different computed input velocity models in the previous section and from various realizations its statistical properties can be calculated. This optimal model will be used to a large extent in the 3D SSH inversions in the following sections.

The influence of the $Pn$-anisotropic correction on the 1D velocity model is mostly concentrated on the lower crust and upper mantle, namely, by showing lower velocities in Layer L3 (lower crust) and slightly higher ones in layer L4 (upper mantle). The results in the upper layers L1 and L2, on the other hand, are rather similar for the two cases, the isotropic and the anisotropic one. This is due to the fact, that these two layers are mainly sampled by $Pg$-phases and are less affected by the $P_n$-phases.

The optimal $V_P/V_S$-ratio found depends on whether the $P_n$-phases are corrected with elliptical anisotropy, or not. For the anisotropic case, the ratio is slightly higher, whereas the over all minimum is astonishingly obtained in the isotropic case. The difference for the RMS between 1.46 to 1.47 is such small that this is not significant to decide whether the isotropic or anisotropic case is the best.

# Chapter 6

# SSH-inversions for 3D velocity models

## 6.1 Introduction, general approach and model-setup

Now, after we have found the optimal 1D vertically inhomogeneous velocity model in the previous chapter, representing the best fit between the measured (observed) and the calculated data, we can use this 1D-model as the optimal input model in the SSH-inversions for 3D laterally heterogeneous velocity models[1]. The results of the 3D SSH-computations can then be interpreted as velocity deviations from the 1D vertically inhomogeneous velocity model, i.e. they represent the lateral velocity perturbations, usually expressed in percent of the 1D layer velocity. Looked upon from the other way, the initial 1D velocity model represents the average of the 3D velocity model.

But before proceeding to the real 3D velocity inversions for the crustal and lithospheric seismic structure underneath Germany, some synthetic tests are done to check the reconstruction (resolution) capabilities of the traveltime data with regard to the model extensions used in the 3D inversions. This will give some nice hints how to interpret the quality of the inverted 3D velocity models, i.e. which areas of the model are perfectly well, fair, or not at all reconstructed or resolved. Also, from these initial synthetic inversion tests, by comparing isotropic with anisotropic inversion solutions, the need for the elliptical anisotropic $P_n$-velocity correction in the real subsequent 3D tomographic study beneath Germany is to be recognized.

For the real 3D tomographic inversions, we start with some simple models with a rather coarse bloc discretization, representing only a crude approximation of the true 3D seismic structure in the crust and upper mantle beneath Germany. These simple models consist of 4 layers, each with a thickness of $10\,km$, and a lateral discretization into $15 \times 15$ blocs. While such a model is rather crude for interpreting local structures it is, on the other hand, rather fine for most tomographic applications. Step by step, the lateral model discretization will be refined up to $35 \times 35$ blocs.

The geographical center for all models is located at $10°$ Longitude and $50°$ Latitude. At this latitude one degree longitude corresponds to approximately $111 * cos(50°) = 72\,km$ in the x-direction (EW), whereas one degree latitude is, as usual, $111\,km$ in the y-direction (NS). The total size of the model has lateral extensions of $650\,km \times 650\,km$, which for the $15 \times 15$ bloc model results in a horizontal size of about $43\,km \times 43\,km$, going down to $18\,km \times 18\,km$ for the $35 \times 35$ model. The reason to chose that model size is related to the ray coverage which is then sufficient enough even to cover the borders of the model area. Choosing huger model sizes would not provide any additional information on the seismic structure because of a lack of ray coverage at the model margins due to a limited distribution

---
[1] Another way to do 3D tomography is to start with any 1D- or 3D- velocity model. But the number of inversions required may be higher and may lead to incomparable results or, even worse, to different inversion solutions qualified by different local minima for the RMS. This is a somewhat unpleasant property of the gradient optimization (minimization) techniques used in the nonlinear inversion approach here which, unlike global optimization techniques (e.g. simulated annealing, genetic algorithms), only search for local minima in the neighborhood of the starting solution. Here we are looking for an average 1D model for routine localizations of earthquakes throughout Germany, our approach to use the optimal 1D velocity model as a starting model for the subsequent 3D inversions makes good sense.

of stations and seismic events. This gap may be filled in succeeding studies by extending the data set with records from adjacent areas like Poland, France, Switzerland, Austria or the Czech Republic, which is left for succeeding studies.

In the following paragraphs we go through the development of the 3D models as they increase in complexity, especially by augmenting the number of blocs in the horizontal layers until a resolution limit is reached whereby the inversion results become unstable as there are not enough rays to cover sufficiently well all the blocs of the model. For each of the models, the inversion solutions for the isotropic, as well as the anisotropic case (i.e. incorporating the anisotropic traveltime correction for the $P_n$-phases) are examined and the sensitivity of the solution to the data is estimated by means of various tests for resolution, covariance and other trade-off characteristics of the data and the model space (see Chapter 4).

For some models, two additional computations for each of the isotropic and the anisotropic case are shown, whereby the hypocenters are fixed to their initial value (set by the parameter IFIX=1 in the SSH-program) as reported in the original BGR-dataset and localized by other institutions during the routine localization process. Such a comparison seems to be necessary to look at the dependency of the inverted velocity models upon the localization of the hypocenters, i.e. the influence of the latter on the former and to see if there are distinct differences between two model approaches when the hypocenters are free (IFIX=0) or fixed (IFIX=1). However, the details regarding the relocation of the hypocenters will be deferred to Chapter 7. In this Chapter, we only concentrate on the inversion results for the seismic velocity structure.

One critical task in 3D tomography and seismic inversion is the parametrization of the models. The term parametrization is meant to be the discretization of the 3D model volume into layers and each of them laterally into blocs. Having not enough blocs in a model will result in a geometrical resolution too low, having too many blocs may provide ambiguous bloc velocities with high covariances, as noise in the data may deliver senseless information about the model space and gaps may occur where too few, or even no rays at all travel through the individual blocs, carrying not enough information with regard to traveltime perturbations. In other words, in such an over-parametrized model, structural differences in the blocs of the model result in the same data fit, i.e. the inversion becomes non-unique.

Based on these facts, the most important goal here is to have sufficient ray coverage for an individual bloc within the model which, necessarily, becomes less as the size of the bloc is decreased, i.e. the number of blocs in the model is increased. This problematic nature is visible in Figures 6.24, 6.29 and 6.31, where the areas with good or sufficient coverage become smaller as the number of blocs is increased[2].

The selection of the optimal inversion solution is done at the first sight by selecting those solutions with minimal RMS or TSS in each external (nonlinear) iteration step of the Levenberg-Marquardt minimization technique (see Chapter 2, for details). Anyways, as it will become obvious during the analysis of the synthetic tests to be performed beforehand of the SSH inversion proper, this is not the only criterion. In fact, additional trade-off curves for various parameters of the data- as well as the model-space will be investigated which will show that the optimal inversion solution will most likely be located within a certain range.

To finally rate and evaluate the resulting models with regard to the influence of the elliptical anisotropic correction for the $P_n$-phases, the F-Test can be used (Seber and Wild, 1989; Koch, 1985b, 1989, 1993a). With the F-Test (see section 6.7), one can decide whether there are statistically significant differences in the variances between two statistical populations, namely, if one particular inverted SSH-model (seismic velocity structure or relocated hypocenters) is significantly better than another one. For details of the application of the F-test in seismic tomography we refer the reader to (Song et al., 2004)

---

[2]To reduce this problematic nature of low or insufficient ray coverage, it would be desirable to have different bloc-sizes within a particular layer of the model, depending on the ray coverage there. However this feature is not yet implemented in the present SSH program.

and (Koch, 1985b, 1989, 1993a). In the present case one would compare the variances of the model fits (the square of the RMS) for the isotropic and anisotropic inversion models, respectively, and, based on the outcome, decide whether the application of the anisotropic $P_n$-traveltime correction has led to a significant improvement or not. It should be noted that the F-Test is not suitable for the synthetic tests, as these are generated with initially anisotropic models, so that, naturally, there is a rather big difference between the data fit variances. However, when using the real, original data set, the structural information is hidden in the arrivaltime data and much more complex and some of the inherent anisotropic structure in the upper mantle underneath Germany may be projected into the velocity structure when inverting the model isotropically. Thus, one cannot expect the differences in the variances to be that large as with the synthetic models.

## 6.2 Synthetic random resolution tests

### 6.2.1 General approach

Before carrying out the real 3D tomography study of the crust and upper mantle beneath Germany, various tests will be performed with synthetically generated seismic velocity anomalies for the purpose of evaluating the capabilities of the data set at hand to properly resolve the model space. This is done for both isotropic and anisotropic model cases. In fact, executing the tests with an anisotropically generated structure (as this can now be taken as a fact for the upper mantle beneath Germany), but inverting the theoretical traveltimes generated in this way without the anisotropic elliptical correction for the $P_n$-phases should show the influence of that correction. On the other hand, one could also use a second approach whereby an isotropic test model for the calculation of the theoretical traveltimes is used. The subsequent inversions should then, in principle, deliver opposite results, namely, the anisotropic inversion should be much more biased than the isotropic one. But, since we accept *a priori* the hypothesis of the existence of anisotropy in the upper mantle, the first test method with the anisotropic correction included in the forward calculation appears to be more appropriate then the latter one. There has been some discussion in the scientific literature on how to best execute such resolution tests. It appears that testing with randomly distributed velocity anomalies may be the best approach and better than the usual checkerboard test (to be carried out in Section 6.5), as an irregular velocity structure reconstruction reacts more sensitive to the traveltime data (Husen, 1999; Haslinger and Kissling, 2001).

For the random synthetic test, starting with a laterally homogeneous, layered model, the seismic velocities are perturbed randomly for several selected blocs within each layer to simulate a realistic 3D seismic structure. This has the effect that the mutual influence of velocity variations in adjacent layers on each other can be recognized. Having two or more neighboring anomalies with opposite signs may end up in a partial canceling of the ensuing traveltime residuals, so that the former cannot be clearly separated by the inversion process. It follows that such a random test may be the most extreme test of the resolution capabilities of the present dataset (Kissling et al. (2001); Husen and Kissling (2001); Husen et al. (2003)). A further advantage of this random test, as compared to the checkerboard test, is, that differences in the resolution capability and possible shifts of the reconstructed bloc anomalies may be more easily detected.

The synthetic tests are done for three different bloc discretization of the model, which are $15 \times 15$, $25 \times 25$ and $35 \times 35$ blocs, similar to the ones used in the subsequent real tomography study. When generating the synthetic dataset, i.e. the theoretical traveltimes for the model to be tested, the same distribution of the recording stations and of the hypocenters as available in the original data set is employed. For easier description of the areas, the following plots are considered to be separated into quadrants and sub quadrants. The quadrants extend from the borders of the model space to the middle lines at 10° Longitude and 45.5° Latitude. Each of them is separated into sub quadrants, approximately following the rectangular areas between the **Longitude** and **Latitude lines** in the plots, shifted downwards by 0.5° in the case of the Latitudes.

## 6.2.2 Synthetic test with 15x15 blocs

The input model is the optimal anisotropic 1D velocity model of the previous Chapter, divided into $15 \times 15$ homogeneous blocs, where to some of them a certain velocity anomaly has been attributed to, to mimic the synthetic structure as shown in Figure 6.1(a). In the following both isotropic and anisotropic inversions of the originally anisotropic model are considered.

The velocity structure of the **isotropic inversion** (Figure 6.1(d)) looks completely different from the original (anisotropic) model. While the reconstruction in the first layer and in the fourth quadrant of layer two is still possible, the reconstruction completely fails for the third and fourth layer. The anomaly in the fourth layer is still visible if one knows its existence, but in reality, it is hard to identify it among the other extremely strong variations of the structure inverted there.

In contrast to the isotropic (Figure 6.1(d)) inversion, the inversions with the **elliptical anisotropic correction** (Figure 6.1(b)) for the $P_n$-phases indicate a much better reconstruction, especially for the third and fourth layer of the model where the anisotropic correction for the $P_n$-phases actually takes effect. Even more, the velocity reconstructions in the first and second layer are better as well, because the anisotropic information in the $P_n$-phases is now correctly interpreted and not projected into these layers, as it appears to be the case for the isotropic inversion before. Indeed, this projection will lead to artifacts in the border areas of these layers, as can be seen in Figure 6.1(d) for the isotropic inversion. Thus we can conclude that for the anisotropic inversion most of the synthetic velocity anomalies are well reconstructed, though sometimes with smaller contrasts than those of the original ones.

While the anisotropic inversion with K = 1, i.e. the smallest damping parameter $\lambda$ delivers rather good results for the reconstruction, the anomalies are more and more suppressed with increasing damping value (Figure 6.2) and the optimal solution of this model case (defined as the solution with the minimal RMS) barely reveals any reconstructed synthetic anomalies (Figure 6.1(c)). With a larger damping parameter the inverted velocity anomalies are more suppressed, so that the final model tends towards the initial 1D homogeneous starting velocity model.

Figures 6.3 shows the development of the velocity solution for the second iteration step (I = 2). One notes here also that for the lowest damping parameter the inverted velocity anomalies are the strongest and resemble the original ones the best, though some artifacts appear at the borders of the model. For the optimal solution (minimal RMS) the velocity anomalies are somewhat dampened and nearly disappear in the first two layers in contrast to the optimal model of the first iteration, whereas the reconstruction in the third and fourth layer is more consistent with the initial model (Figure 6.1(c)). This is an astonishing result which has to be investigated further.

Concerning the simultaneously reconstructed hypocenter locations, the most salient results to that regard are shown in Figures 6.4 and 6.5 for both the isotropic and anisotropic inversion cases. As one can see from Figure 6.4, the horizontal relocation shifts of the epicenters are within a range of only a few hundred meters for the anisotropic case, but within a much larger range of up to 1 $km$ and more for the isotropic case. A similar situation holds for the vertical (depth) shifts of the hypocenters (Figure 6.5) where, again, the anisotropic inversion produces less vertical scatter, i.e. the hypocenters are better constrained to their initial (theoretically true) values, than the isotropic inversion.

In conclusion of this section, one can say that if one assumes an anisotropic upper mantle in the theoretical model, only a true anisotropic SSH inversion (i.e. correcting the $P_n$-phases for elliptical anisotropy) has the potential to deliver a "trustworthy" reconstruction of both the structural velocities and the hypocenters. Neglecting the anisotropy results in a largely biased and unstable model reconstruction. Even so, the results also indicate that the proper choice of the optimal model, i.e. the optimal damping parameter, is not easy and fraught with some intricacies, as there appears to be some discrepancy between the model which best reconstructs the structural anomalies and the one which minimizes the observed traveltime residuals (the RMS). This issue will be discussed further in subsequent sections.

## 6.2. SYNTHETIC RANDOM RESOLUTION TESTS

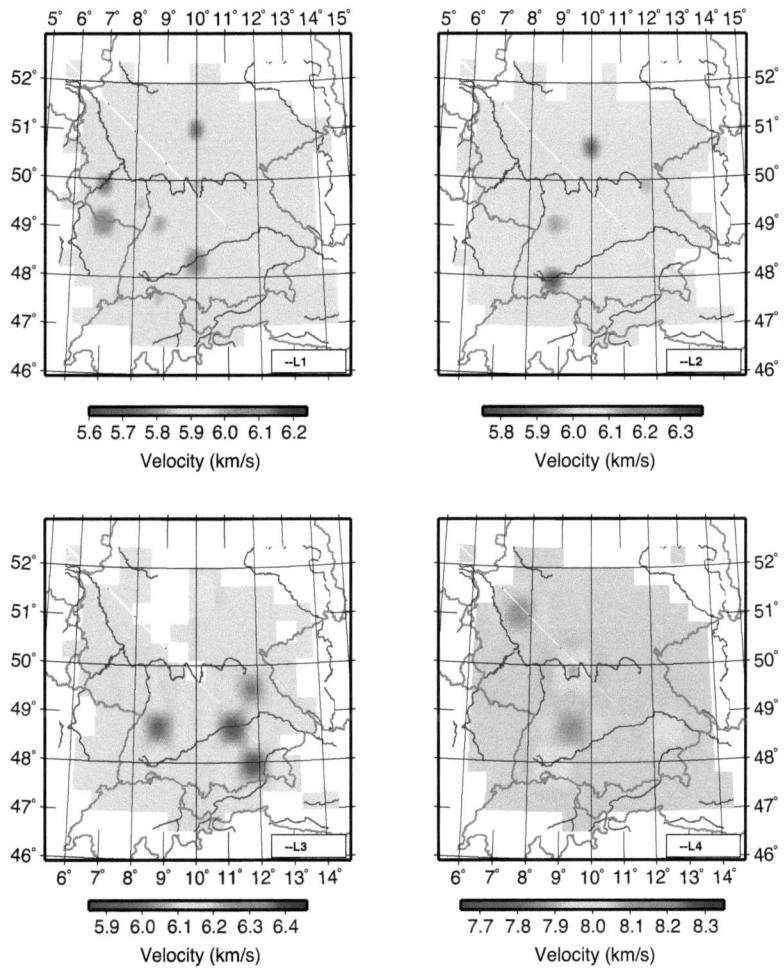

(a) input

106                    CHAPTER 6. SSH-INVERSIONS FOR 3D VELOCITY MODELS

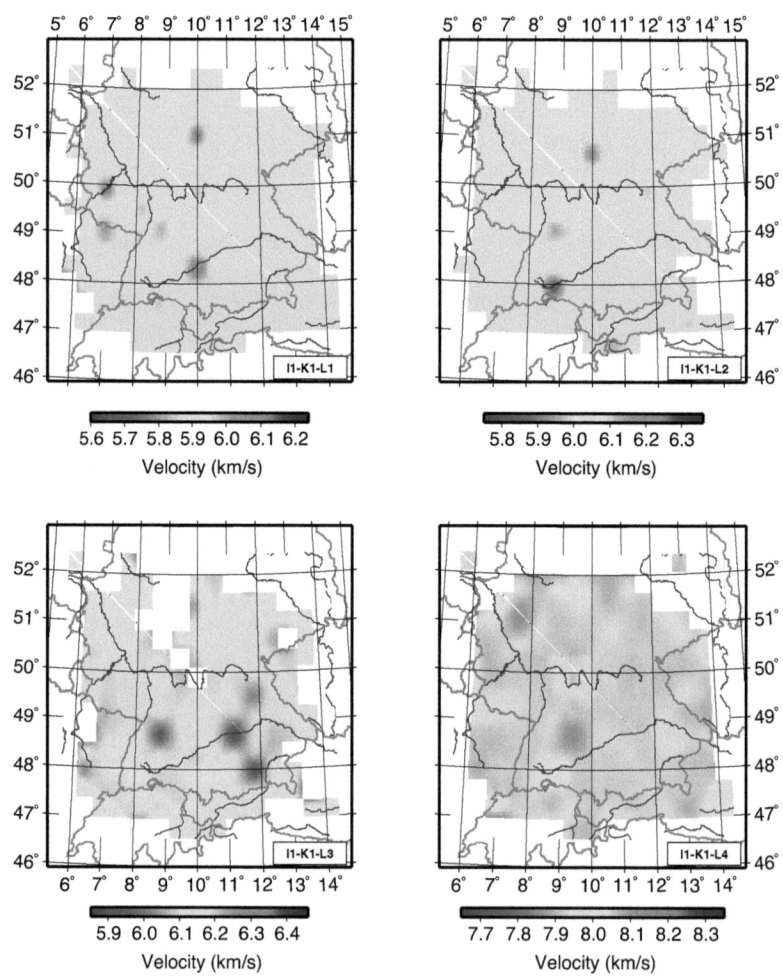

(b) ani-min, I1, K1

## 6.2. SYNTHETIC RANDOM RESOLUTION TESTS

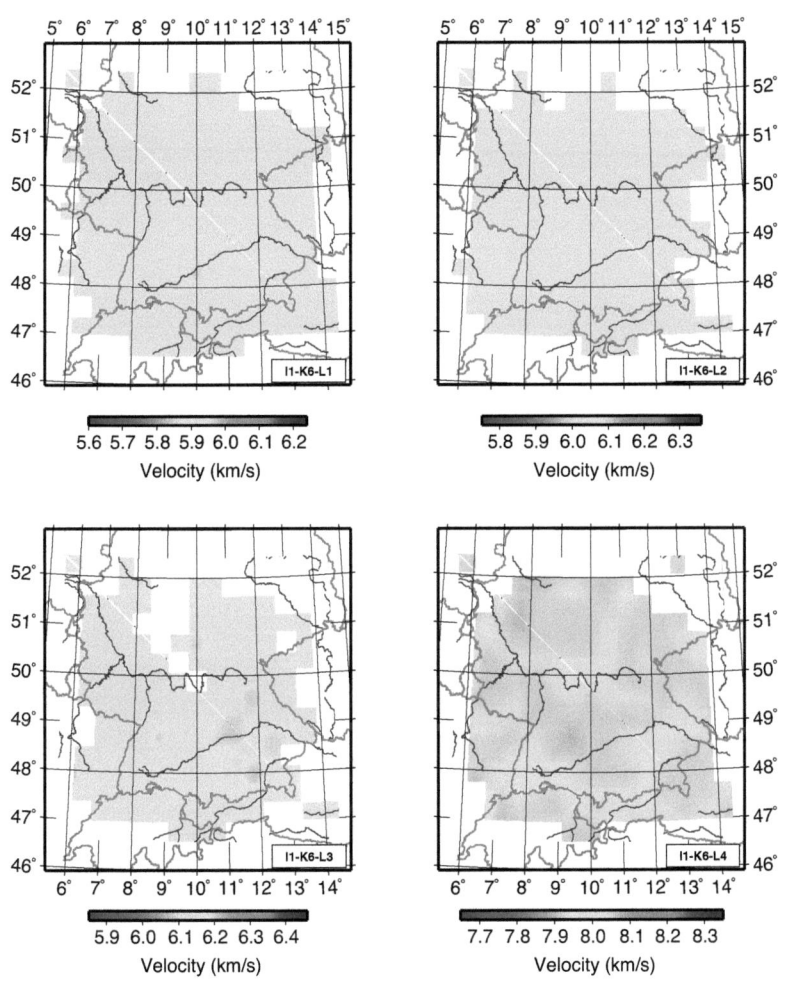

(c) ani-opt, I1, K6

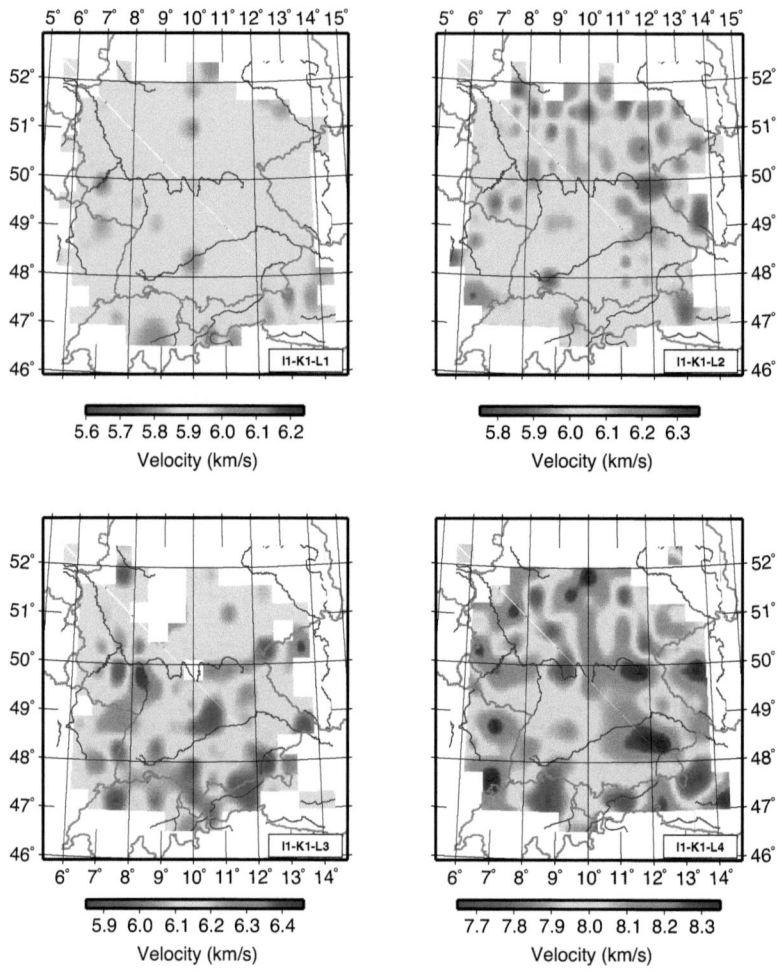

(d) iso, I1, K1

Figure 6.1: Synthetic random test with 15 × 15 blocs. The panels show the input model (first set, Figure 6.1(a)); inversion results for the first iteration with a low damping value (second set, Figure 6.1(b)), the optimal inversion result (minimal RMS) (third set, Figure 6.1(c)) and results for the isotropic inversion with no elliptical anisotropic correction included (last set, Figure 6.1(d)).

## 6.2. SYNTHETIC RANDOM RESOLUTION TESTS

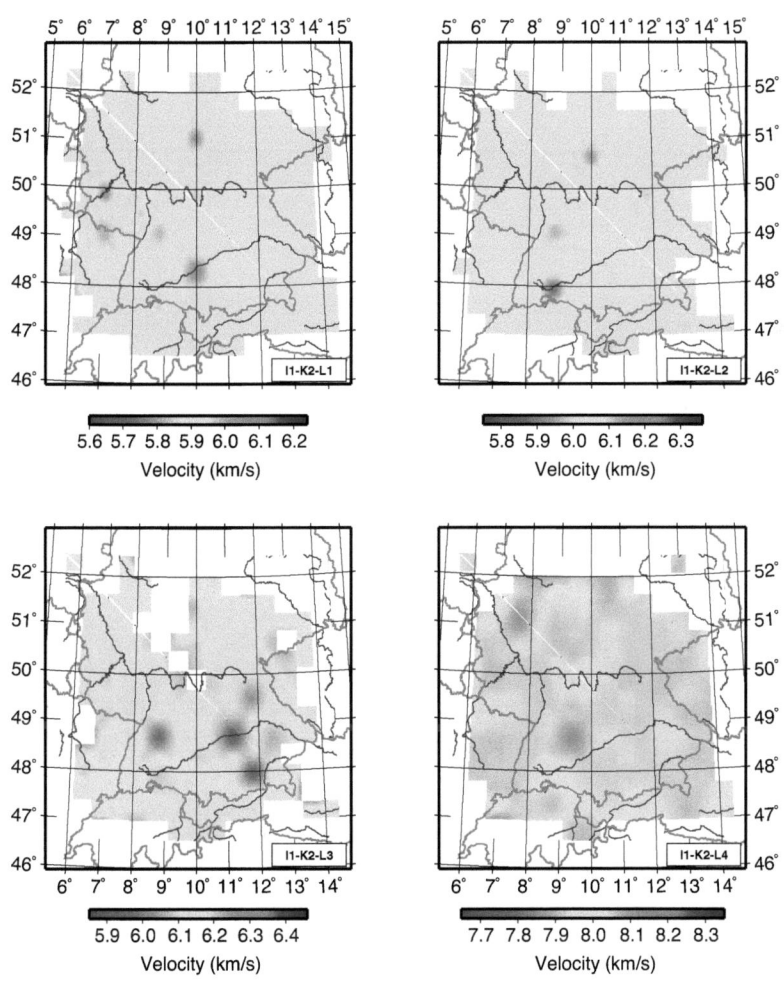

(a) ani, I1, K2

110  CHAPTER 6. SSH-INVERSIONS FOR 3D VELOCITY MODELS

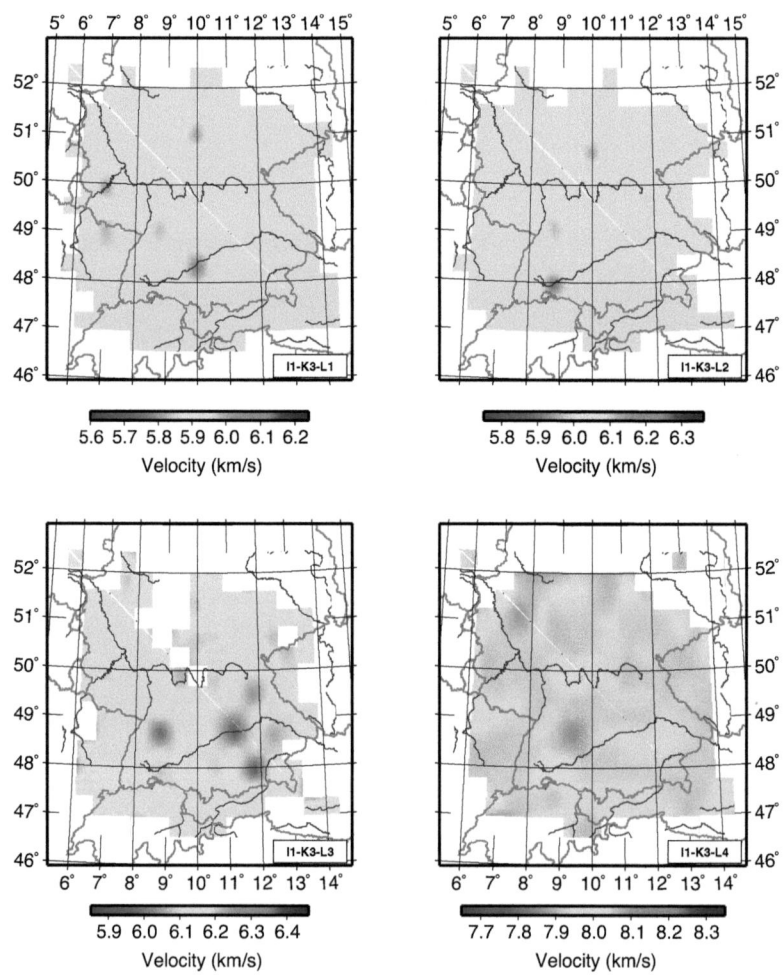

(b) ani-min, I1, K3

## 6.2. SYNTHETIC RANDOM RESOLUTION TESTS

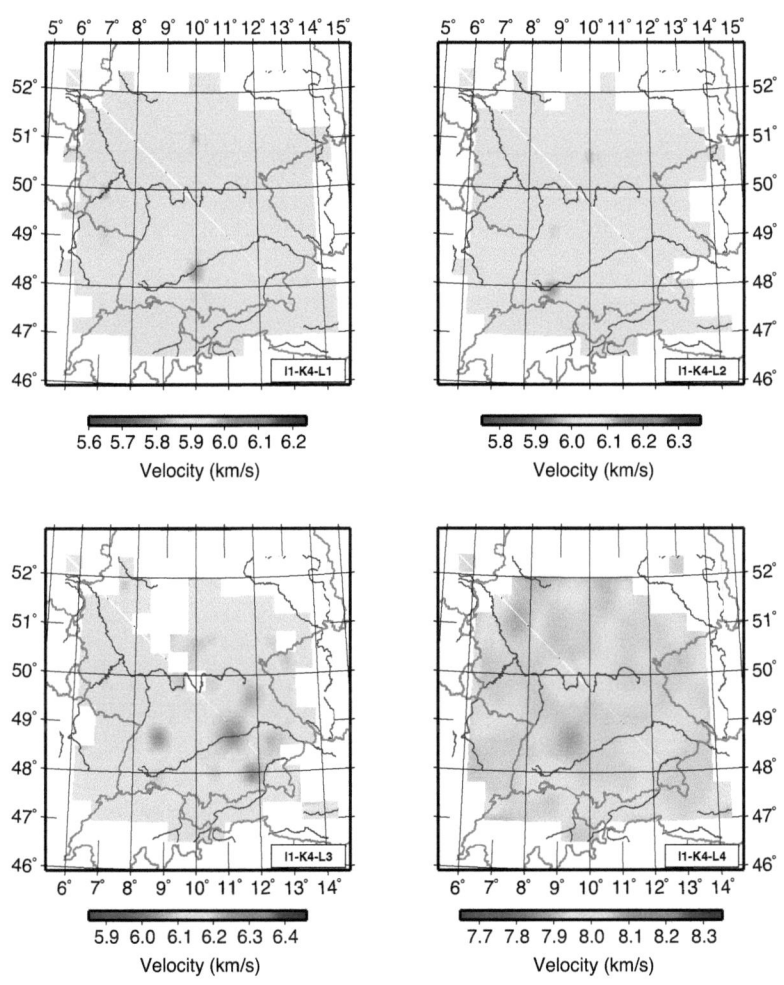

(c) ani-opt, I1, K4

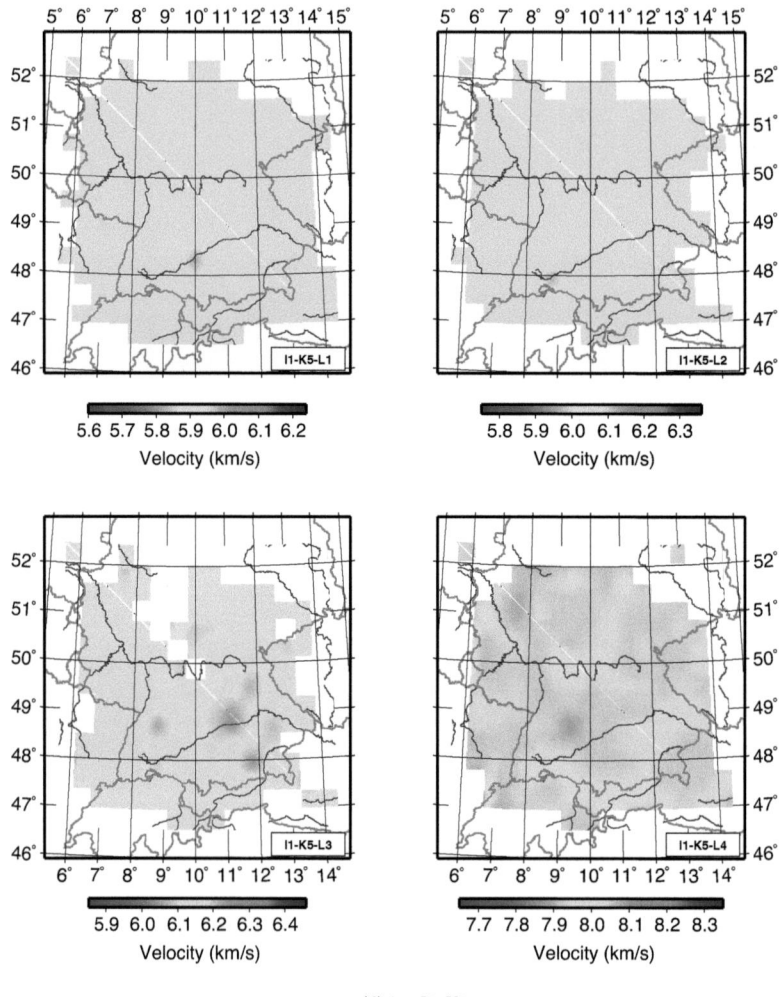

(d) iso, I1, K5

Figure 6.2: Similar to Figure 6.1, but for the inversions with K = 2 to 5 to show the fading of the reconstructed anomalies.

## 6.2. SYNTHETIC RANDOM RESOLUTION TESTS 113

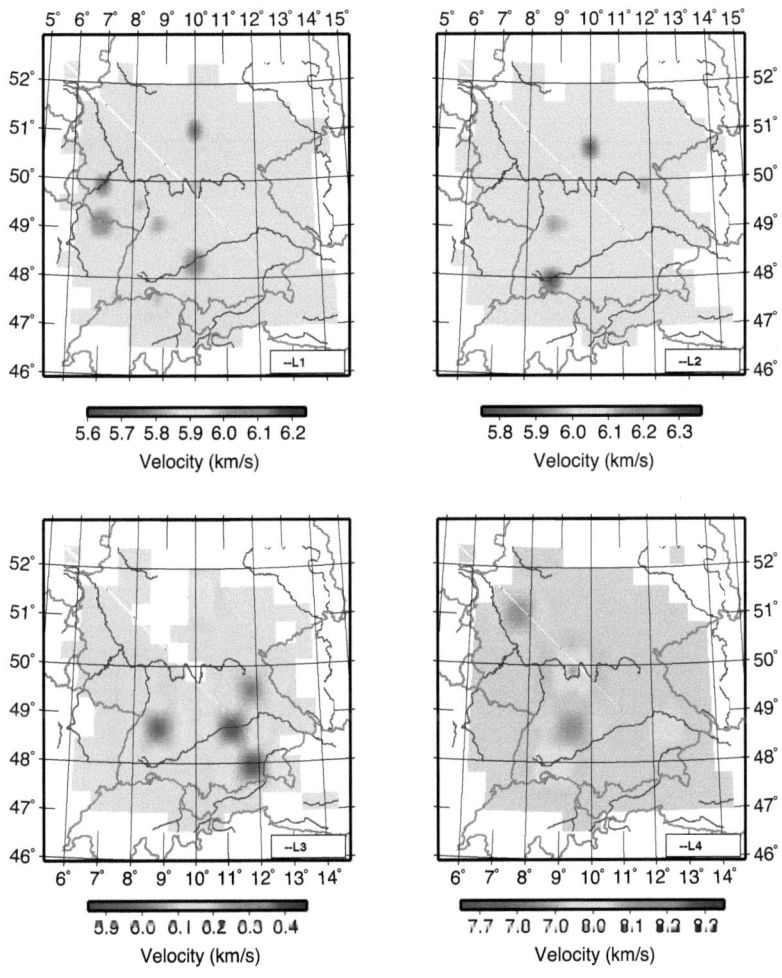

(a) input

114  CHAPTER 6. SSH-INVERSIONS FOR 3D VELOCITY MODELS

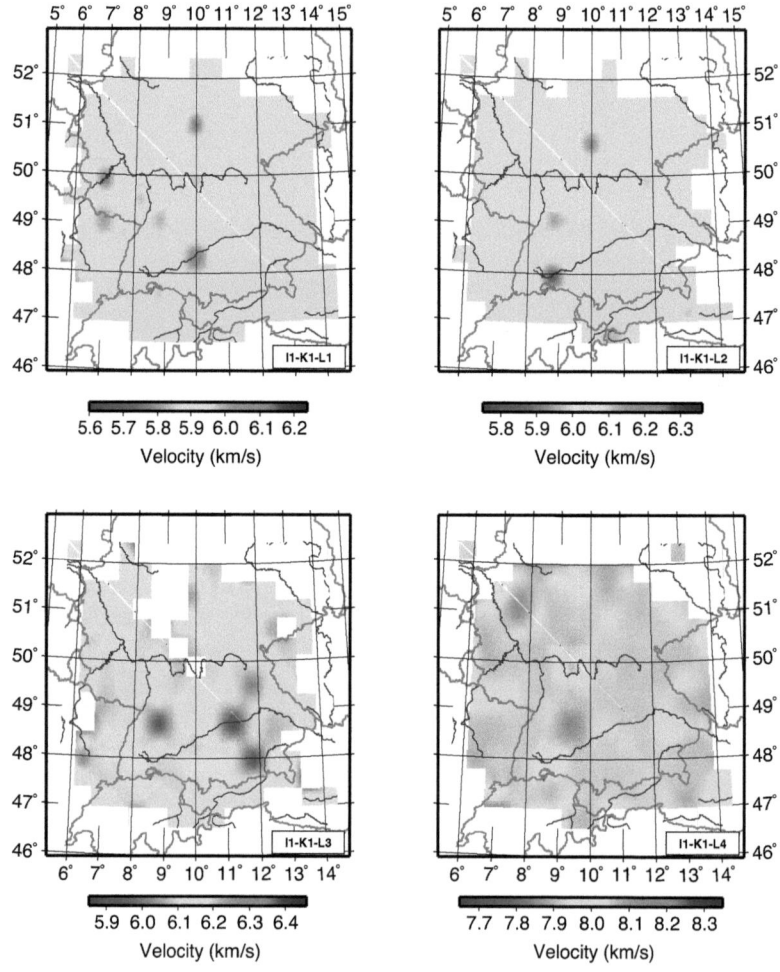

(b) ani-min, I1, K1

## 6.2. SYNTHETIC RANDOM RESOLUTION TESTS

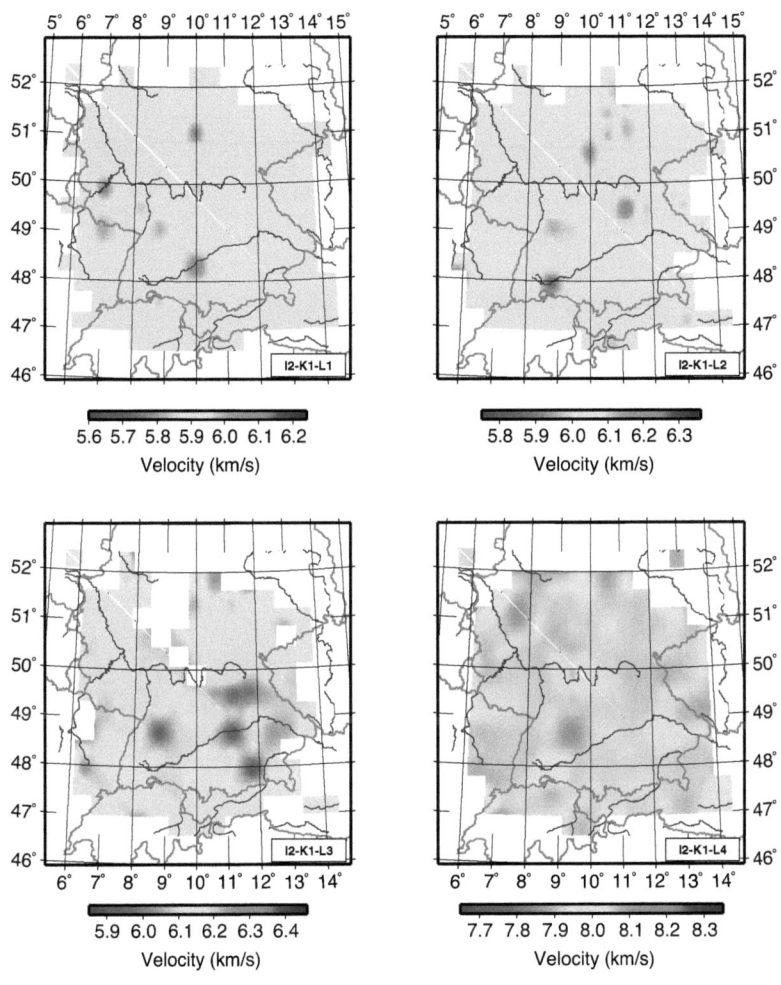

(c) ani-min, I2, K1

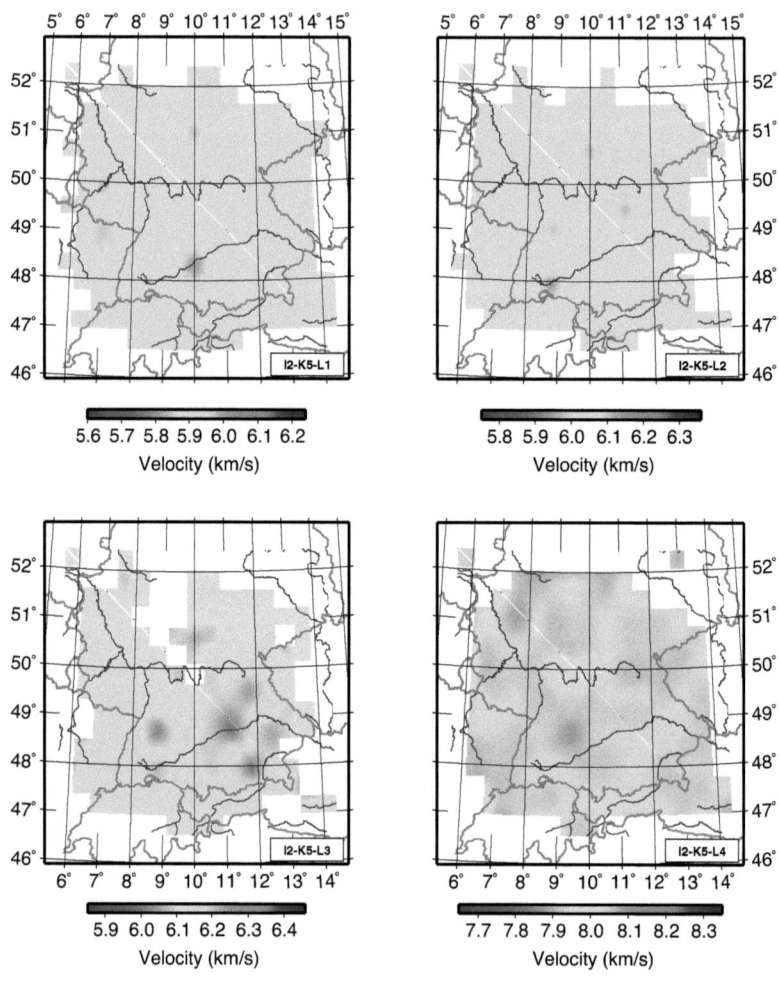

(d) ani-opt, I2, K5

Figure 6.3: Synthetic random test with 15 × 15 blocs, second nonlinear iteration I = 2. The first set of plots (Figure 6.3(a)) shows the input model and the second set (Figure 6.3(b)) the inversion results for the first iteration with a low damping value. The third set of plots shows the inversion results for the second iteration, minimal $\lambda$ at K = 1 (Figure 6.3(c)) and for the optimal RMS at K = 5 (fourth set, Figure 6.3(d)).

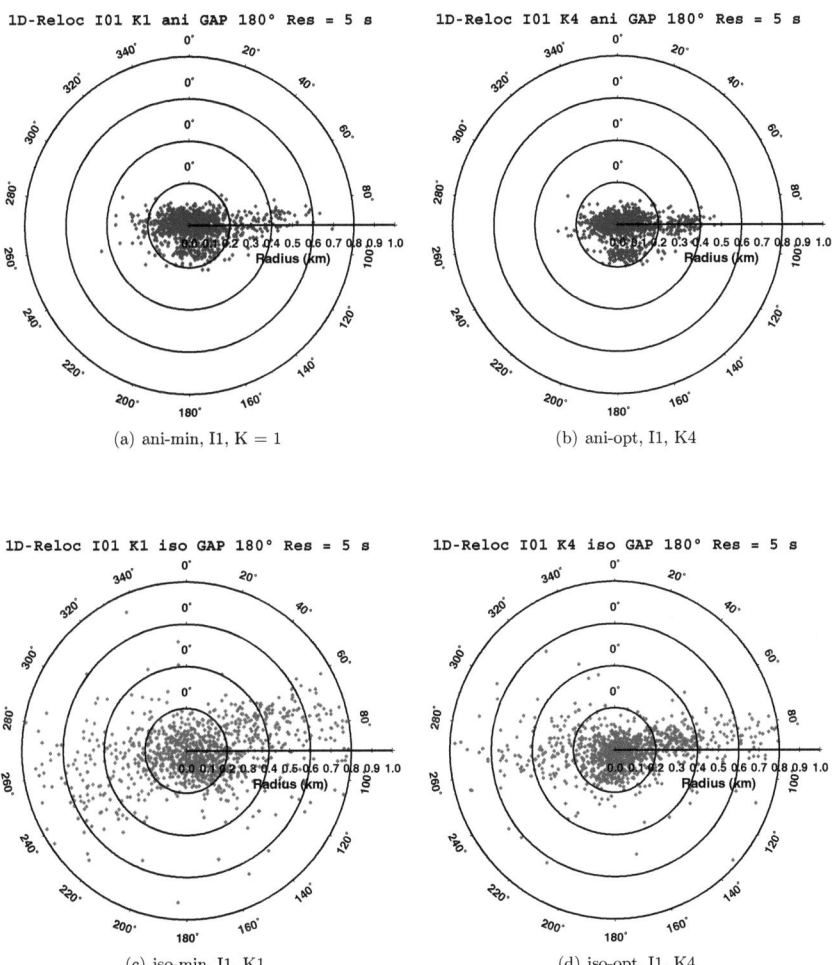

Figure 6.4: The results for the hypocenter shift in the anisotropic and isotropic inversion, both cases with smallest and optimal $\lambda$ for minimal RMS.

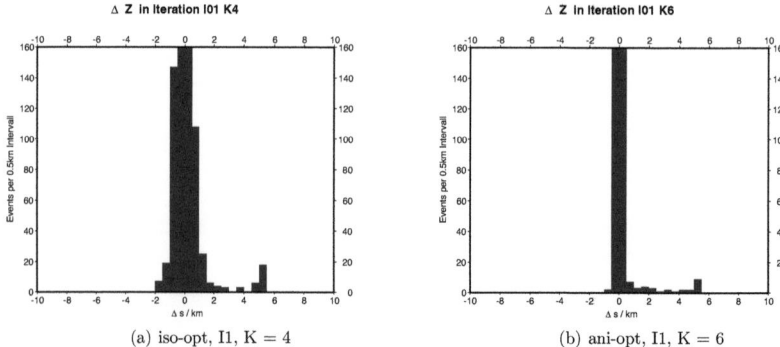

Figure 6.5: The results for the depth shift of the hypocenters in the anisotropic and isotropic inversion.

### 6.2.3 Synthetic test with 25x25 blocs

In this section the synthetic resolution test is applied to models with a finer discretization of $25 \times 25$ blocs. To make the tests more realistic, an additional model with the theoretical traveltimes perturbed by random noise with $\sigma = 0.1\,s$, which is a realistic value for reading errors in the pick-of a seismic phase in a seismogram, has been inverted for.

The results of the model inversions without and with noise added, respectively, are illustrated in Figure 6.6. One notes again that good results are obtained for small $\lambda$, where the RMS is again not at the minimum. From the pictures one notes that the input anomalies are best reconstructed for the first three layers, whereas the fourth layer again has a lot of undulations which make it difficult to identify some structure. Unfortunately, the picture looks different for the optimal model, i.e. the one with the minimal RMS. Here, the structure in the first three layers nearly disappears, but the anomalies in the fourth layer become slightly visible, whereas the rest of the velocity structure in this layer is rather homogeneous in contrast to the result with the lowest damping value. This is a remarkable result, which has to be considered in the later interpretations of the results for the "true" seismic structure underneath Germany.

The reconstruction of the structure is again better for inversions with small $\lambda$'s, where the anomalies are quite good visible, especially in the first two layers. The structure in the 4th layer can not be identified, because the anomalies in the 4th layer become visible, but the results are ambiguous for inversions with low damping values $\lambda$, whereas the structure in the other layers is suppressed.

Adding some noise during the calculation of the new arrivaltimes and inverting these data leads to a rather different result for layer four. Fortunately, the result for layer one is still good, for layer two in the middle of the area and in layer three only in the middle. The undulations in layer four become rather high, so it will be impossible to detect any anomaly except its velocity exceeds the range of the undulations.

## 6.2. SYNTHETIC RANDOM RESOLUTION TESTS

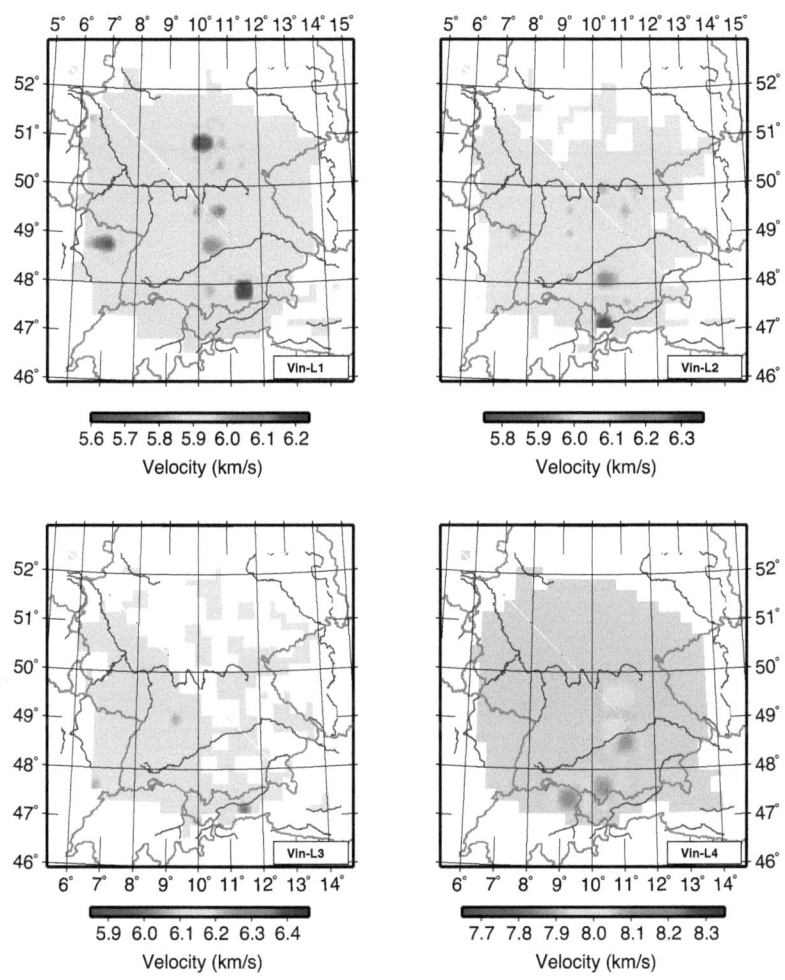

(a) input

120                    CHAPTER 6. SSH-INVERSIONS FOR 3D VELOCITY MODELS

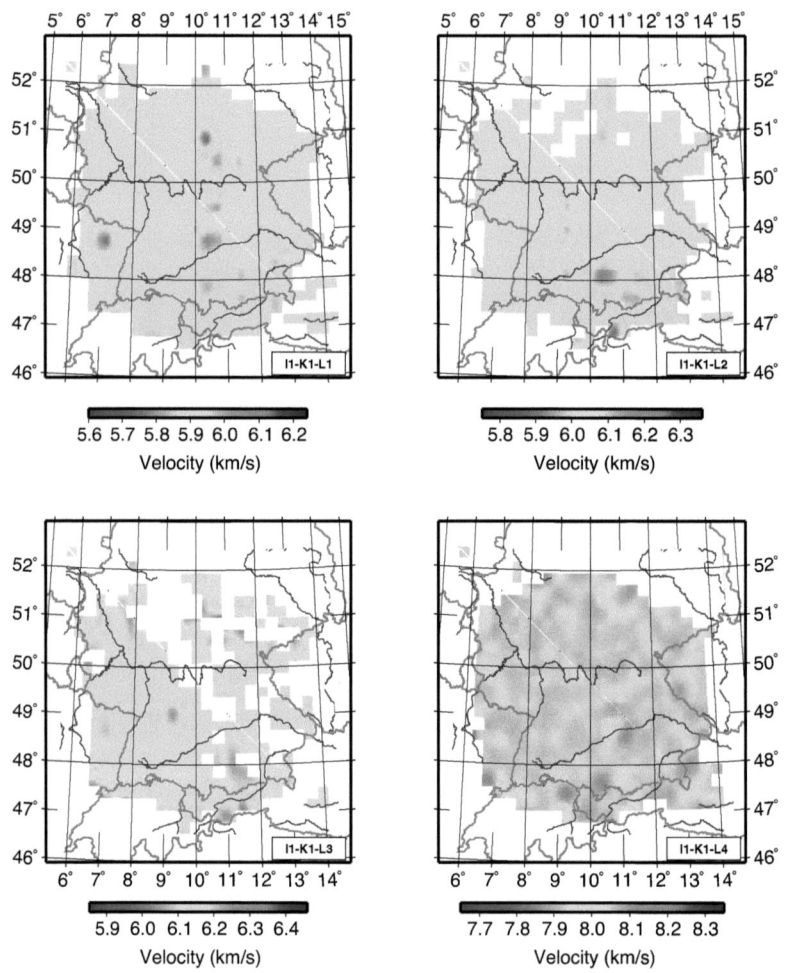

(b) ani-min, K1

## 6.2. SYNTHETIC RANDOM RESOLUTION TESTS

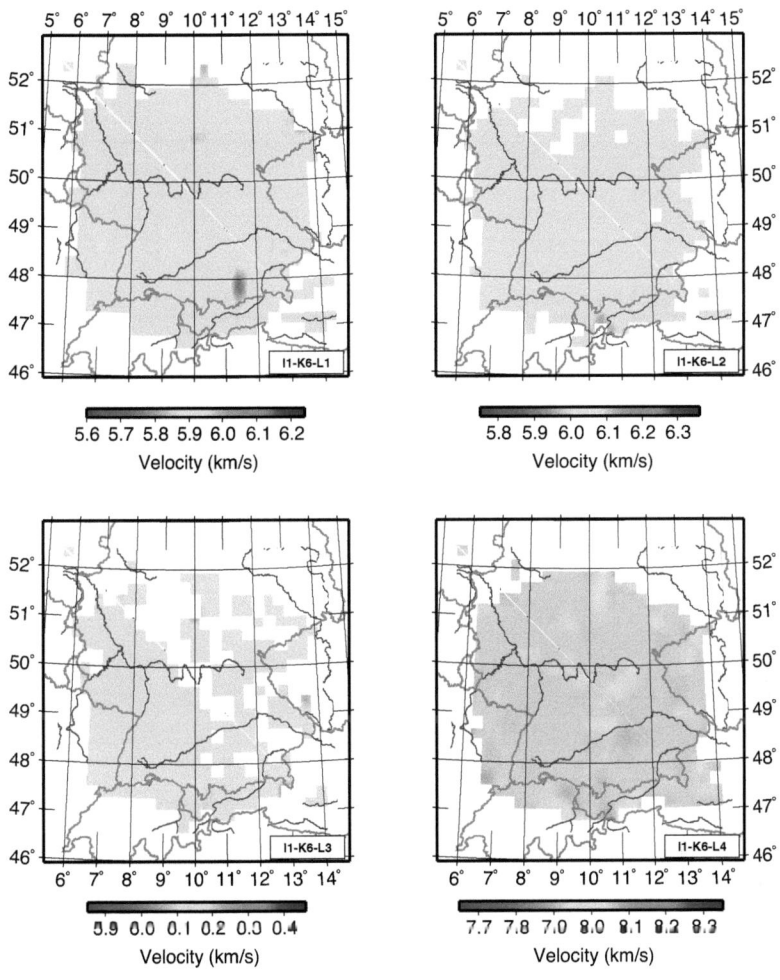

(c) ani-opt, K6

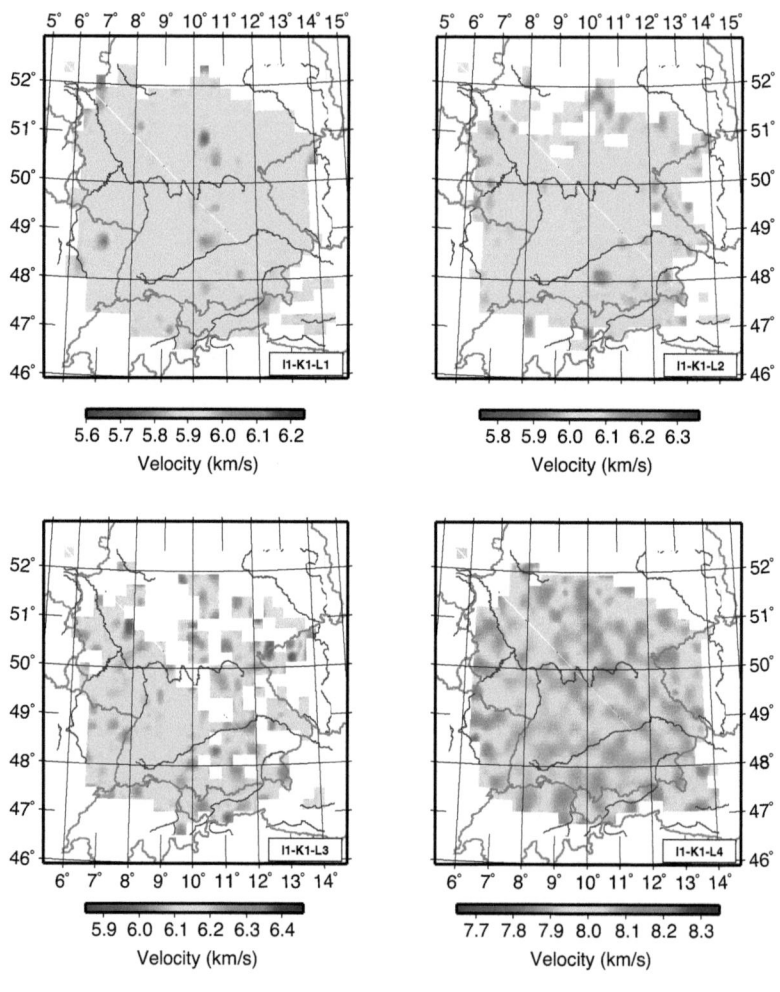

(d) ani-noise, K1, $\sigma = 0.1\,s$

Figure 6.6: Synthetic random test with $25 \times 25$ blocs. The plots show the input model (first set, 6.6(a)), the results for the first iteration with low damping parameter (second set, Figure 6.6(b)), the results for the minimal RMS (third set, Figure 6.6(c)) and the results for noisy traveltimes ($\sigma = 0.1\,s$) and low damping parameter (last set, Figure 6.6(d)).

## 6.2. SYNTHETIC RANDOM RESOLUTION TESTS

Table 6.1: Statistical values for the relocations of the original hypocenters after full SSH inversions of the different synthetic random test models discussed. The first line of the distinct models contain the result for the lowest damping value $\lambda$ and the second one those with the smallest RMS until now considered to be the optimal solution.

| Model | ani | I,K | Variance | | | Standard deviation | | |
|---|---|---|---|---|---|---|---|---|
| | | | x $km^2$ | y $km^2$ | z $km^2$ | x $km$ | y $km$ | z $km$ |
| 15 x 15 | no | 1,1 | 0.1438 | 0.1089 | 1.2706 | 0.380 | 0.331 | 1.127 |
| | no | 1,4 | 0.0733 | 0.0356 | 0.7754 | 0.271 | 0.190 | 0.880 |
| | yes | 1,1 | 0.0129 | 0.0126 | 0.5313 | 0.113 | 0.112 | 0.728 |
| | yes | 1,6 | 0.0036 | 0.0044 | 0.3279 | 0.060 | 0.066 | 0.572 |
| 25 x 25 | yes | 1,1 | 0.0059 | 0.0053 | 0.5829 | 0.078 | 0.073 | 0.755 |
| | yes | 1,6 | 0.0015 | 0.0039 | 0.3513 | 0.038 | 0.062 | 0.592 |
| 35 x 35 | yes | 1,1 | 0.0064 | 0.0057 | 0.5044 | 0.081 | 0.075 | 0.702 |
| | yes | 1,9 | 0.0000 | 0.0000 | 0.0101 | 0.004 | 0.004 | 0.100 |

Table 6.2: Some TSS values along the ridge parameter $\lambda$, named K and for several bloc discretizations.

| K | blocs | ani | 0 | 1 | 2 | 3 | 4 | 5 | 6 | 7 | 8 | 9 |
|---|---|---|---|---|---|---|---|---|---|---|---|---|
| TSS | 15x15 | no | 1599 | 2218 | 2163 | 2063 | 931 | 974 | 1071 | 1235 | 1291 | 1431 |
| | | yes | 215 | 282 | 280 | 252 | 241 | 252 | 97 | 115 | 140 | 167 |
| | 25x25 | yes | 202 | 192 | 160 | 151 | 133 | 131 | 110 | 109 | 131 | 157 |
| | 35x35 | yes | 29 | 130 | 128 | 119 | 91 | 57 | 32 | 32 | 27 | 26 |

### 6.2.4 Synthetic test with 35x35 blocs

The discretization into $35 \times 35$ blocs for each layer is the finest one used in the present study, and also gives the limit for the reconstruction capabilities. This is due to the decreasing ray coverage per bloc, which is at the lower limit for most of the blocs in the volume (see Figure 6.31), which is not anymore homogeneous enough to resolve the whole structure[3].

The interpretation of the results follows the same line as that of the previous $15 \times 15$ and $25 \times 25$ test models. For the noise-free model of 6.7(d), the velocity anomalies are still well reconstructed, though this is achieved only for small damping parameters which does not yet give the supposed optimal model, i.e. the one with the minimal RMS of the nonlinear least-squares model fit to the data. To find this optimal solution the damping parameter has to be increased which has the effect of again wiping out the anomalies completely.

The situation is more detrimental for the complementary model with random noise of $\sigma \pm 0.1\,s$ added to the theoretical traveltimes (Figures in 6.8). The reconstruction becomes now much worse particularly in the second, third and fourth layer. In the second and third layer, there are so many artifacts, that the original velocity anomalies cannot be identified for a low damping parameter (K = 1). For the optimal model with the minimal RMS (at K = 8), again, the anomalies in the first three layers have completely gone, whereas the ones in the fourth layer are still slightly visible. This shows the importance of an accurate phase detection for higher discretized models.

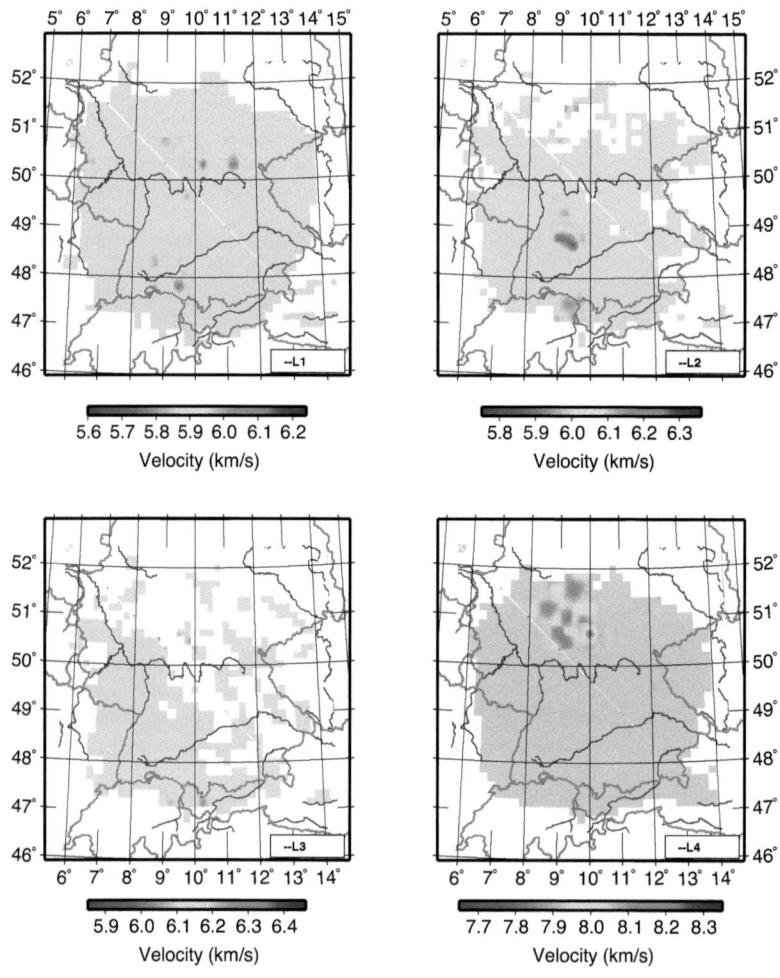

(a) input

## 6.2. SYNTHETIC RANDOM RESOLUTION TESTS

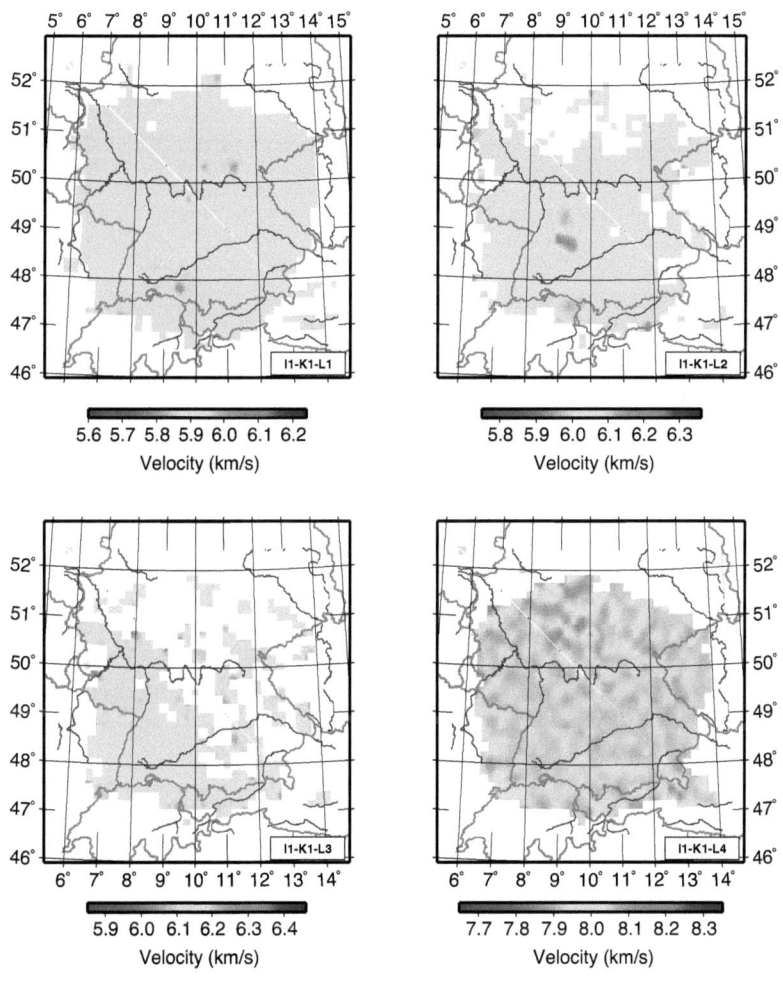

(b) ani-min, K1

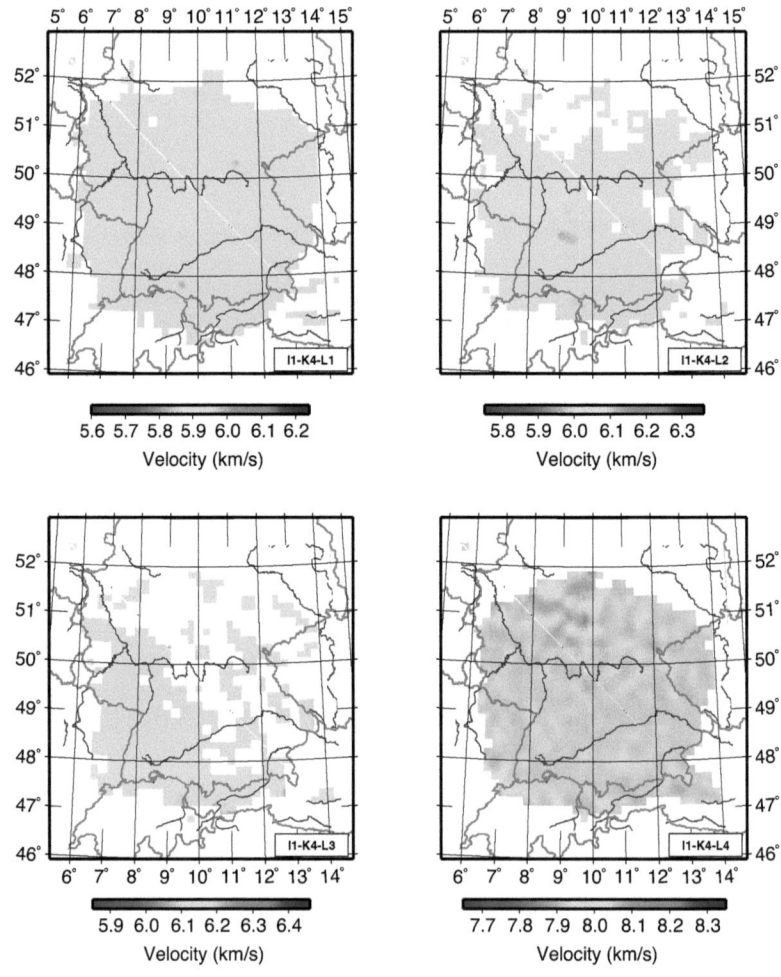

(c) ani, K4

## 6.2. SYNTHETIC RANDOM RESOLUTION TESTS

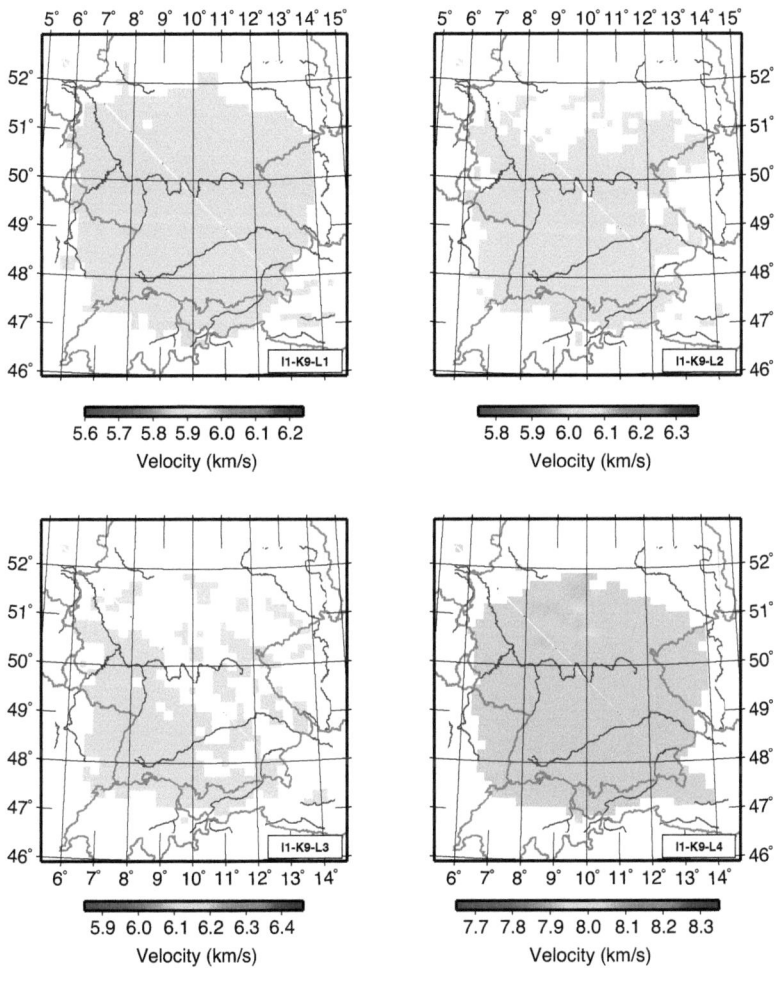

(d) ani-opt, K9

Figure 6.7: Synthetic random test with $35 \times 35$ blocs with the initial model in Figure 6.7(a), results with low damping parameter in Figure 6.7(b), results with a damping value of median size (at K = 4) in Figure 6.7(c) and the optimal model (minimal RMS) in Figure 6.7(d).

128    CHAPTER 6. SSH-INVERSIONS FOR 3D VELOCITY MODELS

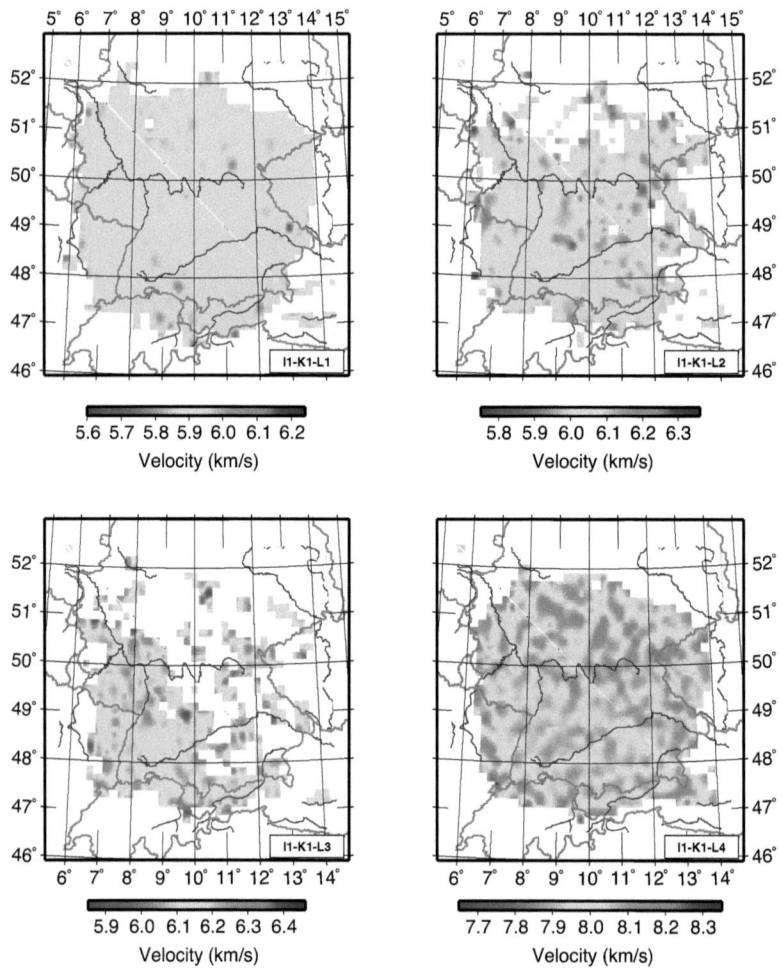

(a) ani-min, K1, noise

## 6.2. SYNTHETIC RANDOM RESOLUTION TESTS

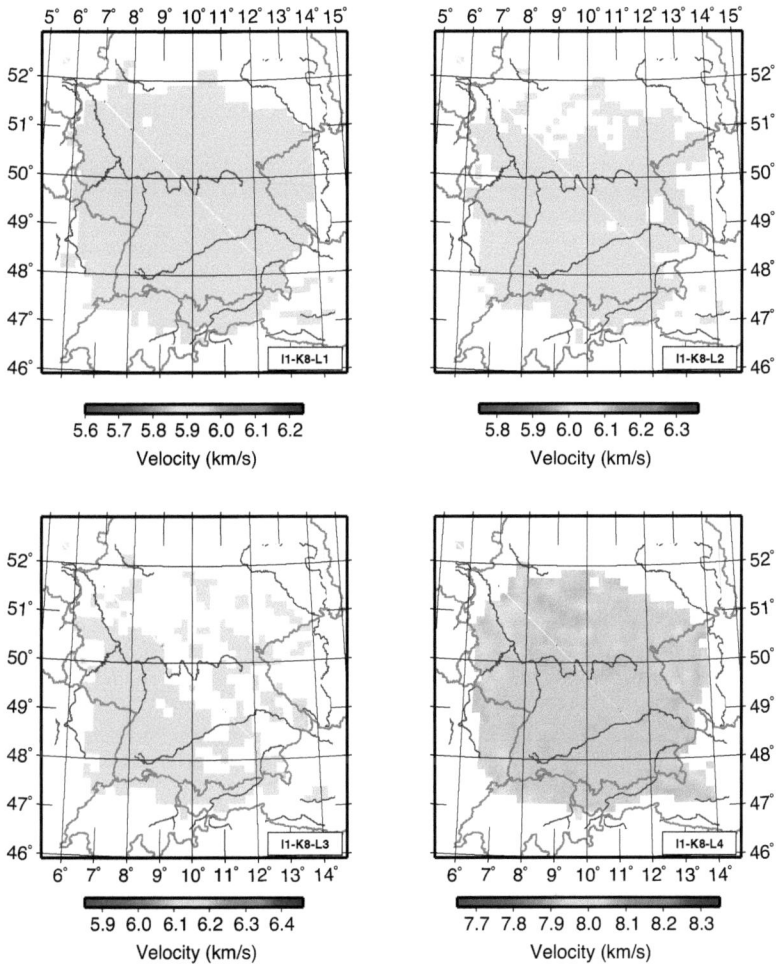

(b) ani-opt, K9, noise

Figure 6.8: Synthetic random test with $35 \times 35$ blocs and noise with $\sigma = 0.1\,s$ in the theoretical data. Shown are the inversions with a small damping parameter in Figure 6.8(a) and the optimal one (minimal RMS, K = 8) in Figure 6.8(b).

### 6.2.5 Final remarks on synthetic random tests

As the investigations for the various synthetic random models show, the identification of the appropriate damping parameter $\lambda$ for the best reconstruction of the original model might be an intrinsic difficulty. Thus, the best solution might be found in the $\lambda$-range that extends from a very small value to the one where the RMS is smallest, i.e. the optimal solution is obtained. As expected, the reconstructed anomalies vanish with increasing $\lambda$ but, at the same time, the instabilities in some of the layers, especially layer four are also reduced. The appropriate damping factor $\lambda$ here is the smallest one for the first three layers because of the best reconstruction of the synthetic anomalies. While further increasing the damping value, the RMS becomes better, but the reconstruction for the structure fades away, leaving only a nearly homogeneous structure.

The important influence of the elliptical anisotropic $P_n$-traveltime correction is clearly shown in Section 6.2.2. While the results for the isotropic reconstruction (Figure 6.1(d)) are still somewhat acceptable in the center of the first layer of the model, they are unsatisfactory for the other layers, unlike the anisotropic inversion, which is able to reconstruct the original velocity anomalies (Figure 6.1(a)) in all four layer to a high degree. This clearly indicates the need for the anisotropic $P_n$-phase correction. The RMS-values for the hypocenter relocation and the structure are in Tables 6.1 and 6.2. Especially in Table 6.2, one can see that even the RMS for the first inversion is sometimes higher than the initial one, the result for the reconstruction is the best one.

Adding noise to the synthetic traveltime data one can observe a decrease in the resolution as the number of blocs in the model is increased. Whereas the reconstruction is still possible for the 25 × 25 bloc-model, things are getting worse for the 35 × 35 bloc models, i.e. areas with false reconstruction increase.

## 6.3 Trade-off characteristics of the synthetic random test models

The results of the previous sections indicate that, in order to select the optimal damping parameter $\lambda$, one has to consider a certain range for the latter. For the synthetic models tested so far there is the tendency to have a better resolution of the model for small $\lambda$, which turn out to be below the values for which the minimal RMS is obtained. At this stage, it is not yet clear whether this discrepancy is a consequence of the particularities of the tested models which only exhibited a few random velocity anomalies, or whether it is due to the linearized optimization technique used.

As discussed in Chapter 2, an important task in seismic tomography is to calculate and analyze trade-off characteristics between several inversion solution parameters within the model- and the data space to judge the inversion results. Eventually one wishes to select the most appropriate inverted model out of a variety of different models computed, which amounts to select the optimal damping parameter $\lambda$. Methods to do this properly are called **Optimal Regularization Techniques** (ORT) and have been enunciated in detail in Chapter 2. Because the seismic inverse problem is both under- and overdetermined[4], normally, ORT is an important tool to analyze the resolution capabilities and the stability of the inversion. So, by varying the damping (also called regularization) parameter $\lambda$, one can analyze the trade-off changes between, for example, the solution norm (hypocenter and velocity norm) and the data fit (residual norm) and, so, find the best compromise, i.e. the optimal solution. Depending which parameters in the model- and the data space are selected in the trade-off pairs, different regularization techniques (RT's) ensue, such as that of Tikhonov, the ridge regression, the stochastic inverse, the method of Backus and Gilbert (Backus and Gilbert, 1967), the method Backus

---

[3] Another limitation is the application of ray theory, which is only possible for bloc discretization much bigger than the used wave-length to disregard any wave effects like selfhealing of interferences

[4] Underdetermined because parts of the model structure are not properly resolved because of insufficient or zero ray distribution in the individual bloc and overdetermined where multiple rays pass a distinct bloc. Underdetermination for the hypocenters does not exist because minimum number of observations per event has been set to seven in the analysis, i.e. there is a minimum of three degrees of freedom per event

## 6.3. TRADE-OFF CHARACTERISTICS OF THE SYNTHETIC MODELS

subjective, described in (Koch, 1993a) and Chapter 2.

As shown before with the synthetic tests in Section 6.2, the choice of the correct $\lambda$ is not easy. It appears to depend on which part of the model space (i.e. which layer) one wants to illuminate. Thus, the results of the synthetic tests show that the best result is obtained for rather small damping values in the first two layers, but not for the third and fourth one, where the solutions become instable for small damping values. In general, one has to say that the value for $\lambda$ can be smaller when the ray coverage (the data, respectively) is good and homogeneous. The less or more uneven the ray coverage of the volume is, the more undetermined or poor determined parameters of the model space remain and the seismic inverse problem has to be damped more and more to avoid instabilities.

Such a lack of information will result in a rather low contrast to resolve the model parameters, especially the velocity structure. This is the clear result of the synthetic tests with the artificially generated anomalies. Out of this, not only a distinct value but more likely a range for $\lambda$ is appropriate and delivers reasonable solutions. Whereas for the first two layers, $\lambda$ has to be selected as a small value, it has to be higher for the third layer, and due to the undulations in the fourth layer (wherever they result from) it has to be set near the optimal value selected along the inversion process or selected out of the plots for the trade-off curves. For details see Chapter 2, Section 2.3.3.8 or Koch (1992, 1993a,b), where the basics for the various tests for trade-off's are described in detail.

To test for the optimal range of the damping (ridge) parameter, some of the most important tests are done and listed below.

- Trade-off between data-fit (explanation of variance) and solution- (velocities and hypocenter) norm.

- Variation of the solution norm as a function of the damping (ridge) parameter.

- Trade-off between (co)-variance and resolution.

Another aspect of the various trade-off tests is the comparison between the isotropic and anisotropic cases, and to find out whether there are differences for the damping values, so, for example, that in the anisotropic case the inversion becomes stable at lower damping values than for the isotropic case.

To evaluate the different trade-off characteristics, here the relationship between, first, the velocity and the ridge parameter (Figures 6.9(a) and 6.9(b)), second, the one between the hypocentral norms and the ridge parameter (Figures 6.9(c) and 6.9(d)) and third, between data-fit and velocity and hypo norm (6.9(e) and 6.9(f)) are analyzed. For brevity, only the 15 × 15 models are considered here. The trade-off between covariance and resolution is considered in the subsequent section.

In Figures 6.9(a), 6.9(b), 6.9(c) and 6.9(d), the trade-off curves between data-fit and velocity or hypocentral norms are plotted for various ridge-parameters $\lambda$ for the individual nonlinear iteration steps, as well for the isotropic as for anisotropic case (fast axis at 27° and 5% velocity distortion).

During the earlier tests with the synthetic models, it was shown that the optimal solution can be found in the range between the smallest damping factor (K = 1) and the one at the optimal RMS (K about 5). So it appears that the traditional idea to select the **damping value at the knee-point is of minor significance**. Indeed, the optimal solution is to be found in the range up to the optimal data fit that goes along with a small damping value[5]. The optimal solution, which is found near the knee-point for iteration step five, marks the end of the range for the optimal reconstruction.

In Figure 6.9(f), the optimal RMS is at the rather high damping value 10.24, according to iteration step 8, but the optimal reconstruction, as selected out of the synthetic tests, is obtained at the minimal damping value at 0.08 for the first iteration.

---

[5]This interesting effect has to be investigated further whether it depends on the special situation according to the synthetic test or whether it is a common feature also for the "true" inversions

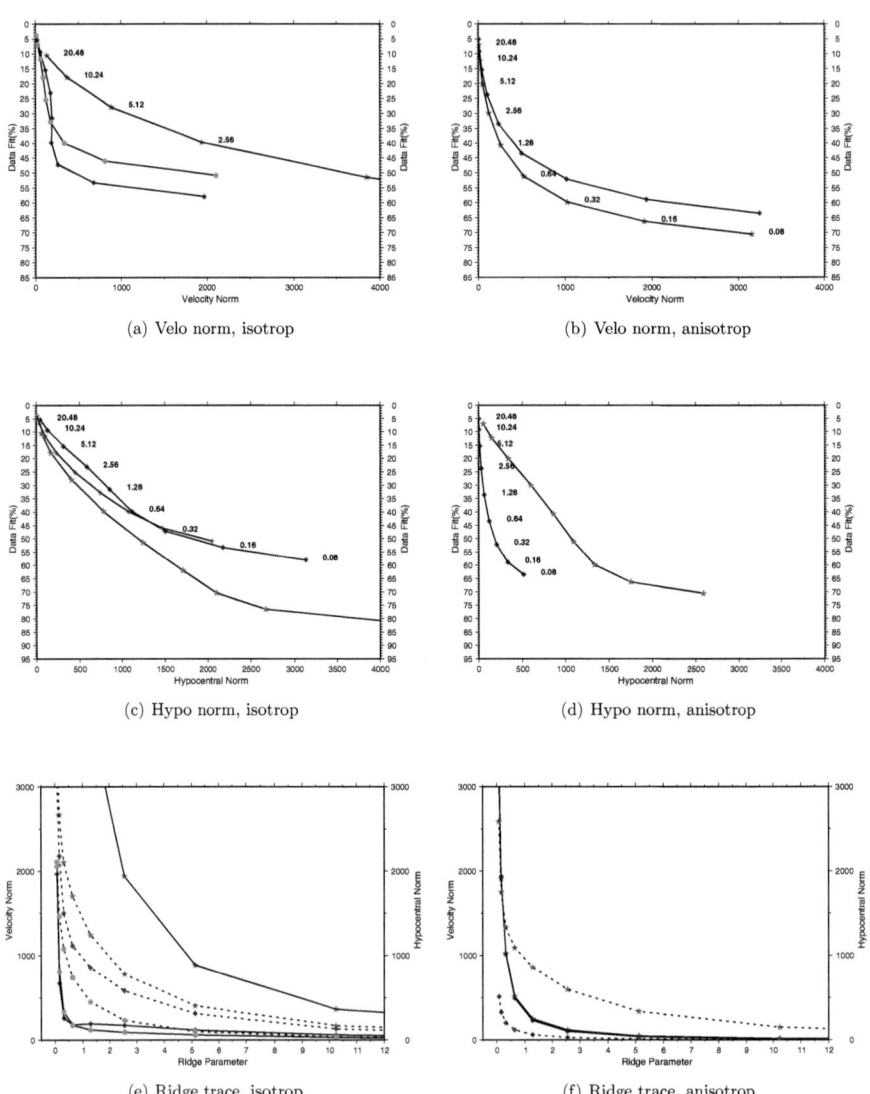

Figure 6.9: Graphical analysis of various trade-off characteristics of the inversion solutions for both the isotropic and anisotropic case, synthetic tests. Upper panels: Trade-off between data-fit and the velocity norm. Middle panels: Trade-off between data-fit and the hypocenter norm. Lower panels: Velocity (dotted lines) and hypocentral (full lines) norms as a function of the ridge parameter. The colors denote the number of the nonlinear iteration step, where red stands for the first, blue for the second, green for the third and yellow for the fourth and last one.

The Data-Fit $||R||^2$, which is computed directly from the equation system[6] for the isotropic inverted case, but anisotropic generated data, astonishingly is better about 10% than that for the anisotropic one after the first inversion in case of the hypocentre norm (Figures 6.9(c) and 6.9(d)). The values become better after the second nonlinear inversion (blue dots). Otherwise, looking at the curvature of the trade-off curves, the former is more pronounced for the anisotropic case, where the knee point is located at smaller damping values.

The trade-off curves for the velocity parameters also show the typical curvature and the knee point as well (Figure 6.9(a) and 6.9(b)). In contrast to the anisotropic case, for the isotropic case the values for the data-fit increase rather fast within a small range of the velocity values for the second nonlinear inversion step. The same holds for the hypocentral norm, where the trade off's are in Figures 6.9(c) and 6.9(d).

The results clearly show that the kneepoint for the different trade-off curves is in the same range as the optimal solution evaluated by the minimal RMS. But just after the first nonlinear inversion, the results are different, the optimal RMS tends to be at higher damping values i.e. higher K-values whereas the kneepoint of the trade-off curves mostly is more distinguished.

## 6.4 Testing for resolution

With a lot of data one can perform numerous tomographic inversion runs and obtain a lot of model results. However, these will eventually only reflect the errors in the data and the set of "tuning" parameters used in the inversion program, but will not represent the full worth of the information the data really contains. To compute and to select now the best models, some important criteria have to be considered which are, for example, the ray coverage or the computations of the resolution and the covariance matrix which, in turn, also depend on the ray coverage as wells as on the discretization of the model space. But with increasing complexity of the model, i.e. its parametrization, the fit of the model to the data will get better, i.e. the residual sum will get smaller, independent of the structural information. In fact, an over-parametrized model will always provide a good data fit, however, there is danger that such a model only fits noise in the data which, in turn, means that such a model may not be the optimal one. Moreover, a consequence of such an over-parametrization is that adjacent model blocs may not be properly resolved and changes in the velocities of the individual blocs may compensate each other. One simple tool to estimate over-parametrization of a model is to compare the bloc discretization with the ray density. Inspection of the "spider web" of the ray coverage will allow to judge, at least qualitatively, the appropriateness of the chosen discretization. Thus, blocs falling between the rays will not be resolved at all (see Chapter 4).

### 6.4.1 Resolution matrix

Determining the regions of the model space in which parameters are well resolved is the key for interpreting inversion results. In linear inverse problems, a straightforward means of investigating the resolution of a model is through the resolution matrix, $R_M$ (Tarantola, 1987; Abt and Fischer, 2008), where the resolution matrix is defined as

$$R_M = (A^T A + \lambda I)^{-1} A^T A \qquad (6.1)$$

The basic meaning of the resolution matrix $R_M$ stems from the following relationship between the true (unknown) model parameter vector $x$ and the inverted solution vector $x_s$ (Koch, 1983b):

$$x_s = R_M * x \qquad (6.2)$$

---
[6]Not to be confounded with the RMS computed from the traveltime residuals using nonlinear ray-tracing

which shows that the model solution $x_s$ is actually a filtered version of the true model $x$, with the filter coefficients given by the row elements of $R_M$. The diagonal elements of the resolution matrix $R_M$ reflect how well individual model parameters are resolved, whereas the trade-off's between different model parameters in different locations of the model are revealed by the off-diagonal terms. If a diagonal element of $R_M$ is not exactly one, then the resolved volume of the corresponding model bloc is actually larger than the given bloc volume, i.e. there is leakage of the resolved volume into adjacent blocks by the amount which is given by the off-diagonal elements of the resolution matrix at the corresponding bloc indices.

The meaning of the resolution matrix $R_M$ is somewhat more complicated in a non-linear inversion problem studied here. Because the partial derivatives may change significantly with each iteration, $R_M$ should be considered as reflecting the resolution of the solution only relative to the model from the previous iteration. In addition, when the large volume constraints are employed, the resolution matrix will reflect the values of a priori variance assigned to the constraint equations, but not the fact that the constraint equations correspond to zero values in the data vector (Appendix B). Nonetheless, the resolution matrix still contains valuable information about how the data samples the anisotropic parameters in a given model. We use the resolution matrix in two ways: (1) to define model blocs that should be combined into larger model volumes since they are not well resolved anyway and, (2) to define which model regions to display.

From the definition of the resolution matrix it follows that the resolution is optimal only for small damping parameters $kI$. With increasing $kI$ the resolution goes monotonically to zero. This becomes immediately obvious by looking at Figure 6.10, 6.11 or 6.12. Overall, one observes from these figures that the resolution for most regions in Germany is near one for the $15 \times 15$ and $25 \times 25$ bloc models. With increasing damping parameters k (here the one's for iteration step 5 in Figures 6.10(b), 6.11(b) and 6.12(b)), the resolution decreases significantly. For the $35 \times 35$ bloc model, good resolution can also be obtained across most of the model area, but now some gaps occur, and the resolution becomes sketchy in the northeastern part of Germany, where it diminishes significantly for all but the upper layer of the model, indicating the limit of resolution.

### 6.4.2 Covariance matrix

Applying the covariance operator $cov(X)$ to the damped least square solution of the linear inverse problem (Chapter 2) results in the following expression for the covariance matrix (Koch, 1992):

$$cov(\delta x_s) = A^+ A^{+T} \sigma^2 \tag{6.3}$$

where $A^+$ is the generalized inverse of $A$ and $\sigma$ is the variance of the data which is either estimated a priori or is evaluated from the *a posteriori* estimated residual sum squared (TSS). Because the general inverse solution is given by $x_s = A^+ y$, it follows from the damped linear (LM) system of equations that $A^+$ can be written as

$$A^+ = (A^T A + \lambda I)^{-1} A^T \tag{6.4}$$

Using this definition of $A^+$ and that of the resolution matrix $R = (A^T A + kI)^{-1} A^T A$ defined in the previous chapter, the covariance matrix can be written in terms of the resolution matrix by the following equation (Koch, 1992):

$$cov(\delta x_s) = (A^T A + \lambda I)^{-1} R \sigma^2 \tag{6.5}$$

which means that the covariance $cov(\delta x_s)$ can be calculated from the solution of the linear system above, with the right hand side given by $R\sigma^2$, which means that the covariance can only be computed once the resolution matrix $R$ has been calculated. Equation 6.5 must be solved once for each block of the model which means that the computation of the covariance for all blocs of the model is prohibitively time-consuming.

## 6.4. TESTING FOR RESOLUTION

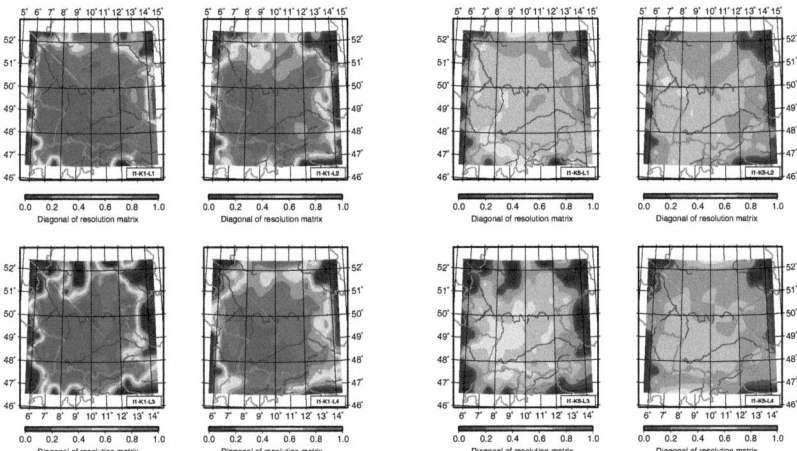

(a) Resolution matrix, small lambda, 15x15 blocs    (b) Resolution matrix, higher lambda, 15x15 blocs

Figure 6.10: Resolution matrices for the $15 \times 15$ bloc model. Resolution values are overall quite good in the central areas of the layers, except for the problematic layer three. Naturally, the resolution values go to zero with increasing damping value $\lambda$.

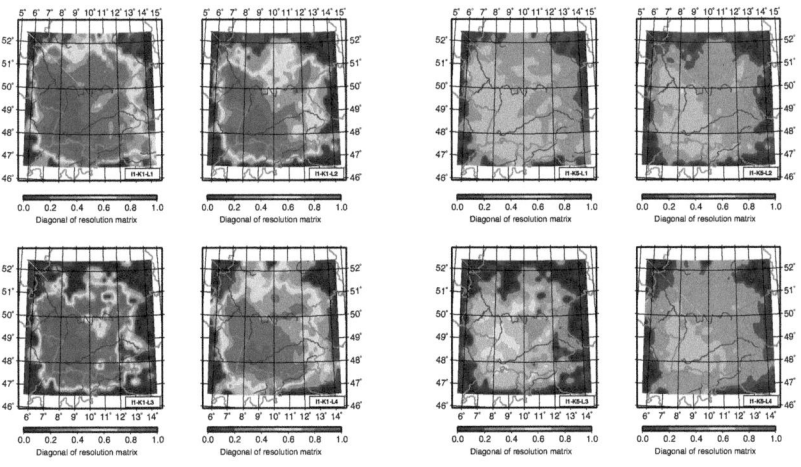

(a) Resolution matrix, small lambda, 25x25 blocs    (b) Resolution matrix, higher lambda, 25x25 blocs

Figure 6.11: Same as in Figure 6.10, but for the $25 \times 25$ bloc model.

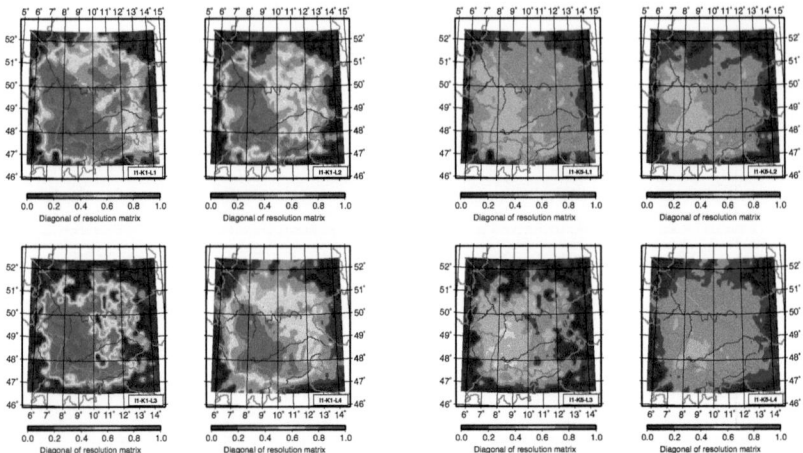

(a) Resolution matrix, small lambda, 35x35 blocs  (b) Resolution matrix, higher lambda, 35x35 blocs

Figure 6.12: Resolution matrices for the $35 \times 35$ bloc model. Compared with the previous two models the resolution deteriorates now in the northwestern part of Germany, but is still quite good in the remainder of the model area, though somewhat less so for the problematic layer three.

Whereas the resolution matrix in some way reflects the ray coverage and the resolution capabilities, the covariance indicates how errors in the data project into the solution. In particular, the diagonal term of the covariance matrix gives the variance of a particular model (velocity) parameter, whereas the off-diagonal elements indicate the amount of dependency of that bloc to adjacent ones, which has to be computed separately.

Figures 6.13, 6.14 and 6.15 show the diagonal elements of the covariance matrices for the $15 \times 15$, $25 \times 25$ and $35 \times 35$ bloc models, respectively. One notes from all figures that with increasing $\lambda$ the diagonal terms of the covariance matrices, i.e. the variances, approach more and more the value zero. So, comparing this with the corresponding behavior of the resolution matrices (see Figures 6.10, 6.11 and 6.12), one notes the typical opposite trend between resolution and covariance, i.e. their trade-off, as epitomized in the Backus-Gilbert formalism. From the inspection of the corresponding trade-off curves it is, in principle, also possible to find the optimal $\lambda$, or the optimal inverse solution.

The visual comparison of the three covariance figures also shows that, as expected, the covariances for individual bloc velocities overall increase for the more finely discretized models. Thus, many areas of the $35 \times 35$ bloc model, particular in the third layer (lower crust) indicate relatively high (co)-variances which, statistically, put the inverted bloc-velocities there into doubt.

## 6.4. TESTING FOR RESOLUTION

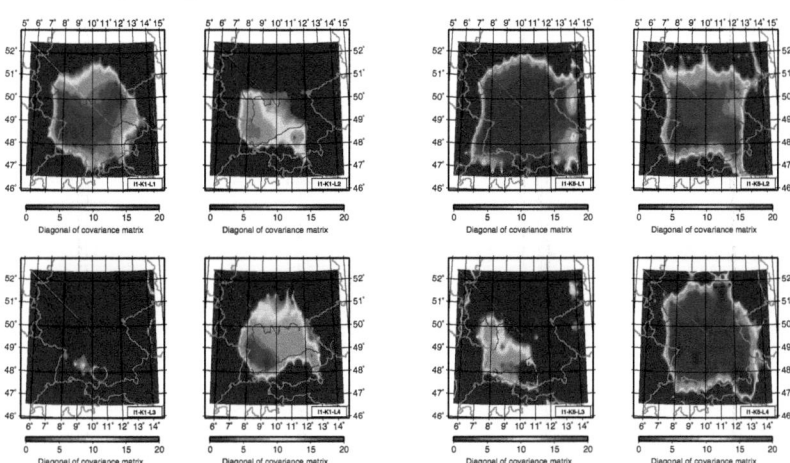

(a) Covariance matrix, small lambda, 15 × 15 blocs  (b) Covariance matrix, higher lambda, 15 × 15 blocs

Figure 6.13: Covariance matrices for the 15 × 15 bloc model. The covariance values are overall quite good in the main areas of all layers, except for the problematic layer three.

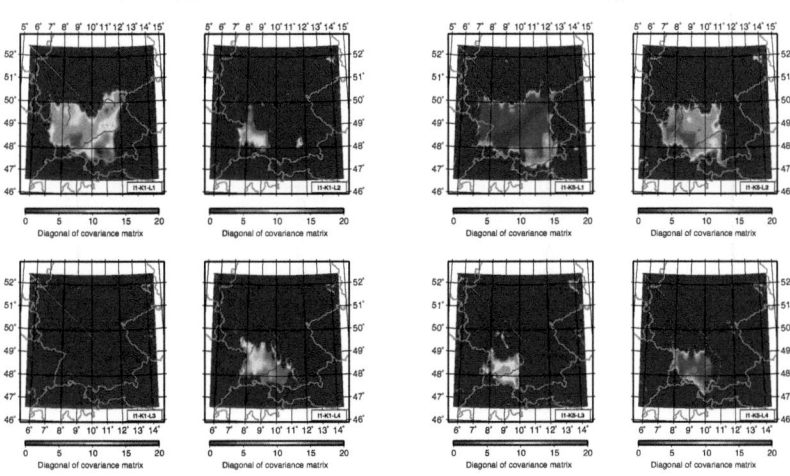

(a) Covariance matrix, small lambda, 25 × 25 blocs  (b) Covariance matrix, higher lambda, 25 × 25 blocs

Figure 6.14: Covariance matrices for the 25 × 25 bloc model. Despite the finer discretization than that of the 15 × 15 bloc model, the covariance values are still satisfactory in the main areas of the layers, except for the problematic layer three.

138                    CHAPTER 6. SSH-INVERSIONS FOR 3D VELOCITY MODELS

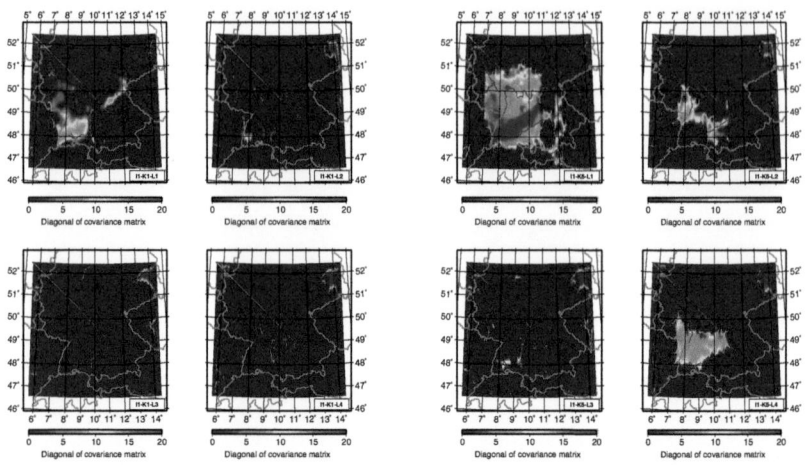

(a) Covariance matrix, small lambda, 35 × 35 blocs  (b) Covariance matrix, higher lambda, 35 × 35 blocs

Figure 6.15: Covariance matrices for the 35 × 35 bloc model. Compared with the coarser 15 × 15 and 25 × 25 bloc models, the covariance values overall become higher, particularly in the problematic layer three.

## 6.5 Checkerboard tests of the resolution capability of the dataset

Another, rather simple way to test the resolution capability of the data set for a selected 3D inversion model is to use a checkerboard-like structure in a synthetic test model. Using this model, new theoretical arrivaltimes are computed for all available hypocenters and recording stations like those listed in the original data set. This means that these theoretical arrivaltimes contain the information on the checkerboard velocity structure, according to the present geometrical hypocenter-station-distribution. In the following checkerboard tests, the theoretical data are also corrupted by random noise of $\sigma = 0.1$, 0.2 and 0.3 $s$ superimposed on the exact arrivaltimes, to mimic typical arrivaltime errors in seismograms, whereby $\sigma$ of 0.2 and 0.3 $s$ represent rather bad data.

With this new synthetic datasets, containing anisotropic influenced $P_n$-phases[7], the inversion runs are carried out each for two simulation cases, namely, with and without anisotropic correction for the $P_n$ phases. Doing so, the importance of including anisotropy into the inversion model will be verified again, this time under the aspect of its effects on the model resolution. Similar to the synthetic tests of the previous sections the checkerboard tests are carried out with three different bloc discretizations (15 × 15, 25 × 25 and 35 × 35, respectively). Naturally, the resolution capability will decrease with a higher number of blocs in each layer and the areas with low resolution will become larger.

In a first set of checkerboard tests, the hypocenters are fixed (IFIX = 1) to avoid that possible relocations may influence the newly computed structure by a compensation or trade-off between calculated hypocentral shifts and velocity anomalies. In the subsequent set of tests the hypocenters are free (IFIX = 0), i.e. full SSH-inversions are performed, to check the dependency of seismic structure and hypocenters. This is an important test to evaluate the regions of good, fair and poor reconstruction of the structure in the later real 3D tomographic analysis. In an ideal situation, with very good ray

---

[7]The real data set used in this study is indeed affected by the anisotropic upper mantle beneath Germany, as proved by the residual analysis in the previous chapter "Data Analysis"

## 6.5. CHECKERBOARD TESTS

Table 6.3: Statistics (Variance and Standard deviation) for the relocalized hypocenters after full SSH inversion and RMS values, including the anisotropic correction.

| Model | Noise | Variance | | | Standard deviation | | | $TSS_{SSH}$ | $TSS_{hypo}$ |
|---|---|---|---|---|---|---|---|---|---|
| | | x | y | z | x | y | z | | |
| | $s$ | $km^2$ | $km^2$ | $km^2$ | $km$ | $km$ | $km$ | $s^2$ | $s^2$ |
| Optimal solution in Iteration I = 2, isotropic case | | | | | | | | | |
| 15 x 15 | 0.0 | 0.286 | 0.023 | 0.987 | 0.535 | 0.151 | 0.993 | 1396 | 0.0 |
| Optimal solutions in Iteration I = 2, anisotropic cases | | | | | | | | | |
| 15 x 15 | 0.0 | 0.036 | 0.021 | 0.452 | 0.190 | 0.144 | 0.672 | 189 | 0.0 |
| | 0.1 | 0.038 | 0.034 | 0.527 | 0.196 | 0.183 | 0.726 | 382 | 240 |
| | 0.2 | 0.049 | 0.051 | 0.704 | 0.222 | 0.225 | 0.839 | 953 | 898 |
| | 0.3 | 0.179 | 0.111 | 1.234 | 0.423 | 0.334 | 1.111 | 2017 | 2006 |
| 25 x 25 | 0.0 | 0.012 | 0.010 | 0.216 | 0.108 | 0.464 | 0.464 | 128 | 0.0 |
| | 0.1 | 0.022 | 0.021 | 0.489 | 0.147 | 0.145 | 0.699 | 353 | 240 |
| | 0.2 | 0.942 | 0.145 | 3.927 | 0.970 | 0.381 | 1.981 | 874 | 898 |
| | 0.3 | 0.047 | 0.047 | 0.919 | 0.217 | 0.217 | 0.958 | 1779 | 2006 |
| 35 x 35 | 0.0 | 0.003 | 0.004 | 0.147 | 0.055 | 0.065 | 0.383 | 118 | 0.0 |
| | 0.1 | 0.027 | 0.043 | 0.658 | 0.165 | 0.206 | 0.811 | 297 | 240 |
| | 0.2 | 0.052 | 0.061 | 1.169 | 0.228 | 0.246 | 1.081 | 845 | 898 |
| | 0.3 | 0.025 | 0.029 | 0.841 | 0.159 | 0.170 | 0.917 | 1700 | 2006 |

coverage and no errors in the arrivaltime data and optimal determined hypocenters, no differences between the two cases (IFIX = 1 and IFIX = 0) should arise. In a real situation, however, due to an inhomogeneous ray distribution and gaps where the blocs are not illuminated by any rays and errors in the data, the hypocenters are shifted within some range for the case "IFIX = 0", which may result in a different computed velocity structure. So, comparing the two cases, this will allow in some way to investigate the stability of the inversion. Also, since full SSH-inversions with free hypocenters are computationally much more burdensome, a seismologist might sometimes be inclined to just invert for the seismic structure (IFIX=1) in routine investigations. Our comparison of the two cases will indicate whether such an approach is vindicated.

Another interesting feature of the tests consists in the comparison of the influence of two different damping parameters $\lambda$ (the lowest one at K = 1 and the one for the optimal solution with minimal RMS) within a nonlinear inversion step, i. e. the damping sequence. This is done according to the results of the synthetic tests of the previous section where the best velocity reconstruction has been obtained for K = 1, i.e. for the smallest damping parameter. To evaluate the different trade-off characteristics as done in Section 6.3, but now for the checkerboard tests, the same correlations are plotted in Figure 6.27, also only for the $15 \times 15$ models. The trade-off between covariance and resolution is considered in the subsequent section.

The results for the optimal damping parameters are similar to the previous one, but now according to the results with the optimal RMS.

### 6.5.1 Checkerboard test for $15 \times 15$ bloc models

For the $15 \times 15$ input bloc models with hypocenters fixed, the checkerboard structure is well resolved across the whole area of the model, as can be seen from Figure 6.16, where the optimal results with minimal RMS are shown. Only at the borders of the model, some discrepancies compared to the input model occur in each layer. This is most obvious in the third layer representing the lower crust which has been shown throughout this study to be the problematic one due to poor illumination by seismic rays.

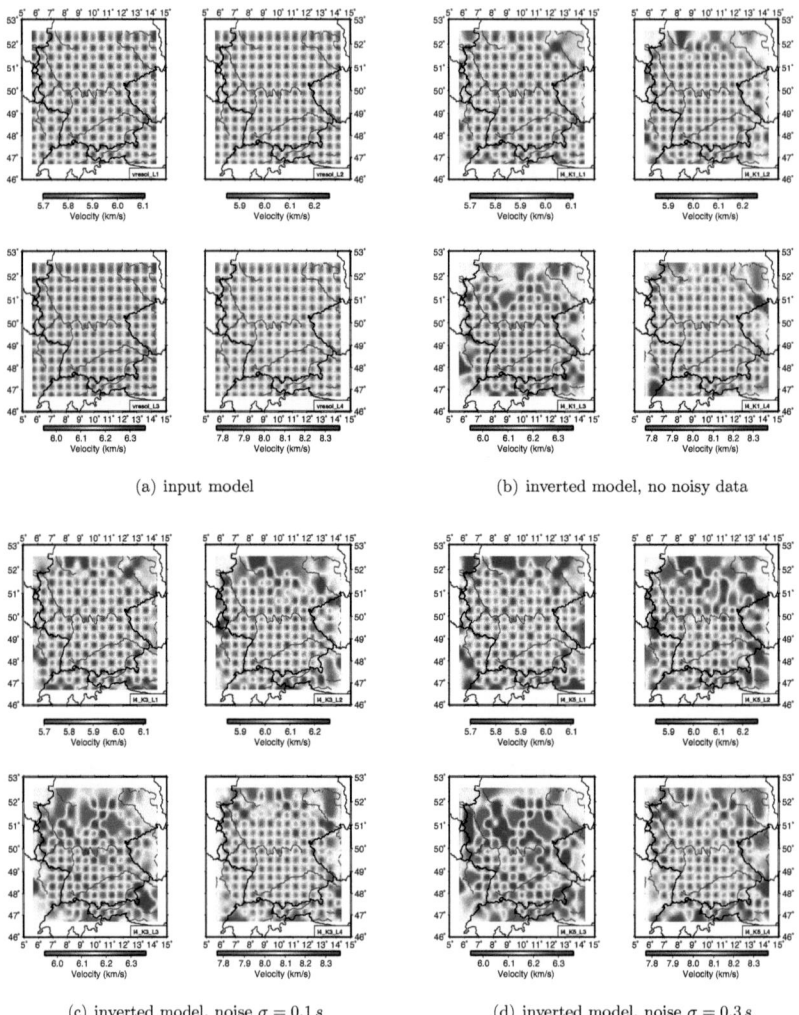

Figure 6.16: Checkerboard test for $15 \times 15$ bloc model with hypocenters **fixed** (without and with noise as indicated by the values of $\sigma$). Shown are the final results after the fourth nonlinear iteration. Most of the areas of the model are clearly reconstructed though the quality is decreasing with increasing $\sigma$.

## 6.5. CHECKERBOARD TESTS

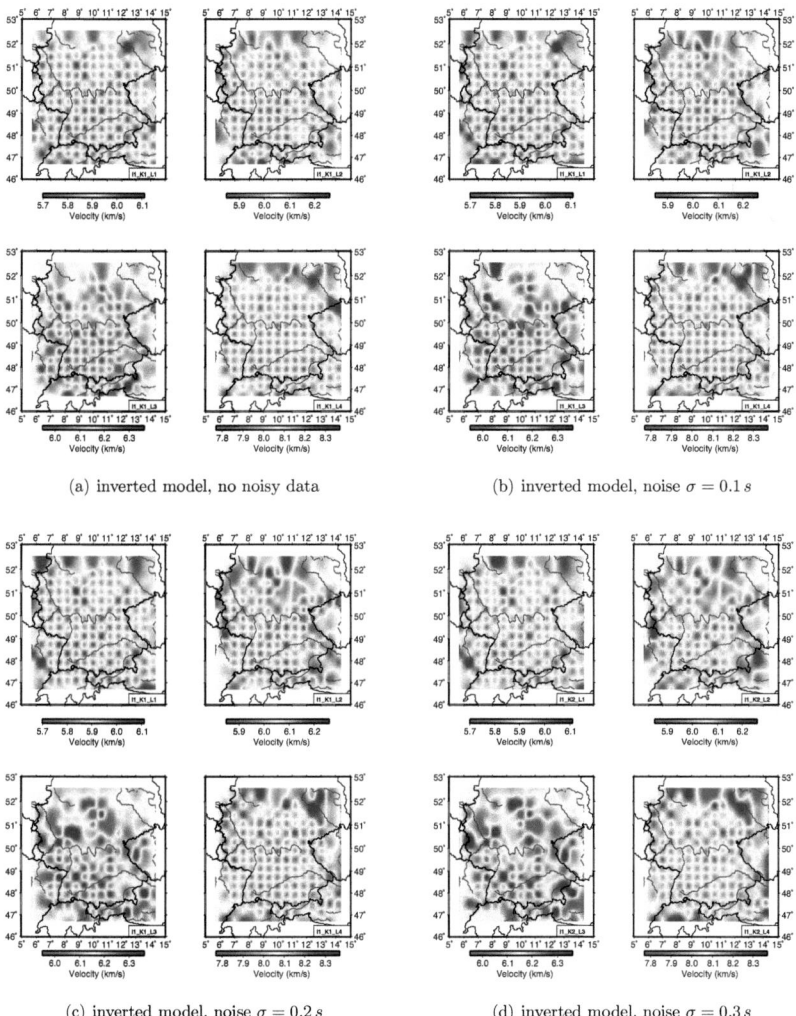

Figure 6.17: Checkerboard test for $15 \times 15$ bloc model with hypocenters **free**, i.e. full inversion. Compared to the solutions with fixed hypocenters, the structure is not as well resolved.

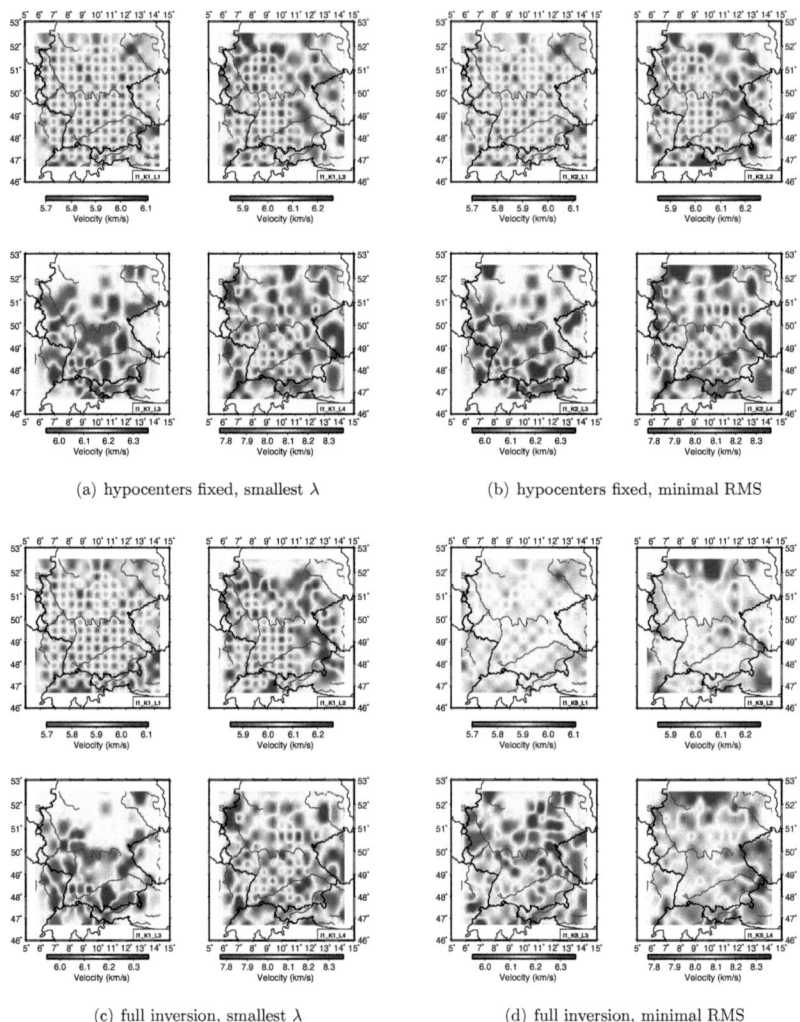

Figure 6.18: Checkerboard test for $15 \times 15$ bloc model with isotropic inversions of anisotropic generated data with hypocenters fixed as well as with hypocenters free (full inversion). The first layer is somewhat reconstructed for small $\lambda$, but the other three layers are hardly resolved.

## 6.5. CHECKERBOARD TESTS

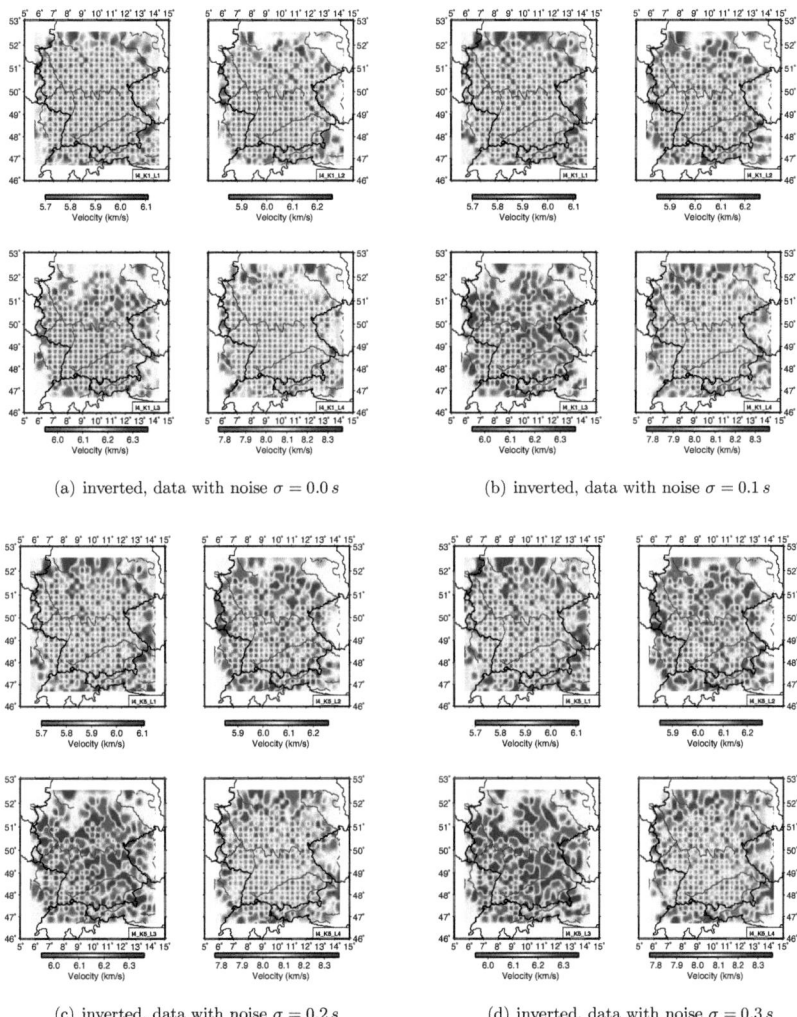

Figure 6.19: Checkerboard test for the 25 × 25 bloc model, hypocenters **fixed**. Although there occur now wide areas with no clear reconstruction of the input model, especially in the third layer, the other parts of the input model are still well resolved.

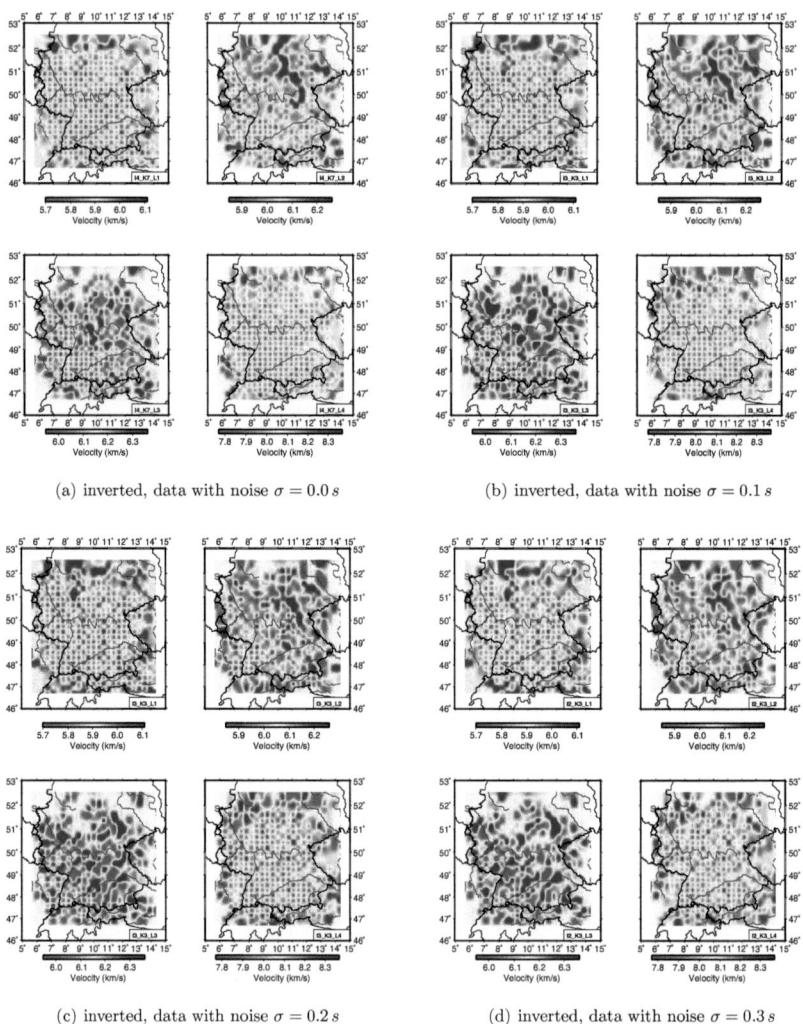

Figure 6.20: Checkerboard test for 25 × 25 bloc model, hypocenters **free**. Although wide areas with no clear reconstruction of the input model occur, especially in the third layer, the reconstruction is still good. In case of the inversions with noise at 0.1 and 0.3 s, the inversion process stops after the third and second nonlinear iteration, respectively.

## 6.5. CHECKERBOARD TESTS

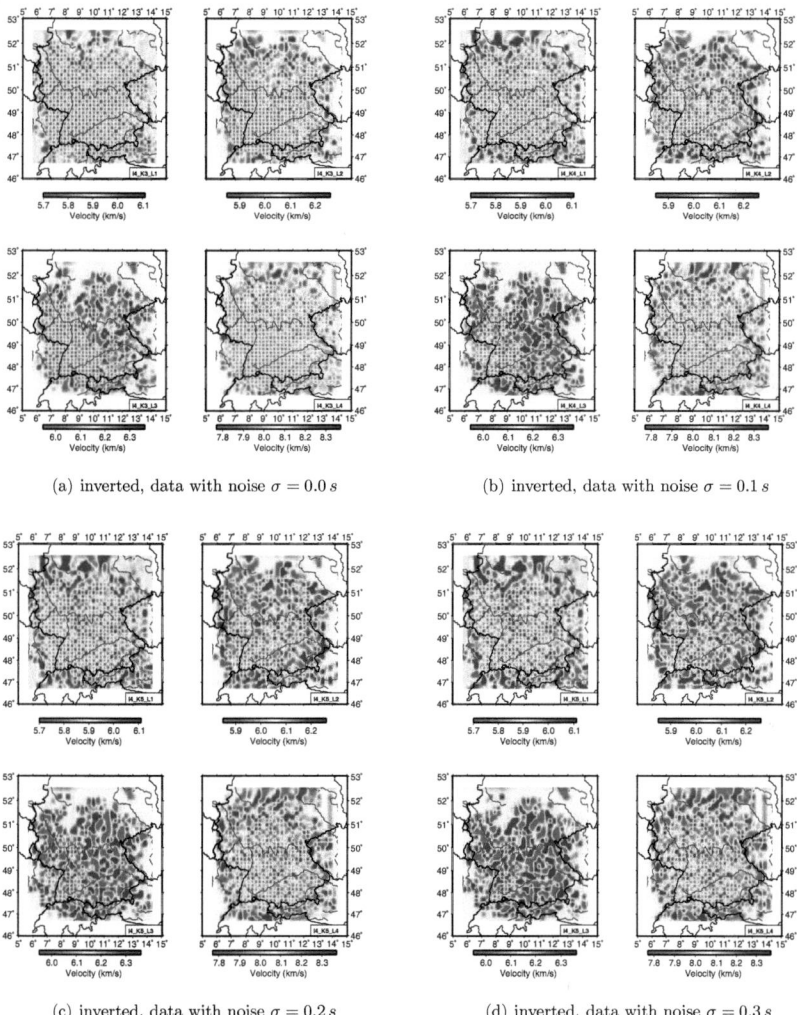

Figure 6.21: Checkerboard test for the 35 × 35 bloc model, hypocenters **fixed**. That rather fine discretization now shows the limits of the possible resolution. There are now wide areas with no clear reconstruction of the input model, especially in the third layer, which is due to blocs with zero or only small ray coverage.

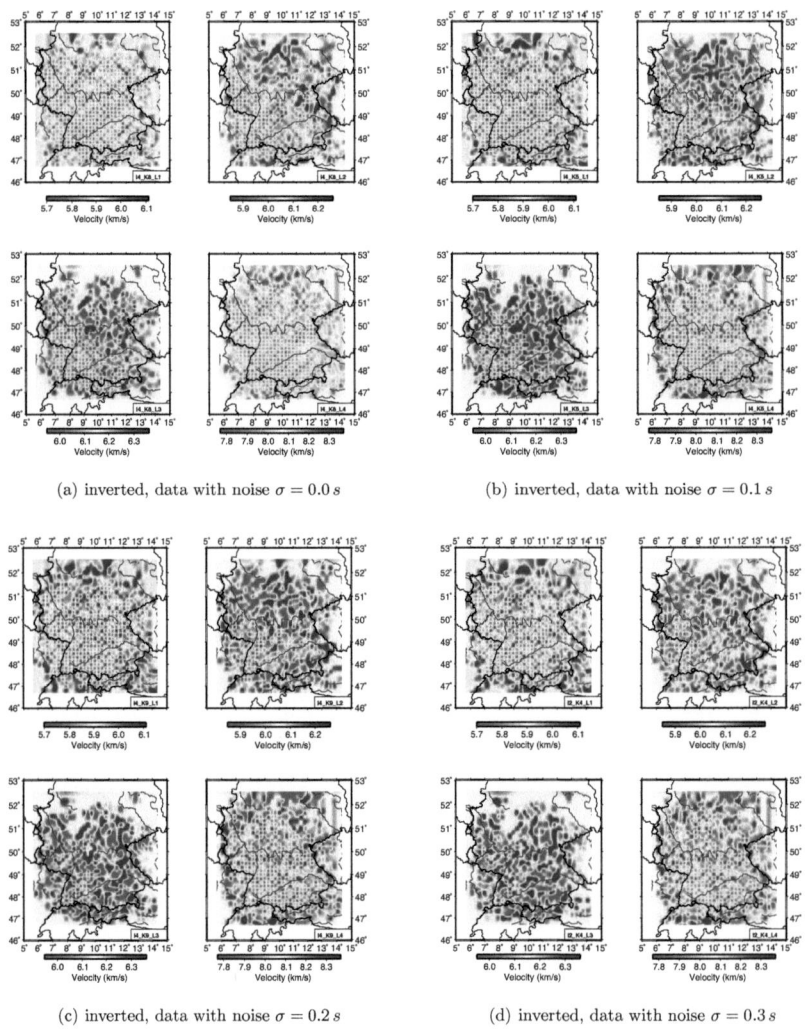

Figure 6.22: Checkerboard test for 35×35 bloc model, hypocenters **free**. That rather fine discretization now shows the limits of the possible resolution. There are wide areas with no clear reconstruction of the initial model, especially in the third layer which is due to blocs with zero or small ray coverage.

## 6.5. CHECKERBOARD TESTS

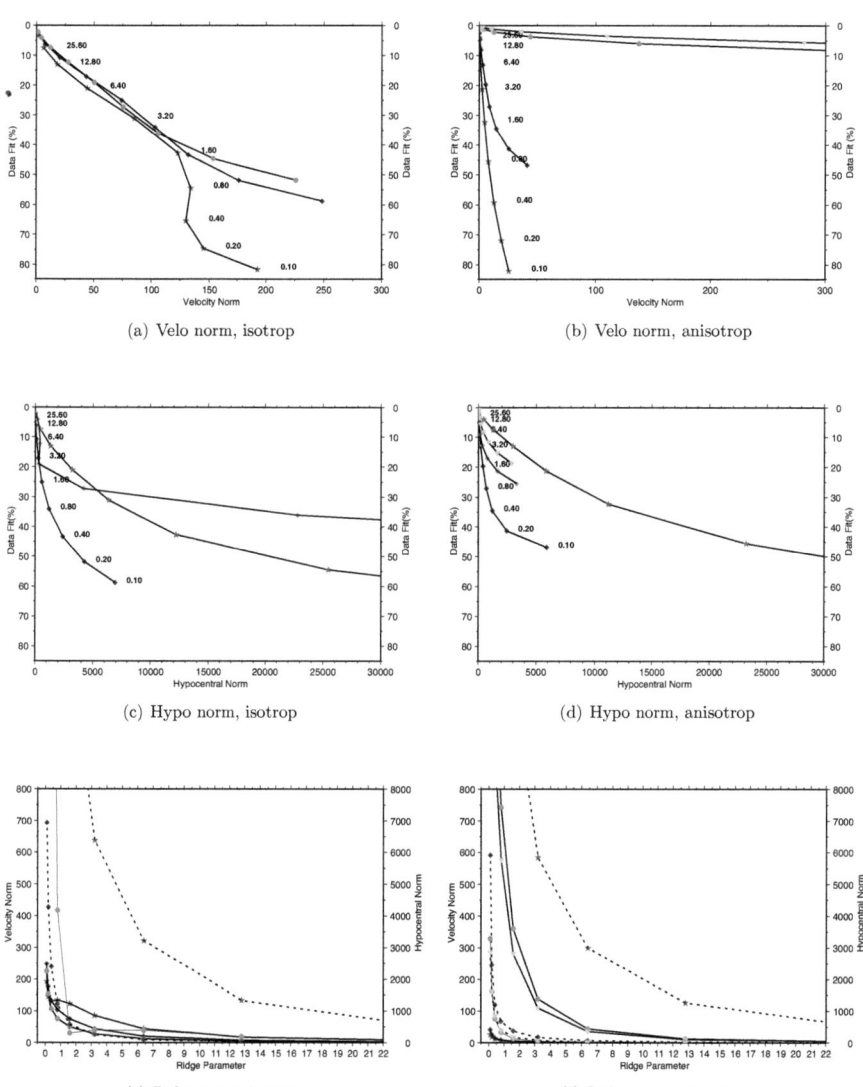

Figure 6.23: Graphical analysis of various trade-off characteristics of the inversion solutions for the 15 × 15 blocs checkerboard model. Upper panels: Trade-off between data-fit and the velocity norm. Middle panels: Trade-off between data-fit and the hypocenter norm. Lower panels: Velocity and hypocentral norms as a function of the ridge parameter. The colors denote the number of the nonlinear iteration step: Red stands for the first, blue for the second, green for the third and yellow for the fourth and last one.

Adding noise to the arrivaltimes makes the situation more realistic, as this will take into account errors in the recorded data owing, for example, to mispics of the incoming waveforms, as well as to other model errors. A noise level with $\sigma = \pm 0.1\,s$ is the most appropriate, whereas higher levels of $\sigma = 0.2$ or $0.3\,s$ are less suitable for a reliable 3D tomography. This immediately becomes obvious from Figure 6.16(d) which shows that, compared with Figure 6.16(c), the reconstruction in layer one, two and four is more uncertain and the areas, where the reconstruction is adequate, dwindle.

The situation becomes slightly different when doing a full inversion for both structure and hypocenters (IFIX = 0), the reconstruction of the input checkerboard model does not get better (Figure 6.17). The velocity contrasts of the inverted checkerboard structure are not as clear as in the case with hypocenters fixed (IFIX = 1) and the areas where the original checkerboard structure is resolved dwindle again. This shows the influence of the hypocentral relocations onto the reconstruction of the seismic structure. Table 6.3, where the relocations of the hypocenters are listed, indicates that even small shifts have such an influence onto the quality of the inversion that the well reconstructed model areas are reduced compared to the inversions with fixed hypocenters. This is an intrinsic problematic of the full coupled seismic structure-hypocenter inverse problem where the number of unknowns is now much higher, i.e. the number of the degrees of freedom smaller than for the "hypocenters fixed" problem, and major part of the information contained in the data set is now projected into the hypocentral relocations.

The inversions without anisotropic correction (Figure 6.18) of an initially anisotropically computed checkerboard model finally show a vast difference to the original checkerboard structure, whereby the areas of reasonable reconstruction shrinks to rather small areas in all, but the first layer. In fact, for layers three and four the reconstruction fails completely. Comparing the results of the inversions with hypocenters fixed and hypocenters free again, the later ones are slightly worse again. With regard to the real 3D tomography carried out in the subsequent chapter, this means that using the real data set with anisotropically influenced arrivaltimes in an inversion which does not account for this anisotropy will deliver highly biased results and is, therefore, not acceptable. Moreover, the analysis above shows also that the influence of the shifts of the hypocenters on the seismic structure is of only minor importance and only takes effect in areas which are not clearly identified by the data.

### 6.5.2 Checkerboard test for $25 \times 25$ bloc models

Naturally, the results for the checkerboard tests with $25 \times 25$ blocs give a more detailed overview of the resolution capability of the data. Compared to the checkerboard test with $15 \times 15$ blocs, the ray coverage per bloc will decrease, so there may occur some areas where no information about the structure is available. Considering this, the result of the checkerboard test (Figure 6.19 with hypocenters fixed and 6.20 with hypocenters free) can be regarded as test for the areas where the ray coverage is good enough to provide a stable result of the inversion and to select these areas[8] in the real 3D tomographic inversions done below.

The reconstruction for the data with $\sigma = 0.0\,s$ is fair and nearly the same as in Figure 6.16(b), so it is acceptable to use this bloc discretization in the true 3D tomography performed later. Also, the reconstruction of the models with $\sigma = 0.1\,s$ is still good, but shows velocity variations in the individual blocs. Furthermore, the areas where the reconstruction is possible, only ambiguous or impossible one's are detected more clearly. The results in Figures 6.19(c) and 6.19(d) are much more ambiguous in the second and third layer, the initial checkerboard is only crudely approximated. Things are getting much worse in the case with hypocenters free, where only in the first and fourth layer, the reconstruction is possible in some way. The increasing levels of noise mainly affect the result in layer three, where it becomes just worse at a noise level at $0.1\,s$.

---

[8]This is indirectly done by shading out the areas with low ray coverage in the tomographic results, which coincides rather good with the results obtained here.

## 6.5.3 Checkerboard test for 35 × 35 bloc models

Finally, the possible resolution for the 35 × 35 bloc model is tested. Here, due to the poor or not available ray coverage in some blocs, the reconstruction is more difficult in some areas. So, the areas with good or bad reconstruction clearly can be identified as can be seen in Figure 6.21 for hypocenters fixed and 6.22 for hypocenters free (full inversion). As recognized in the tests before, for the models with bigger blocs, the best reconstruction is in the first layer and overall in the southwestern part of the model area where also most of the seismic events occur with a corresponding good ray coverage there. The worst resolution is in the third layer where the reconstruction at the borders fails completely, although the reconstruction is still acceptable in the middle part of the model area.

Things are getting worse when noise is added to the original synthetic arrivaltimes, simulating real data. The data are disturbed with noise at $\sigma = 0.1\,s$ (Figure 6.21(b)), $\sigma = 0.2\,s$ (Figure 6.21(c)) and $0.3\,s$ (Figure 6.21(d)) where noise with $\sigma = 0.1\,s$ represents the upper limit for rather good data in seismic inversion studies, but $\sigma = 0.3\,s$ represents data of rather poor quality with hard-to-define seismogram phases and other systematic model errors. [9]

Whereas the reconstructing for data with noise at $\sigma = 0.1\,s$ is still good for local earthquake tomography with the inhomogeneous distributed seismic sources and stations, the reconstruction for $\sigma = 0.3\,s$ fails. Only in the first and fourth layer the reconstruction is possible, however, with large unresolved border areas. The differences between the reconstructions with noise at $\sigma = 0.1\,s$ and no noise are mainly in the velocity contrasts for the individual blocs, but basically no changes in the checkerboard structure itself, except in some areas of the third layer.

The situation for the case with noisy data at $\sigma = 0.3\,s$ is different. Now, there are not only strong variations of the velocity contrasts, but completely new structures arise in the reconstructed model. Concerning the noisy data, one always has to remember, that the noise is constructed using a Gaussian distribution. So, in the case with noise at $\sigma = 0.3\,s$, 68% of the data have arrivaltime disturbances within the $\pm 0.3\,s$ interval and 95.4% are in the $\pm 0.6\,s$ range. The rest, 32 % or 4.6 % respectively, of the data have even bigger noise values, too bad for a reliable tomographic inversion, and this is probably the major cause of the failed checkerboard reconstruction. Consequently, applying noise with $\sigma = 0.1\,s$, reflects the situation for a realistic, well analyzed data set, though with some accidentally mispicked phases.

## 6.5.4 Conclusive remarks on the checkerboard tests

Overall it can be stated that the resolution capability for the available dataset is rather good for all three bloc discretizations investigated with, expectedly, slightly decreasing resolution along the borders of the model area when the number of model blocs as well as the noise in the data is increased. Despite this problematic, the resolution is still fair, except for the models with noise at $\pm 0.3\,s$, which results in a completely unresolvable structure for the 35 × 35 bloc discretization. For the 15 × 15 bloc models, the velocity structure is much clearer than for those with the finer bloc discretization which exhibit more areas with a lack of resolution. Nevertheless, the advantage of these finer models is the better differentiation between the areas with good and bad resolution and a refinement of the borders of the resolvable area.

## 6.6  3D laterally heterogeneous seismic velocity models

Here, the real 3D laterally heterogeneous seismic velocity models for the crust and upper mantle beneath Germany are computed for several SSH-inversion cases, namely with different bloc sizes (15 × 15, 25 × 25 and 35 × 35) with hypocenters fixed (IFIX=0) as well as free (IFIX=0, i.e. full

---

[9]Actually, because of the Gaussian distribution of the random noise, the absolute size of the arrivaltime perturbations is much bigger, as only 68% of the data are contained in the $\pm \sigma$ interval.

SSH inversion). Also, these models are executed without (isotropic) and with the anisotropic $P_n$-traveltime correction included. Further parameters to be varied within the models are the GAP- and the RES-criterion (see Chapter 4), where, for example, the use of the GAP criterion of 120° improves the stability for lateral changes in the hypocenter estimate, as will be shown later in Chapter 7. The later models with increased bloc discretization are only computed for the less restrictive criteria GAP = 180° to avoid low ray coverage. Additionally, after processing the only $P$-phases, in case of the 15 × 15 models, $S$-phases are used to compute a separate $S$-velocity model and compared to each other in Section 6.8. From these models local $V_P/V_S$ ratios [10] are derived by comparing the optimal solutions for the $P$- and $S$-inversions and interpreted seismologically as well as petrologically.

### 6.6.1  15 × 15 bloc models

The models consist, like the previous 1D models, of three layers for the crust, with vertical thickness of $10\,km$ each, and a fourth layer representing the upper mantle below $30\,km$. Each layer is divided into 15 × 15 blocs which, with an overall NS- and EW- extension of $650\,km$ for the model, results in a lateral extension of about $43\,km$ for one bloc, which is big enough to have a good ray distribution in each bloc (see below), but also small enough to resolve major lateral seismic anomalies underneath Germany.

The calculations are done for the different cases enumerated above (only P-phases; only S-phases; without and with elliptical anisotropic correction, hypocenters fixed, hypocenters free, two values of the GAP: 120° and 180°, and RES = $5\,s$). From these models the influence of the correction for anisotropic influenced $P_n$-phases onto the computed structure and, in Chapter 7, Relocation, onto the relocation of the hypocenters is explored.

#### 6.6.1.1  Ray coverage for the 15 × 15 bloc models

The ray coverage is one important criteria to get a first idea on the reconstruction capabilities of the tomographic inversion models. Figure 6.24 shows that the ray coverage for the 15 × 15 bloc model is, as expected, rather good for the first and second layer, especially in the center of the model, and very good in SW-Germany, where the seismic active region of the Rhinegraben is situated. Also, as illustrated by the panels representing the fourth layer of the model, i.e. the upper mantle, the $P_n$-ray coverage is very good and homogeneous here which is very fortunate as this should allow for an unbiased evaluation of the effects of the $P_n$ anisotropy on the tomographic models.

The influence of the GAP criterion, which is more restrictive to the ray coverage than the residual limit RES, is also visible in the upper panels of Figure 6.24. The ray coverage is diminished the strongest in the second and third layer when the azimuthal gap is reduced from 180° to 120°. In layer three, the ray coverage becomes less than 50 rays per bloc in some parts of the area. Even so, the fourth layer (upper mantle) is still fairly well covered because of the large number of $P_n$-phases in the data set. On the other hand, limiting the maximum residual RES to $5\,s$ has not much influence on the overall ray coverage as this criterion is most likely overruled by the GAP criterion.

#### 6.6.1.2  15 × 15 P-wave models

Here, only $P$-phases are used in the SSH-inversion runs which are carried out for the various case combinations, discussed. It turns out that for cases of pure SSH inversion for seismic structure (hypocenters fixed, IFIX=1) differences in the crustal structure between the isotropic and anisotropic models are small with only slight variations in the absolute velocity anomalies there. For the upper mantle, on the other, where the influence of the anisotropic correction comes to bear, structural difference do occur.

---

[10] To compute the individual $V_P/V_S$-ratios for each bloc separately, the inversion algorithm has to be extended for this, that will be another challenge for further development.

## 6.6. 3D LATERALLY HETEROGENEOUS SEISMIC VELOCITY MODELS

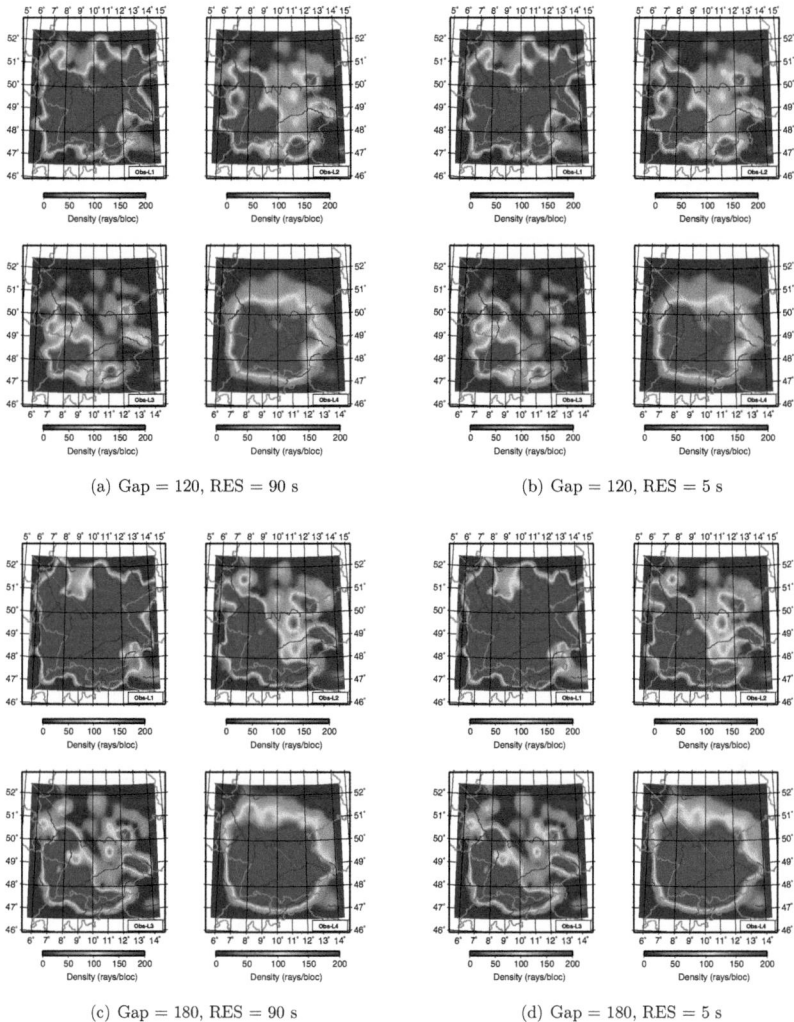

(a) Gap = 120, RES = 90 s  (b) Gap = 120, RES = 5 s

(c) Gap = 180, RES = 90 s  (d) Gap = 180, RES = 5 s

Figure 6.24: Ray coverage for the models with $15 \times 15$ blocs and RES = $90\,s$ (left side) and RES = $5\,s$ (right side). The scale is between 0 (blue) and 200 (red) rays per bloc. Blocs covered with more than 200 rays still are in red. The most dominant area is in the south-west around the Rhinegraben area with its dominant seismic activity. The ray coverage in the third layer is less compared to the others. This coincides with the distribution of most of the hypocenters at depth between zero and twenty $km$.

In the following paragraphs we only talk about the pure tomographic seismic structure inferred from the various SSH inversion cases whereby isotropic and anisotropic model cases are discussed in tandem. The effects of the 3D seismic models on the relocation of the hypocenters (for the full SSH-inversion) are described in the subsequent Chapter (7), "Localization of Hypocenters".

- **Results for GAP = 120°, hypocenters fixed**
  The results (Figures 6.25(a) and 6.25(b)) and for **layer one** and **two** are nearly the same for the anisotropic and isotropic inversion, what is to be expected because, as these two layers are mainly affected by rays with direct phases $P_g$, which are not subjected to an anisotropic correction. The very small differences are due to small contributions of deep $P_n$ phases, which mix up to the sum of the arrivaltimes at a particular station.

  The most prominent structural differences are located in **layer four**, where the blue (high velocity) area extending from south-west to north-east is more clearly separated between the basement of the Vosges and the Black Forest Mountains along the Rhinegraben for the anisotropic than for the isotropic case. The border of that blue area together with the red one in the fore-alpine region more exactly follows the line of the river Danube, which is more realistic with regard to the tectonic situation. Also the abnormal structure in **layer four** between 50° and 51° Latitude and 8° and 10° Longitude, the blue area, probably representing the Vogelsberg area, is more dominant and shifted southwards and the Intraplate-Volcanites in the region of Bonn west of it also become visible. The red area south of the Vogelsberg area starting at 50° Latitude and 8.5° at the entrance of the river Main into the river Rhine is more pronounced in the anisotropic model case and more exactly following the sedimentary structure along the River Rhine.

- **Results for GAP = 120°, hypocenters free**
  Compared to the case with hypocenters fixed, the plots of the "hypocenters free" case show a lot more differences in the seismic velocity distribution for all layers (Figures 6.25(c) and 6.25(d)). This clearly indicates the importance of the elliptical anisotropic correction for the $P_n$-phases. While the hypocenters are now free, i.e. relocated during the SSH-inversion, there are a lot more degrees of freedom in terms of model parameters to satisfy the anisotropic traveltime information as the latter is now projected not only into the seismic structure but into the relocation of the hypocenters as well. Applying the anisotropic correction results in a reconstruction of the structure which is again similar to that of "hypocenters fixed" case.

  For the **first layer** the seismic structure is nearly the same in both cases, owing to the fact again that this layer is mainly only probed by $P_g$-phases. But starting with **layer two**, the differences become more and more visible. The blueish area in the middle east of the study area, starting with the deep blue spot at 11° Longitude and 50.5° Latitude with a $V_P$ velocity of 6.2 $km/s$ follows the shape of the Tepla-Barrandium, the Erzgebirge and the Böhmerwald, which mainly consist of volcanic and granitoidic rocks. The reddish area between 48.5° and 49.5° Latitude and 9° and 11° Longitude again follows in some way the extent of the Saxo Thuringikum which mainly consists of sedimentary rocks. The major anomalies in **layer two** are still existent in both inversion cases, but show some difference. So, for example, in the middle area of Figure 6.25(c), the low velocity anomalies conflate to a ring-like structure for the isotropic case which disappears in the anisotropic one.

  The blueish, high velocity anomaly in the region west of "Nördlingen", where some million years ago a meteoritic impact happened, again appears in the anisotropic and the isotropic case, but being more distinct in the former on. East of that region, Song et al. (2004) detected a major high velocity anomaly in the region of "Ingolstadt", which is about 100 $km$ east of that anomaly, and corresponds to the blue anomaly in **layer one**.

  The **third layer** representing the lower crust which, as discussed several times earlier, is overall the least well-resolved layer of the SSH-model, except in the southwest of the model area. Even so, the structural differences are only minor for the isotropic and anisotropic cases. Some small

## 6.6. 3D LATERALLY HETEROGENEOUS SEISMIC VELOCITY MODELS

differences occur in the western part. But due to the overall problematic seismic interpretation of this layer, this has no further consequence.

In the **fourth layer**, the upper mantle, for the anisotropic model the huge blue area extending from the southwest to the northeast along the Danube river again shows up two blue anomalies, likewise in the "hypocenters fixed" case, though with a slightly different shape. However, as there are some major discrepancies with the isotropic model, one is lead to the conclusion that the anisotropic traveltime correction is indispensable for a proper 3D seismic tomography of the upper mantle beneath Germany.

- **Results for GAP $= 180°$, hypocenters fixed**
  SSH-models computed with the larger GAP of $180°$, instead of $120°$ before, naturally, have a better and more extended ray coverage of the model area. Furthermore, as now many more phases (22491 compared with 12325 phases earlier) included in the inversion, the model parameters should, statistically, be better determined. Otherwise, comparing the results of the corresponding subfigures in Figures 6.26(a) and 6.26(b), only minor differences are visible for all, except the problematic **layer three** and the northern part of the model and especially in the middle of layer two where a blue high velocity anomaly occurs. Differences are mainly only in the absolute contrasts of the velocity anomalies leading sometimes to a "melting together" of individual bloc structures, such as, for example, for the red low-velocity structure in the second layer in the subfigures in Figures 6.26(a) and 6.26(b). There is also some evidence that the reconstruction is clearer for the model inversions with the more restrictive GAP limit of $120°$ than that of $180°$.

  Concerning the major topic of this thesis, the incorporation of anisotropy into the SSH-inversion, differences between the isotropic and anisotropic models occur again, though are somewhat more accentuated than for the GAP $= 120°$ case above. For example, the huge blue area between $40°$ and $50°$ Latitude and $8°$ and $12°$ Longitude spreads out more to the southeast for the isotropic inversion case. The red continuous low velocity area in the third subquadrant at $50°$ and $51°$ Latitude and $7°$ and $9°$ Longitude is split by two blueish high velocity zones at $50.5°$ Latitude and $7.5°$ Longitude and $51°$ Latitude and $9°$ Longitude.

- **Results for GAP $= 180°$, hypocenters free**
  For this "hypocenters-free" (full inversion) case (Figures 6.26(c) and 6.26(d)), the differences between isotropic and anisotropic models are more pronounced than for the "hypocenters fixed" case which are more similar to the anisotropic case with hypocenters free. Thus, looking at the results for **layer one** and **two**, these are again rather identical for the two cases and similar to those with the hypocenters fixed. Only a few distinct anomalies are of different shape, like the red area in the southeast following the Molasse area north of the forealps, following more the real geological situation in the "hypocenters free" case. Also the reddish low velocity anomaly in **layer one** starting southwards at $50°$ Latitude and $9°$ Longitude is more emphasized for this case, what again coincides with the known geological situation there (see Figure 1.2 in Chapter 1). The reddish area lying between $48.5°$ Latitude, $9°$ Longitude and $49.5°$ Latitude, $11°$ Longitude again follows roughly the extent of the Saxo Thuringikum inferred from the "GAP $= 120°$, hypocenters free" case above, though not as to the east. For details see Berthelsen (1992).

  Fore the anisotropic model the huge blue area in the **fourth layer** in the southeast again is more linear without the spreading to the southeast as for the isotropic case and similar to the case with hypocenters fixed. The blue area west of the Rhine disappears compared to the "hypocenters fixed" case, together with the huge red area in the southeast, south of the Danube river. One major difference occurs in the region of Karlsruhe, where a low velocity anomaly arises which cannot be seen neither in the isotropic nor in both "hypocenters fixed" cases.

The trade-off curves which have been used to retrieve some of the 3D seismic inversion models discussed above are illustrated in Figure 6.27(b), where the optimal settings for the damping (ridge) parameter

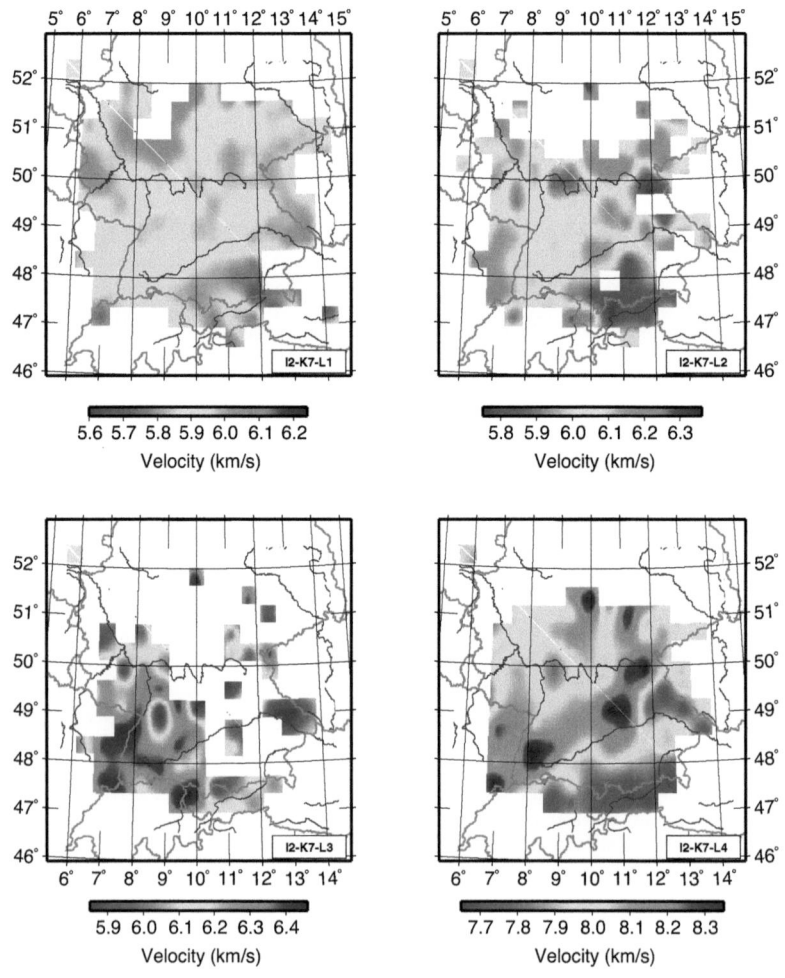

(a) isotropic, Gap = 120, ifix = 1

## 6.6. 3D LATERALLY HETEROGENEOUS SEISMIC VELOCITY MODELS   155

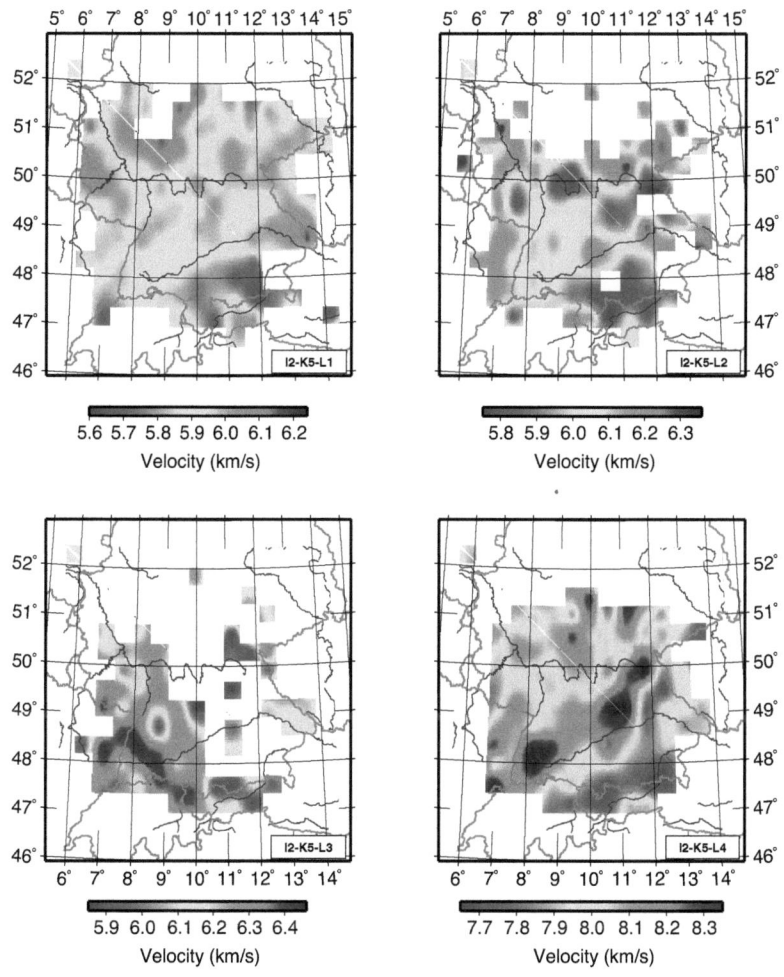

(b) anisotropic, Gap = 120, ifix = 1

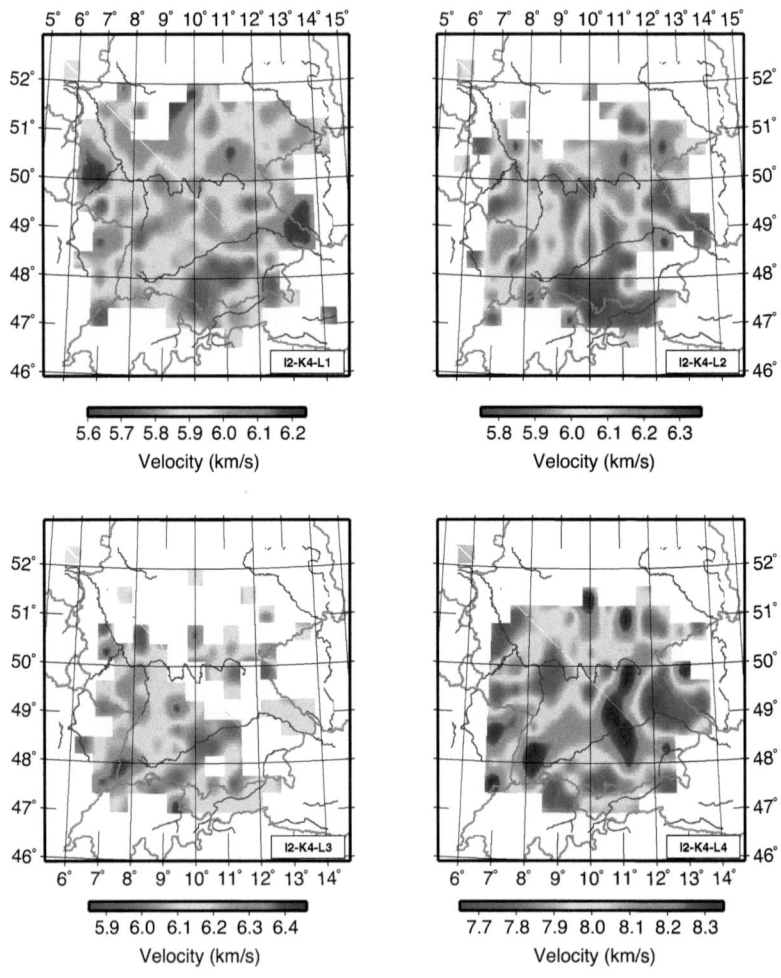

(c) isotropic, Gap = 120, ifix = 0

## 6.6. 3D LATERALLY HETEROGENEOUS SEISMIC VELOCITY MODELS

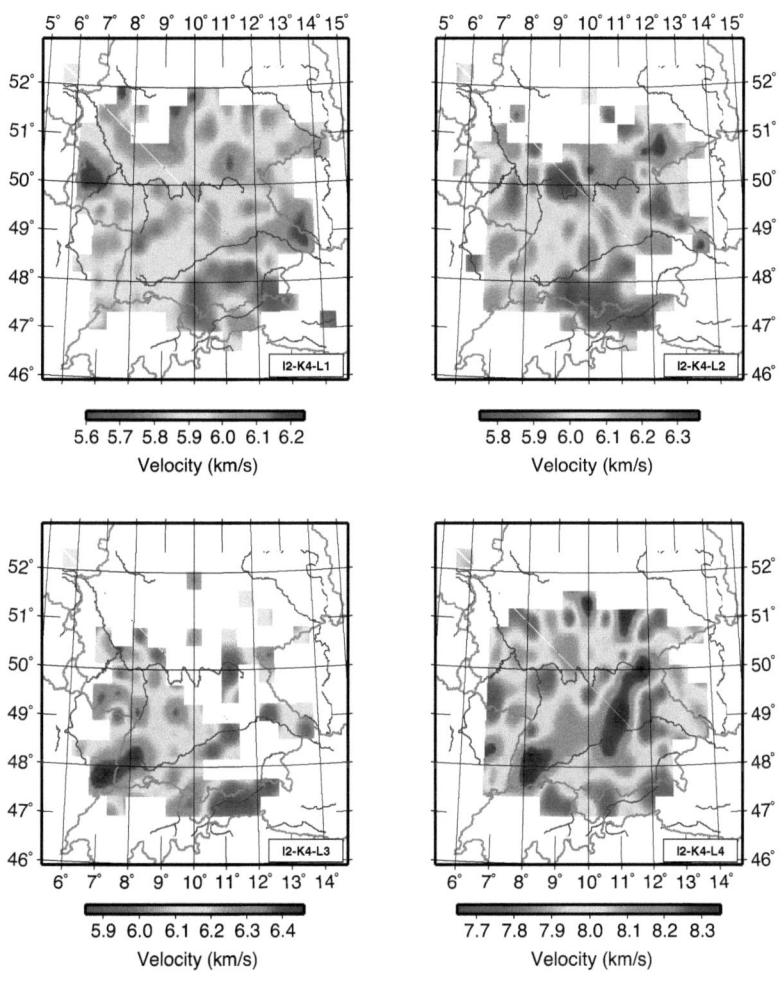

(d) anisotropic, Gap = 120, ifix = 0

Figure 6.25: 15 × 15 3D isotropic and anisotropic velocity models with GAP = 120°, optimal results after the second nonlinear iteration. Upper panel set (6.25(a), 6.25(b)) with hypocenters fixed, denoted by ifix = 1. Lower panel set (6.25(c), 6.25(d)) with hypocenters free, denoted by ifix = 0.

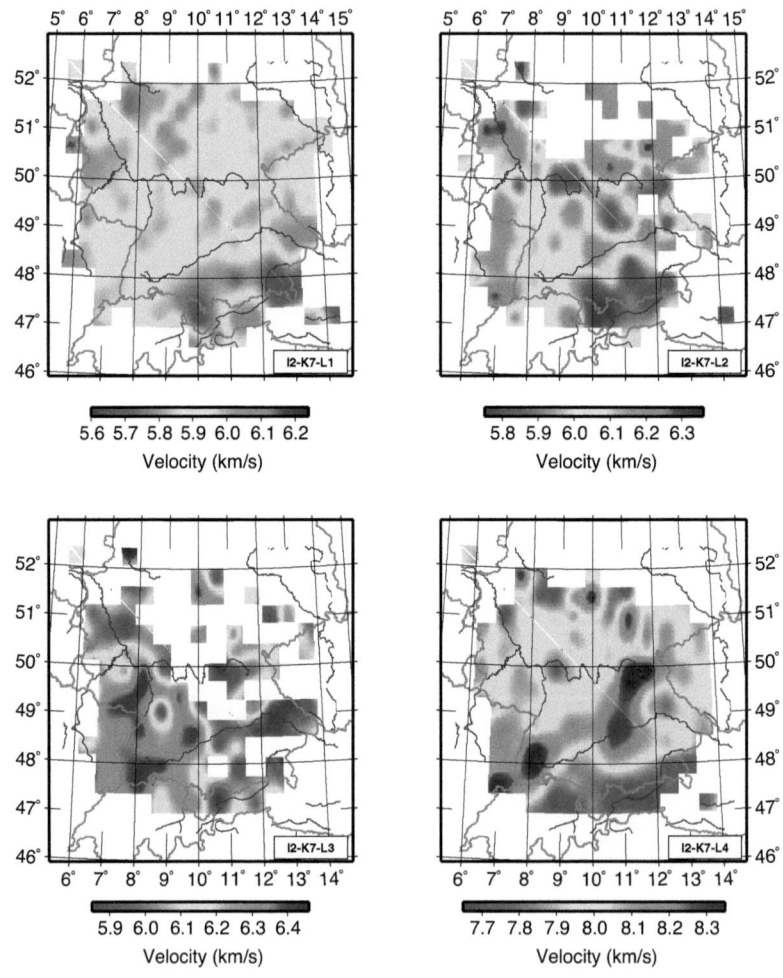

(a) isotropic, Gap = 180, ifix = 1

## 6.6. 3D LATERALLY HETEROGENEOUS SEISMIC VELOCITY MODELS 159

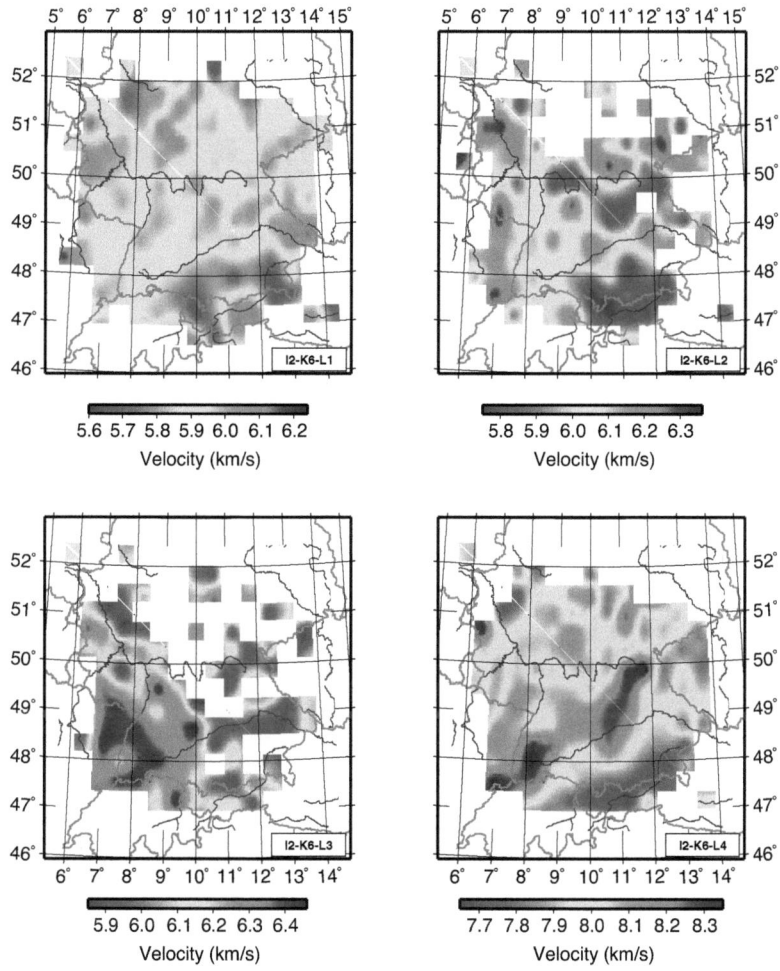

(b) anisotropic, Gap = 180, ifix = 1

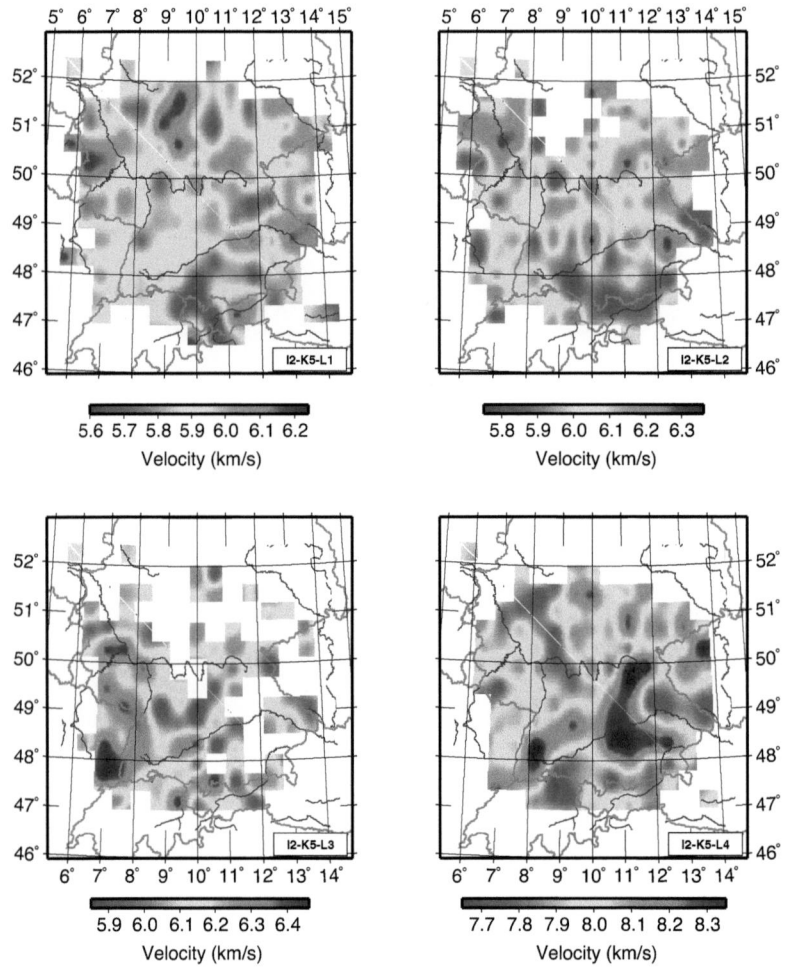

(c) isotropic, Gap = 180, ifix = 0

## 6.6. 3D LATERALLY HETEROGENEOUS SEISMIC VELOCITY MODELS

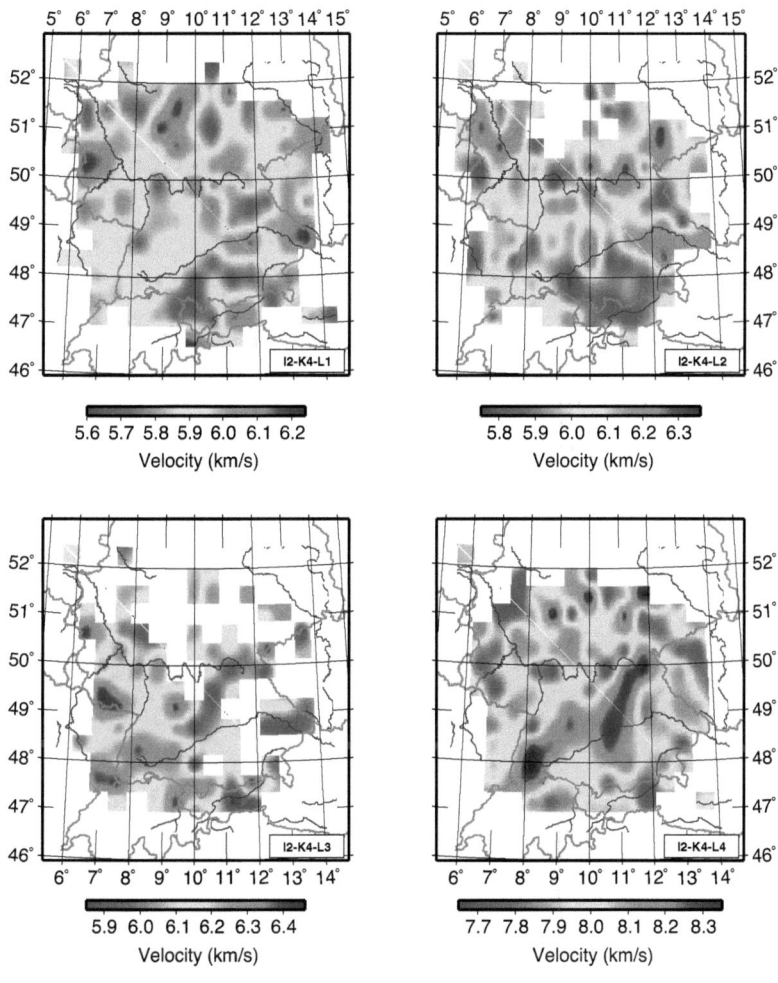

(d) anisotropic, Gap = 180, ifix = 0

Figure 6.26: 15 × 15 3D isotropic and anisotropic velocity models with GAP = 180°, optimal results after the second nonlinear iteration.

$\lambda$ can be chosen by selecting the ridge parameter in the range of the strongest change between the x- and y- values in the appropriate diagrams. Similar to the results of the synthetic tests of an earlier section, the optimal values are located here also somewhere between the one for the minimal RMS and the smallest $\lambda$ at K = 1, that does not correspond the usual concept of choosing the optimal ridge-parameter at the knee-point of the distinct curves. The value for the minimal RMS is between K = 7 and 5 for the "hypocenters fixed" cases and about 1 to 4 for the "hypocenters free" cases after the first non-linear iteration (see Table 6.4) showing that for the first case the stabilization by damping the solution needs to be higher. According to the selection of the optimal solution for the next non-linear iteration step by the minimal RMS and the application of the LM procedure, for further iterations the ridge parameter increases to deliver the optimal result with minimal RMS or TSS, because the optimal result is just reached in the previous step.

### 6.6.1.3   $15 \times 15$ $S$-wave models and comparison with $P$-wave models

In this section anisotropic $S$-wave models are inverted for and compared to the $P$-wave models above. All of these inversions are carried out with "hypocenters free" (full SSH-inversion). Naturally, the results for $S$-waves are not similar due to the different rock and tectonic arrangement, which will be reflected by a different $V_P/V_S$ ratio.

Astonishing at all is the result, that the isotropic and anisotropic results do not differ much from each other. This will lead to a different distribution and different values for the $V_P/V_S$-ratio between the anisotropic and isotropic case as it will be computed in detail in Section 6.8.

The inversion results for the $V_S$-structure in **layer one** and **two** (Figures 6.28(b) and 6.28(d)) are rather different from those for $V_P$. So, for example, in the middle of **layer one**, along 49° Latitude, in a band between 8° and 14° Longitude, a huge, greenish and coherent high $V_S$-velocity area occurs, with some blue maximums inside. Due to the rather small resolved area in the **fourth layer**, the anisotropic correction for the $S_n$-phases is nearly not visible, the seismic structure is nearly the same.

But some of the anomalies, occurring in the plots for the $V_P$ velocities can be recognized again in the $V_S$ velocity plots. So, the low velocity anomaly ($V_P = 5.7\,km/s$) south of the river Main at it's entrance to the river Rhine has its counterpart with a higher $V_S = 3.6\,km/s$ velocity.

Here it is only exemplarily to show the different structure obtained by the separate inversions. A complete comparison between the $P$- and $S$-wave models is in Section 6.8, where the $V_P/V_S$-ratio is computed only for the case with bloc discretization at $15 \times 15$, so in the following computations for the higher bloc discretization, they are not considered at all. As proved earlier, the limit of $5\,s$ for the residuals is a little more than the influence of anisotropy on the original residual, i.e. the variation from the isotropic model, so all computations are done with that limit.

### 6.6.1.4   Summary of statistical results of the $15 \times 15$ bloc models

Although the previous figures have already illustrated qualitatively and subjectively that the 3D inversion models with the anisotropic $P_n$ - traveltime correction included are better and appear to be more consistent with the overall geological picture of the model space than the isotropic models, a more objective, quantitative measure of the inversion quality is needed. As discussed in a previous chapter this is the TSS or the RMS which measures the fit of the model to the data. Table 6.4 indicates, indeed, that the TSS and the RMS are lower for the anisotropic than for the isotropic models. This holds for both cases "hypocenters fixed" and "hypocenters free", i.e. full inversion which is, again, an indicator of the necessity to apply the elliptical anisotropic correction for 3D seismic tomography underneath Germany. It should be noted that the significance of these statistical improvements could also be tested by means of an F-test, as done by Song et al. (2004) for various 2D anisotropic $P_n$-velocity tomographic models underneath Germany (see Section 6.7). In Table 6.5 the average layer velocities, as well as their standard deviations, computed for the different model cases discussed, i.e. ifix = 1 (hypocenters fixed), ifix = 0 (hypocenters free), ani = 1 (anisotropic), ani = 0 (isotropic) are

## 6.6. 3D LATERALLY HETEROGENEOUS SEISMIC VELOCITY MODELS

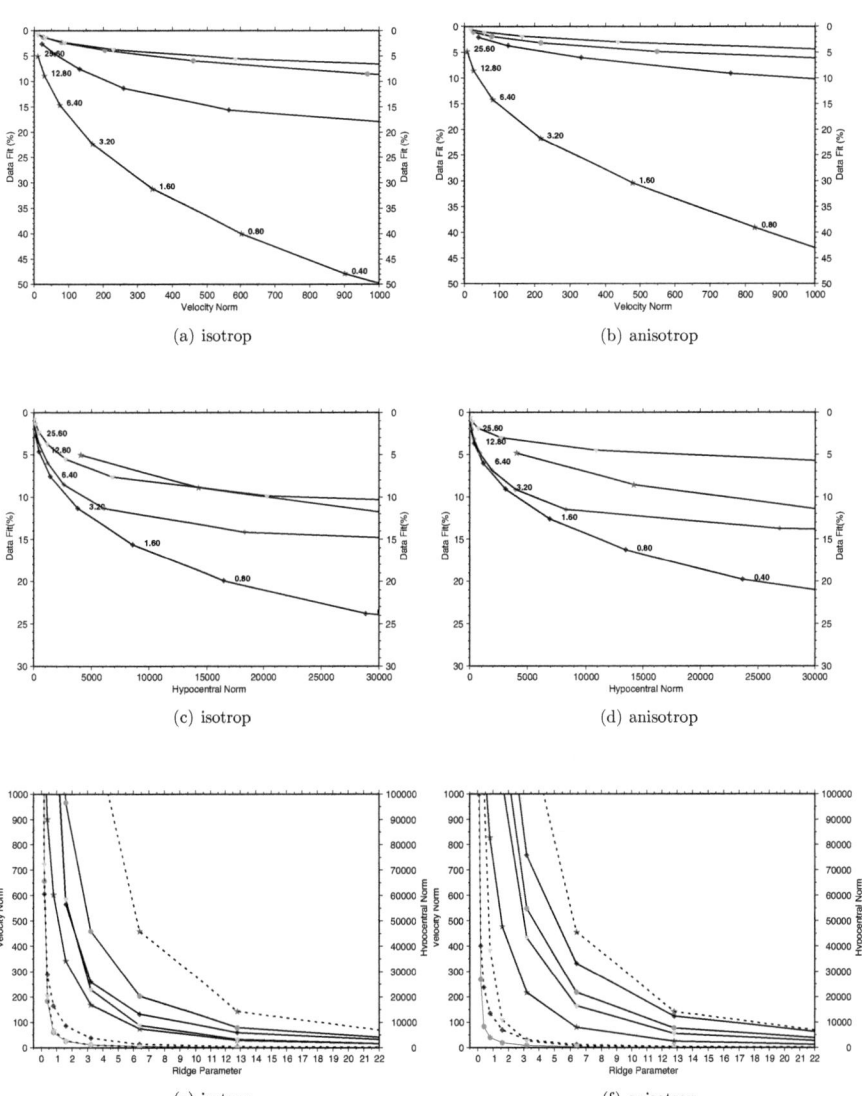

Figure 6.27: Upper and middle set are Trade-off curves between data-fit and the squared sum of hypocentral parameter differences or velocity parameters. The lower set shows the trade off between velocity or hypocentral norm and the ridge parameter, measured by the absolute damping value. The trade off between velocity norm (full line), hypocentral norm (dotted line) and the ridge parameter characterize the optimal ridge parameter in the region of the strongest change of the hyperbolic like curve. The colors denote the nonlinear iteration steps, where red stands for the first, blue for the second, green for the third and yellow for the fourth and last one.

Table 6.4: Statistical results of all previously computed 15 × 15 inversion models as quantified by the TSS and the RMS, after the first nonlinear iteration step.

| GAP degree | K | ani | ifix | res | $TSS_{SSH}$ $s^2$ | $RMS_{SSH}$ $s$ | $TSS_{hypo}$ $s^2$ | $RMS_{hypo}$ $s$ |
|---|---|---|---|---|---|---|---|---|
| 120 | 6 | no  | yes | 5 | 12444 | 1.0048 | 14943 | 1.1011 |
| 120 | 6 | yes | yes | 5 | 11508 | 0.9663 | 14943 | 1.1011 |
| 120 | 1 | no  | no  | 5 | 8693  | 0.8399 | 14943 | 1.1011 |
| 120 | 3 | yes | no  | 5 | 7665  | 0.7886 | 14943 | 1.1011 |
| 180 | 5 | no  | yes | 5 | 26173 | 1.0788 | 29311 | 1.1416 |
| 180 | 4 | yes | yes | 5 | 24847 | 1.0511 | 29311 | 1.1416 |
| 180 | 1 | no  | no  | 5 | 16636 | 0.8600 | 29311 | 1.1416 |
| 180 | 2 | yes | no  | 5 | 14879 | 0.8134 | 29311 | 1.1416 |

Table 6.5: Layer averages of the computed bloc velocities for the various 15 × 15 model cases.

| model | ani | ifix | V1 | V2 | V3 | V4 | Sta1 | Sta2 | Sta3 | Sta4 |
|---|---|---|---|---|---|---|---|---|---|---|
| 15x15 | no  | yes | 5.90 | 6.05 | 6.14 | 8.02 | 0.010 | 0.020 | 0.024 | 0.036 |
|       | yes | yes | 5.90 | 6.05 | 6.15 | 8.02 | 0.013 | 0.023 | 0.025 | 0.032 |
|       | no  | no  | 5.92 | 6.04 | 6.15 | 8.02 | 0.019 | 0.018 | 0.015 | 0.033 |
|       | yes | no  | 5.91 | 6.04 | 6.15 | 8.01 | 0.019 | 0.019 | 0.015 | 0.033 |

listed, together with the standard deviations. The latter are a little bigger for the inversion case with fixed hypocenters where the traveltime information is not absorbed by hypocentral relocations but fully projected into the seismic structure. Surprisingly, no significant differences in the average (1D) layer velocities for the four 3D inversion cases are encountered and this average model is very close to the optimal 1D velocity model retrieved in Chapter 5, except that the velocity V3 in the - problematic - third layer is slightly smaller for the averaged 3D model.

#### 6.6.1.5 Summary of structural results of the 15 × 15 bloc models

The discussion of the different 15 × 15 bloc model variants in the previous sections show that such a model discretization is fine enough to resolve major tectonic and geological structures. Here it appears appropriate to recapitulate the common and most salient seismic structural features retrieved for the crust and upper mantle beneath Germany from these tomographic models.

Before going to the structural results, it has to be mentioned, that they coarsely are the same for the models with hypocenters fixed (IFIX = 1) and the simultaneous inversion for both, hypocenters and structure (IFIX = 0) and the anisotropic correction for $P_n$-phases in contrast to the inversion with hypocenters free and no anisotropic correction. This is an indication of the high quality of the hypocenter localization, which is mostly done by local earthquake data centers (LED), so the starting solution for the complete inversion is near the optimal solution, resulting in an optimal solution for the structure.

In **layer one** major tectonic and geological structures are visible south of the Vogelsberg and the Erzgebirge and the Bohemian Mountain. Also, in the Upper Rhinegraben area the basement of the Black Forest and the Vosges, with the sedimentary basin between them (the Rhinegraben proper) appear to be reconstructed as blueish areas east and west of the Rhinegraben[11]. Geologically, the Black

---
[11] Later results with finer bloc discretization will resolve the structure there more clearly.

## 6.6. 3D LATERALLY HETEROGENEOUS SEISMIC VELOCITY MODELS

Forest consists of a cover of sediments, mainly sandstone ($3.5\,km/s < V_P < 4.5\,km/s$) down to about $4\,km$ depth on top of a core of gneiss ($V_P \approx 6\,km/s$ down to $10\,km$) (Illies, 1977; Konrad and Nairn, 1972; Franzke et al., 2003). The basements of these two mountains become visible in **layer one** and **layer four**, in both, the isotropic and anisotropic models, as well as for both, "hypocenters free" and "fixed" model cases, but with a better separation applying GAP = 120°. Also the two basements appear to be separated more clearly for the anisotropic (Figure 6.26(d)) than for the isotropic case (Figure 6.26(d)). The sedimentary in-filling of the Rhinegraben is visible in **layer one**, Figure 6.25(d) and, with lower contrast, in Figure 6.26(d) as well. The red area parallel to the River Rhine until the entrance of the River Main further follows the sedimentary fillings of the Graben proper. The blue area north west of this region rather good follows the center of the Rhenish Slate Mountains (Rheinisches Schiefergebirge).

Another major anomaly occurs in Figure 6.25(d) north of the region of the Nördlinger Ries, which might depend on the former meteoritic impact (Stöffler et al., 2002) indicated by blueish color. Near that region, Song et al. (2004) detected a major high velocity anomaly in the region of "Ingolstadt", which is about $100\,km$ south-east of that anomaly, whether there might be a dependency which can not clearly detected now. In **the second layer**, the color changes to deeper blue indicating a high velocity area, which might result out of the compressional and melting effects of that impact onto this deeper structure. The high velocity anomaly mentioned by Song et al. (2004) between the Harz and the Vogelsberg at the northern border of the study area can not be clearly detected, in contrast to the models with GAP = 180°.

The blue area in Figure 6.25(d) follows the border between Germany and Czechia, what is equivalent to the Erzgebirge and the Bohemian massif with the big blue dot at 50.5° Latitude and 11° Longitude marking the tip of that mountainous region with the Thüringer Wald.

The Vogelsberg area with its tertiary volcanism is represented by a blue area (higher velocities as they are for basalts) in the **first** and **fourth** layer around Latitude = 50.5° and Longitude = 9.5° and Latitude = 51° and Longitude = 8.5°. Its shape is much more dominant in the cases with hypocenters free and GAP = 180° whereas for GAP = 120°, the area is partially shaded out due to the lower ray coverage. In the isotropic case with hypocenters free, these anomalies can not be detected.

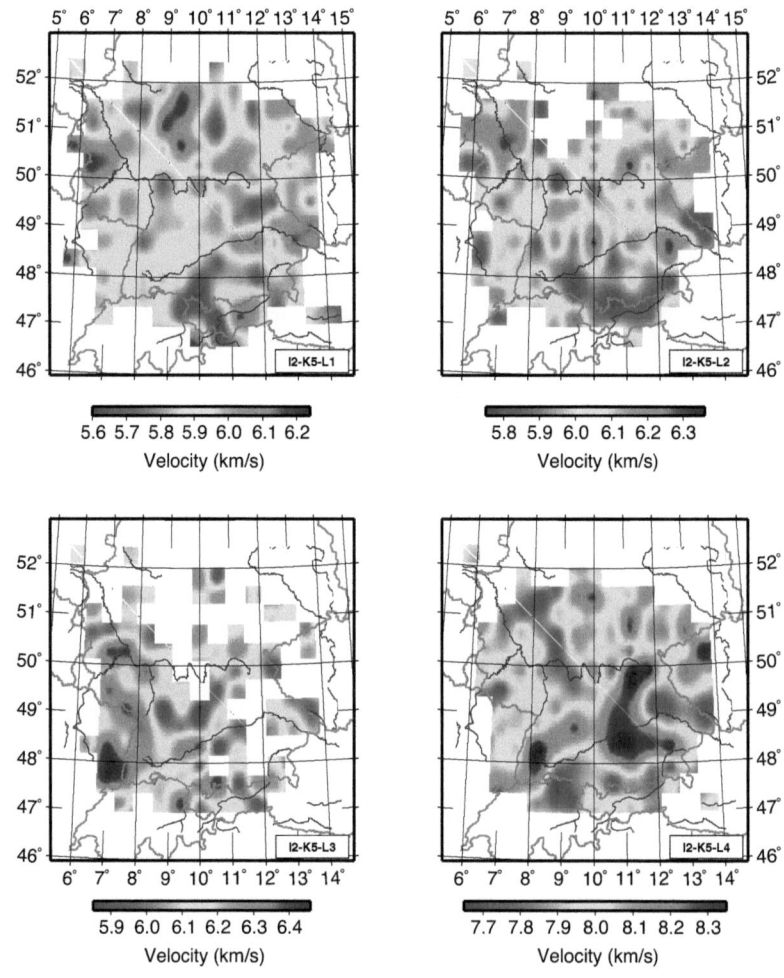

(a) P-phases, Gap = 180, RES = 5 s, iso

## 6.6. 3D LATERALLY HETEROGENEOUS SEISMIC VELOCITY MODELS

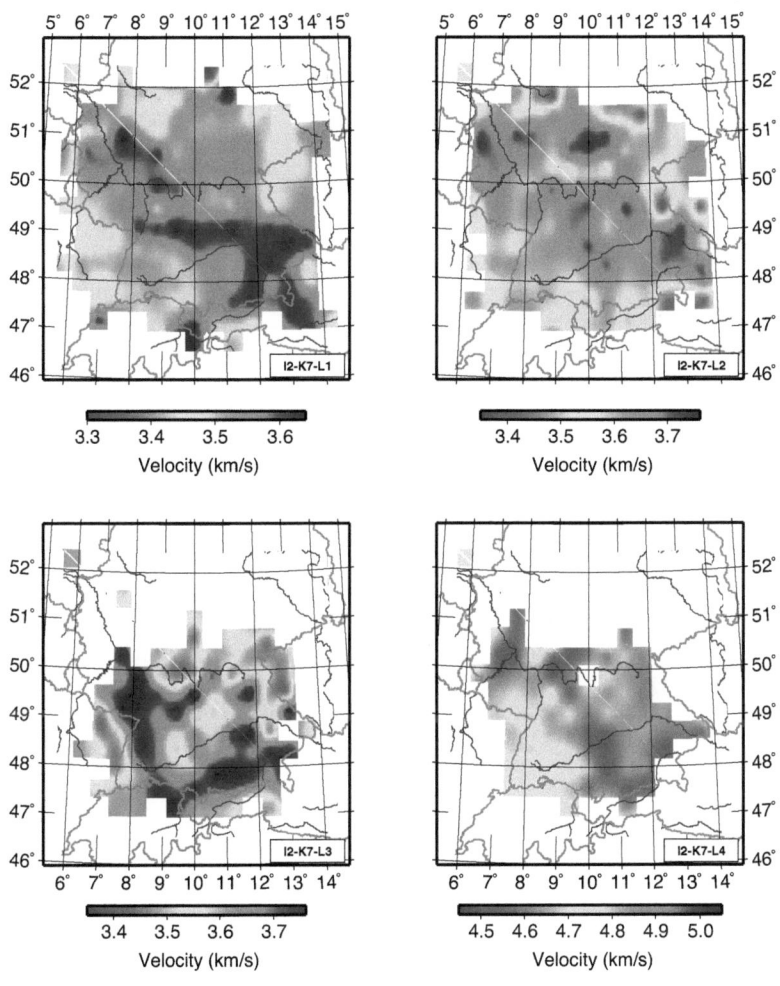

(b) S-phases, Gap = 180, RES = 5 s, iso

168  CHAPTER 6. SSH-INVERSIONS FOR 3D VELOCITY MODELS

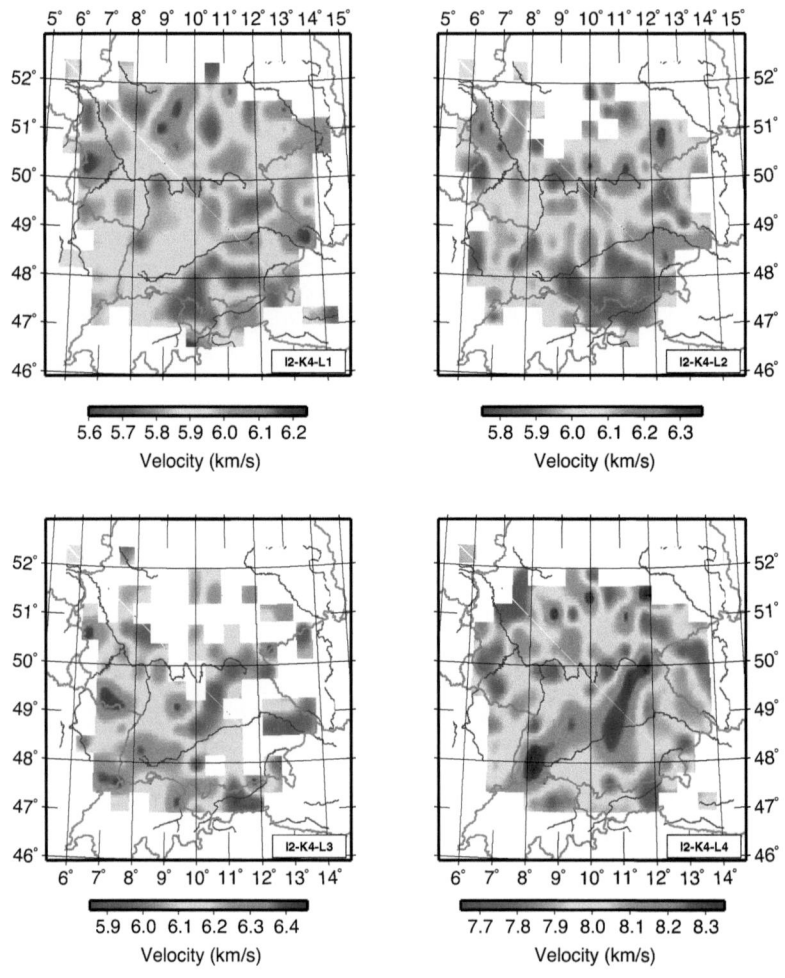

(c) P-phases, Gap = 180, RES = 5 s, ani

## 6.6. 3D LATERALLY HETEROGENEOUS SEISMIC VELOCITY MODELS

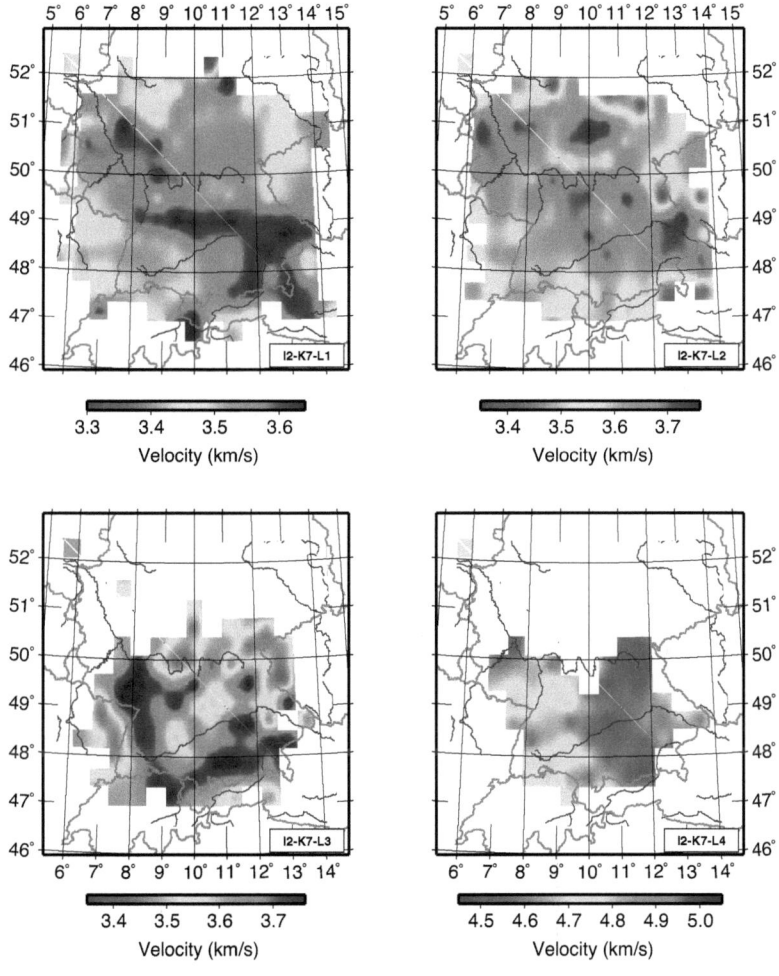

(d) S-phases, Gap = 180, RES = 5 s, ani

Figure 6.28: Comparison of isotropic (upper two sets, Figures 6.28(a) and 6.28(b)) and anisotropic (lower two sets, Figures 6.28(c) and 6.28(d)) $P$- and $S$-wave $15 \times 15$ models.

170  CHAPTER 6.  SSH-INVERSIONS FOR 3D VELOCITY MODELS

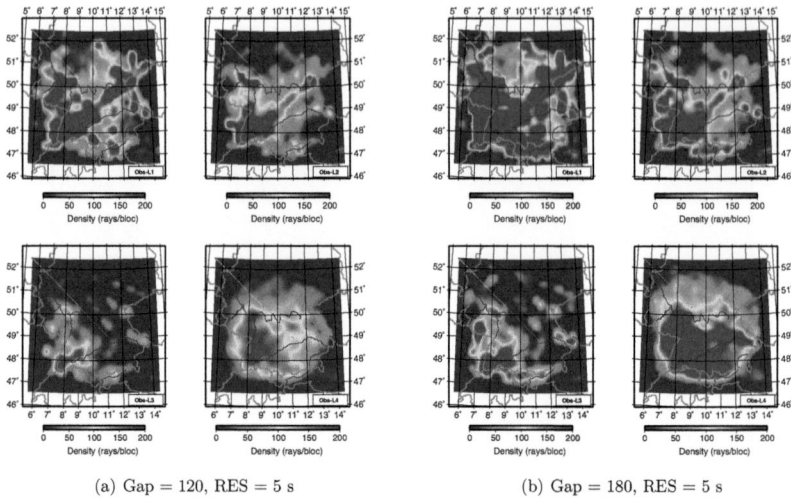

(a) Gap = 120, RES = 5 s    (b) Gap = 180, RES = 5 s

Figure 6.29: Ray coverage for the models with 25 × 25 blocs and GAP = 120° and 180°.

### 6.6.2  25 × 25 blocs models

In the next step of the tomographic model refinement 25 × 25 bloc models are investigated. This discretization reduces the bloc size to $26\,km \times 26\,km$ bloc size, i.e. nearly to half the lateral size of the 15 × 15 bloc model above.

#### 6.6.2.1  Ray coverage for 25 × 25 bloc models

With such a fine subdivision one has to look carefully at the ray coverage and the resolution of the model space, as the probability to have uncovered areas or volumes increases. Figure 6.29 illustrates the ray coverage for the two model cases GAP = 180° and = 120°. As expected, the ray coverage for the former is quantitatively better than for the latter case, but for both cases it is good in layers one and two and, with some reduction, also in layer four. For the problematic layer three, adequate ray coverage is limited to the southwest portion of the model area.

Because of the higher number of model parameters in the 25 × 25 bloc models which, requires a higher number of observations for a reliable inversion, and the definitely better ray coverage with GAP = 180° model, all subsequent model inversions will only be executed with this option which results in a total number of 1216 events with a total of 22491 observations.

#### 6.6.2.2  25 × 25 P-wave models

Figure 6.30 shows the tomographic P-wave structures obtained for this model discretization for the crust and upper mantle for both anisotropic and isotropic inversions. A comparison with the seismic structure obtained for the earlier 15 × 15 bloc model shows that the major geological and tectonic features are, overall, more or less similar giving some additional confidence on the quality of the 3D tomographic inversions carried out here. We thus forgo another discussion of the structural properties of this 25 × 25 model. It is sufficient to mention that the low velocity anomaly at 50° Latitude and 8.7° Longitude is now slightly shifted northwards, closer to the Vogelsberg region and that the sediments in the upper Rhinegraben between Basel and Karlsruhe are more pronounced. Over all, the major

## 6.6. 3D LATERALLY HETEROGENEOUS SEISMIC VELOCITY MODELS

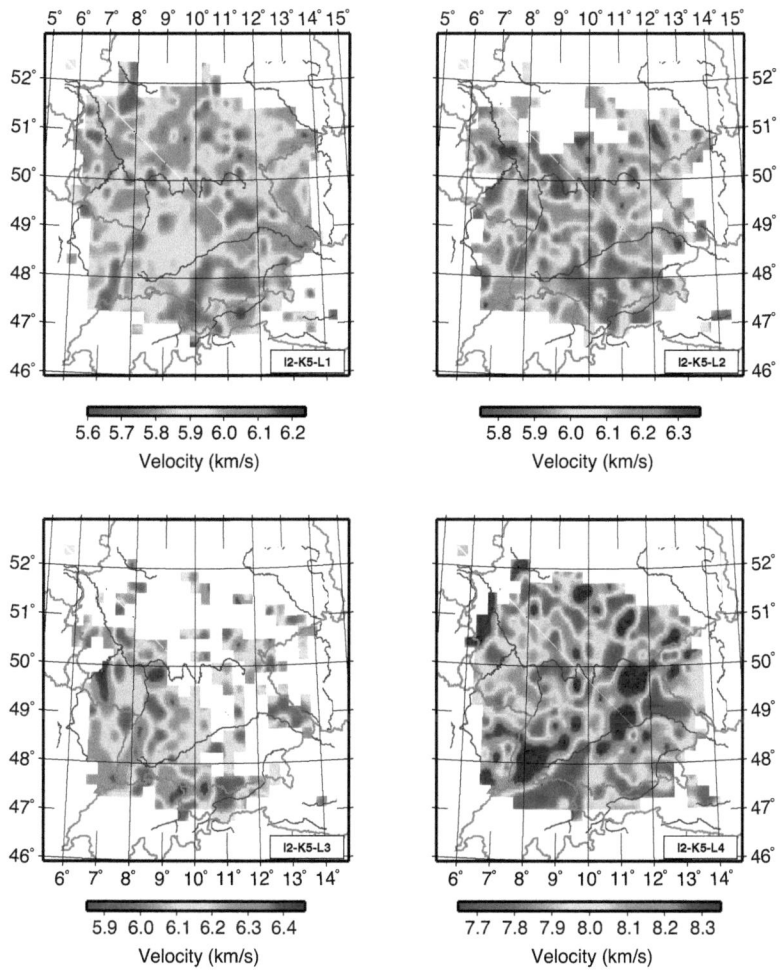

(a) isotropic, Gap = 180

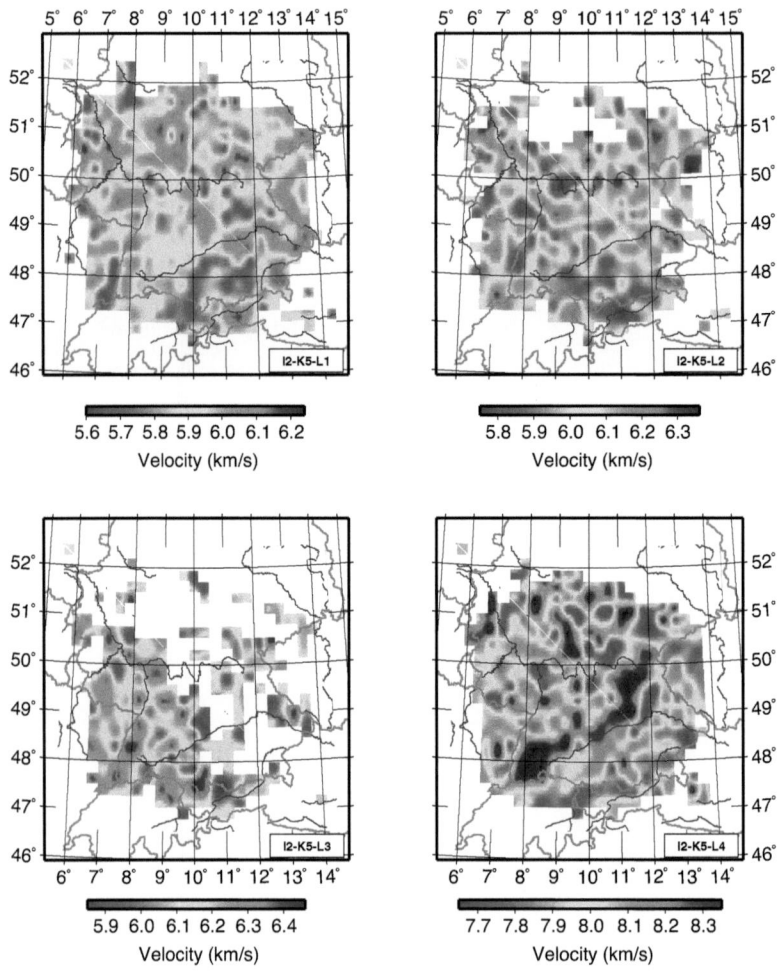

(b) anisotropic, Gap = 180

Figure 6.30: The high resolution models with discretization into 25x25 blocs. Figure 6.30(a) and 6.30(b) are the results for the isotropic and anisotropic inversion with GAP = 180°.

## 6.6. 3D LATERALLY HETEROGENEOUS SEISMIC VELOCITY MODELS

geological and tectonic structures are the same as in the 15 × 15 bloc models, as expected. Also the differences between the anisotropic and isotropic inversion remain the same, showing again a rather different picture for the velocity distribution in layer four. So for example, the area at the entrance of the River Main into the River Rhine has a different distribution of the velocities with opposite values (low velocities instead of high velocities in the anisotropic case).

### 6.6.3  35 × 35 bloc models

Finally, some inversions are done for a 35 × 35 bloc discretization which results in x- and y- widths of 19 $km$ for the blocs. Such a small block size is now at the limit of the validity to apply ray theory as the dimensions of the blocs are in the range of the wavelengths of the seismic signals (see Chapter 2). Going beyond that discretization one comes into the range, that so-called self-healing of wave fronts or Fresnel zones will take affect. Wave fronts traveling through anomalies or traveling through structural inhomogeneities will forget about that disturbance in a sufficient distance of about twice, three times the size of the anomaly. This phenomenon can be explained by Huygens principle of elementary waves which results in the theory of Fresnel zones (Williamson and Worthington, 1993).

Having frequencies of about 1Hz and seismic velocities between 6 $km/s$ and 8 $km/s$ and distances of about 100 $km$ to 50 $km$ to the station, the radius $r_{max}$ of the wavefront becomes about 12 $km$ to 9 $km$ for $V_P = 6\,km/s$ and 14 to 10 $km$ for $V_P = 8\,km/s$.

Another effect is the limitation of the available data, which is shown in Section 6.5, where the areas of reasonable resolution is clearly reduced and more and more gaps occur in the ray covered area. Only the south-western part is resolved with sufficient probability, especially in **layer three**. The best resolution is achieved in the first layer, where the ray-density is the best. With increasing depth, the resolution immensely decreases, and gets slightly better in the fourth layer, where the $P_n$ phases resolve the layer again. Comparing the resolution with the ray-coverage in section 6.6.3.1, the resolution directly follows the ray coverage.

From this, it becomes obvious, that the database has to be increased for the eastern part, to receive a better resolution for the whole area of Germany. Here, the registrations of the Czech network have to be mentioned, that did register a lot of seismic events, also occurring in the western part.

Another complication is the enormous increase of free parameters for the model, compared to the 15 × 15 bloc model, together with decreasing ray coverage. Whereas the structure for the 35 × 35 model has 1225 possible parameters for each of the four layers, the 15 × 15 bloc model only has 225, what is just less than a fifth of the former. A greater number of parameters automatically leads to a better fit of the residuals together with ambiguous solutions for the model space.

#### 6.6.3.1  Ray coverage for the 35 × 35 bloc models

According to the high discretization of the 35 × 35 model, its individual blocs are now crossed by less rays than the previous coarser models. Figure 6.31 shows the ray coverage for the two gap cases, and one immediately notes that only the GAP = 180° case may have ray coverage good enough for a reasonable tomographic inversion, as it provides sufficient phases for most of the blocs, i.e. a sufficient number of degrees of freedom $p = n - m$ ($n$ = number of observations, $m$ = number of model parameters).

One may also compare these ray coverage plots with the corresponding figures of the results for the checkerboard tests carried out in Section 6.5, to get a further grip on the resolution power of this 35 × 35 model. There one can see that only the southwestern part of the model is resolved with sufficient quality, especially in **layer three**. This indicates that there is a need for an increase of the database for the eastern part of Germany - which could be realized by, for example, using the registrations of the Czech network. The best resolution is achieved in the first layer, where the ray-density is also the highest. With increasing depth, the resolution decreases and gets slightly better in

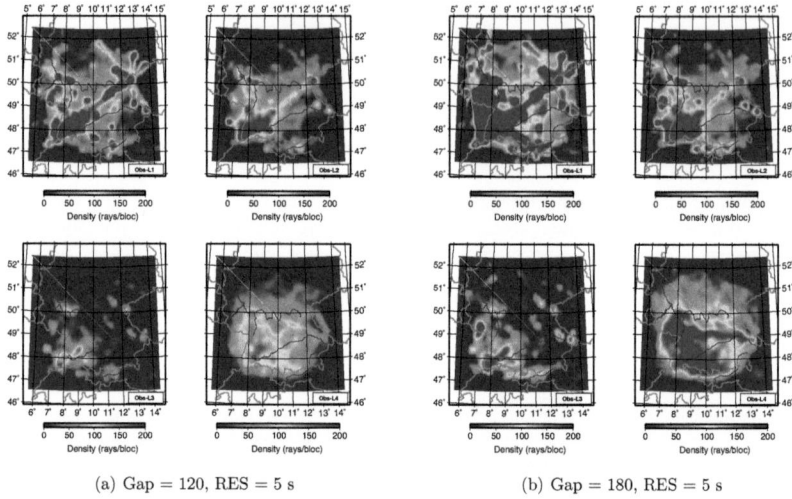

(a) Gap = 120, RES = 5 s  (b) Gap = 180, RES = 5 s

Figure 6.31: Ray coverage for the $35 \times 35$ bloc model for GAP = $120°$ and GAP = $180°$.

the fourth layer, where the $P_n$ phases come to bear. Overall, the checkerboard resolution pictures go in parallel with those of the ray coverage.

#### 6.6.3.2  $35 \times 35$ $P$-wave model

Figure 6.32 shows the tomographic $P$-wave structures obtained for $35 \times 35$ model discretization for the crust and upper mantle for both anisotropic and isotropic inversions. A comparison with the seismic structure obtained for the earlier $15 \times 15$ and $25 \times 25$ bloc models illustrates that, despite the small discretization of this model, which results in more inhomogeneous structural features, major pattern, which were present in the previous models, but not resolved there in detail, are also recognizable here. Other small-scale features of this $35 \times 35$ model may be difficult to be identified in detail.

## 6.7  F-Test statistics for the 3D Models

For testing whether there are significant improvements along the different bloc discretizations and also between the isotropic and anisotropic case, the well-known F-test (Koch, 1985b, 1989, 1993a; Song et al., 2004) is applied to test the various models. Basically, the F-test answers the question whether a decrease in the TSS (data-fit) is worth the increase of parameters, which here are basically the number of velocity parameters. A second important feature is to compare the isotropic and the anisotropic inversion results for a significant improvement. The way to proceed the F-test (equation 6.6) is to compare a model $m_1$ with $p_1$ parameters with a model $m_2$ with $p_2$ parameters (i.e. number of blocs and number of the four hypocenter parameters) and n, the number of data wherefrom one has to calculate the degree of freedom n - p for the special case (N.Draper and Smith, 1981).

$$\frac{TSS_{m_1}/(n-p1) - TSS_{m_2}/(n-p2)}{TSS_{m_2}/(n-p2)} = F(p_2 - p_1, n - p_2) \qquad (6.6)$$

Using the F test the $H_0$ hypothesis that the model $m_2$ results in a statistical improvement of the model fit over the model $m_1$ can be tested.

## 6.7. F-TEST STATISTICS FOR THE 3D MODELS

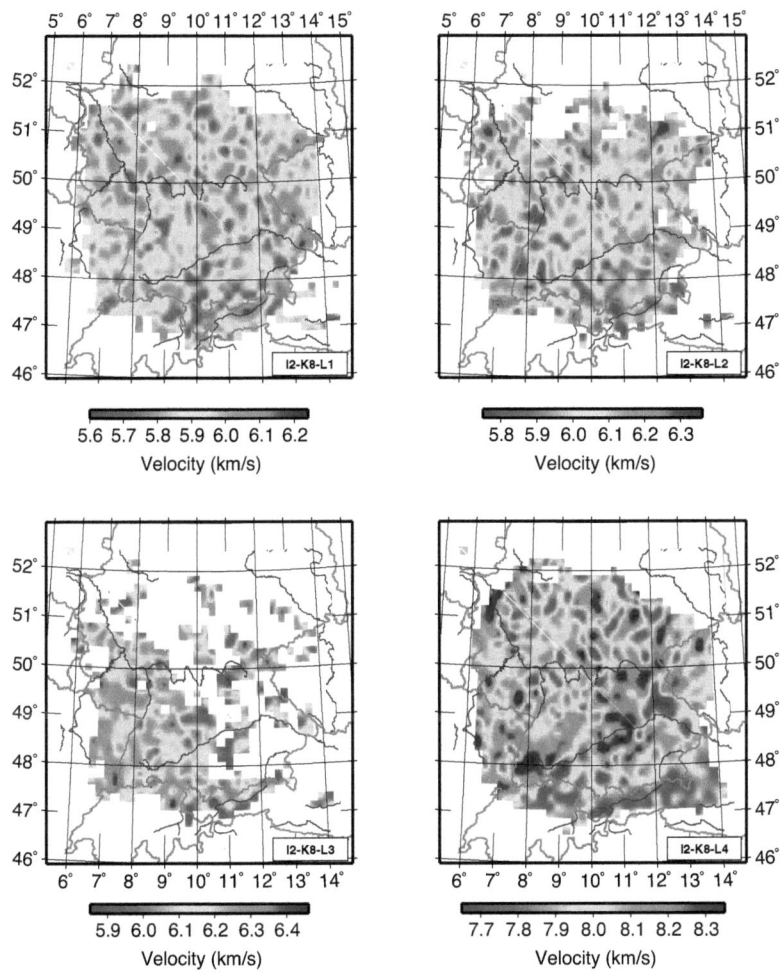

(a) isotropic, Gap = 180

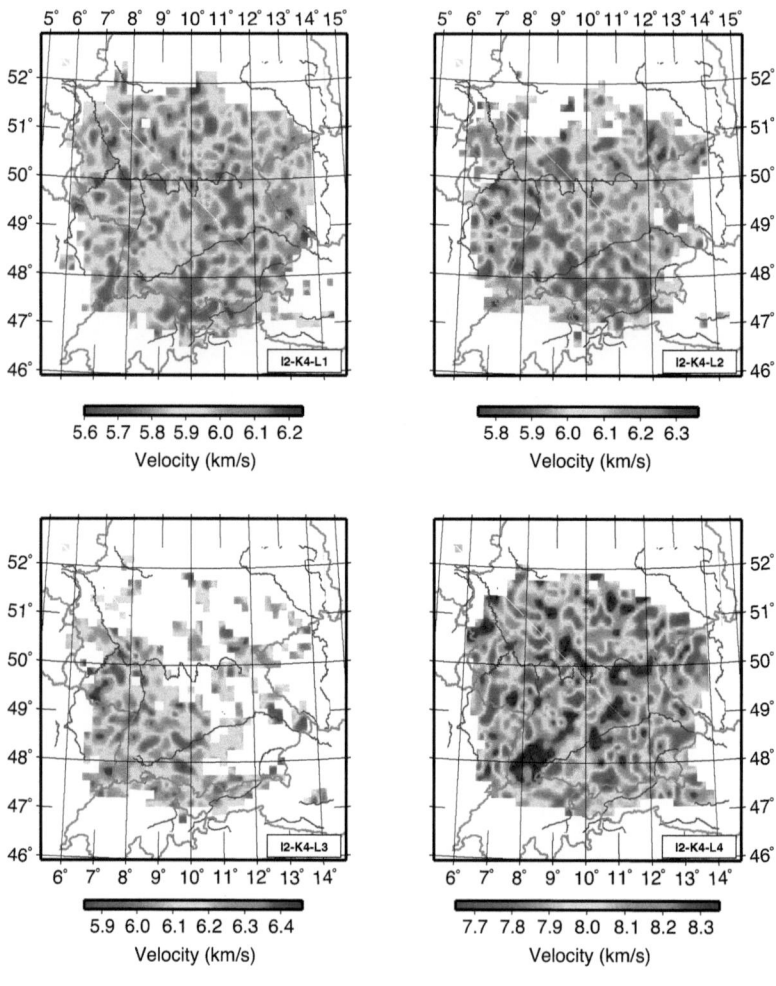

(b) anisotropic, Gap = 180

Figure 6.32: 35 × 35 3D isotropic and anisotropic velocity models with GAP = 180°.

Table 6.6: Layer averages of the computed bloc velocities for the various 15 × 15 model cases.

| Compared models | TSS($s^2$) | n | p | DF | $TSS/(DF)$ | $F_{calc}$ | $F_{table}$ | $H_0$ |
|---|---|---|---|---|---|---|---|---|
| 3D-iso-15x15 (baseline) | 16636 | 22491 | 5764 | 16727 | 0.995 | - | - | - |
| 3D-ani-15x15 | 14879 | 22491 | 5764 | 16727 | 0.890 | 0.18 | 1.0000 | no |
| 3D-iso-25x25 | 22310 | 21709 | 7308 | 14401 | 1.549 | 0.358 | 1.0000 | no |
| 3D-ani-25x25 | 15113 | 22491 | 7364 | 15127 | 0.999 | 0.0001 | 1.0000 | no |
| 3D-iso-35x35 | 16995 | 22491 | 9764 | 12727 | 1.335 | 0.255 | 1.0000 | no |
| 3D-ani-35x35 | 15767 | 22491 | 9764 | 12727 | 1.239 | 0.197 | 1.0000 | no |

The $H_0$ is accepted at the $\alpha = (100 - 95)$ percent level = 5 percent rejection level, whenever $F_{calc} > F_{0.95}(p_2 - p_1, n - p_2)$. As one can see in Table 6.6, for all models, the hypothesis $H_0$ does not hold, so there is no significant improvement between the distinct models. Only the RMS, which is our main criteria, shows an improvement.

## 6.8 Calculation and interpretation of $V_P/V_S$-ratios

In this section 3D laterally inhomogeneous $V_P/V_S$-ratios are computed for both isotropic and anisotropic inversion cases, i.e. by inverting for pure P- and S- wave models without and with correction for the corresponding upper mantle phases. From the inversion for pure S-wave models, it became obvious that different $V_P/V_S$ ratios have to be used for further combined P- and S-wave inversions. Compared to the pure P-wave inversion with a minimal RMS of $0.3412\,s$, the inversion with only S-waves leads to a slightly higher RMS of $0.3511\,s$ but also to a different structure as can be seen in Figure 6.33. Concerning layer four, it can be seen, that the area for an effective calculation of the $V_P/V_S$-ratio is rather small. The local $V_P/V_S$ ratios are then derived directly from the optimal solutions for the P- and S-inversions[12].

### 6.8.1 $V_P/V_S$-ratio and its relation to the composition of the lithosphere

As mentioned earlier the $V_P/V_S$-ratio, which is directly related to the Poisson-ratio $\sigma$, is extremely useful for characterizing various rock properties, namely, of the presence of fluids in the crust. (Koch, 1992; Sherburn et al., 2006). As such it provides additional information beyond of what is available from the P-wave velocities alone. For example, large sedimentary basins normally are characterized by low $V_P \leq 4\,km/s$ and high $V_P/V_S \geq 1.9$, which can be attributed to unconsolidated, water-saturated sediments, whereas the basement rocks underneath have usually $V_P \geq 5\,km/s$ and $V_P/V_S \leq 1.7$.

In the following paragraphs a short overview on the major petrological features and the composition of the crust and the upper mantle as they relate mainly with the $V_P/V_S$-ratio are provided. For more details we refer the reader to (Anderson, 2007), page 94 ff.

- Petrological and geophysical properties of the crust
  The continental crust was formed during the cooling process of the earth, when rock material with lower melting temperatures was separated from the rest, higher temperature melting rocks, and leaving them behind in deeper layers of the earth, i.e. the mantle and the core. Due to this process, the crust has a different composition than the mantle. Furthermore, the tectonics (driven by mantle convection) and geological processes (sedimentary processes, volcanic activity, metamorphic and tectonic processes) occurring in the crust have led to the rather heterogeneous structure that seismologists endeavor to determine by the methods as the ones presented in this thesis.

---

[12]To include the direct computation of the $V_P/V_S$-ratio in the inversion process might be of major interest for further studies with more detailed $V_P/V_S$-ratios.

178  CHAPTER 6. SSH-INVERSIONS FOR 3D VELOCITY MODELS

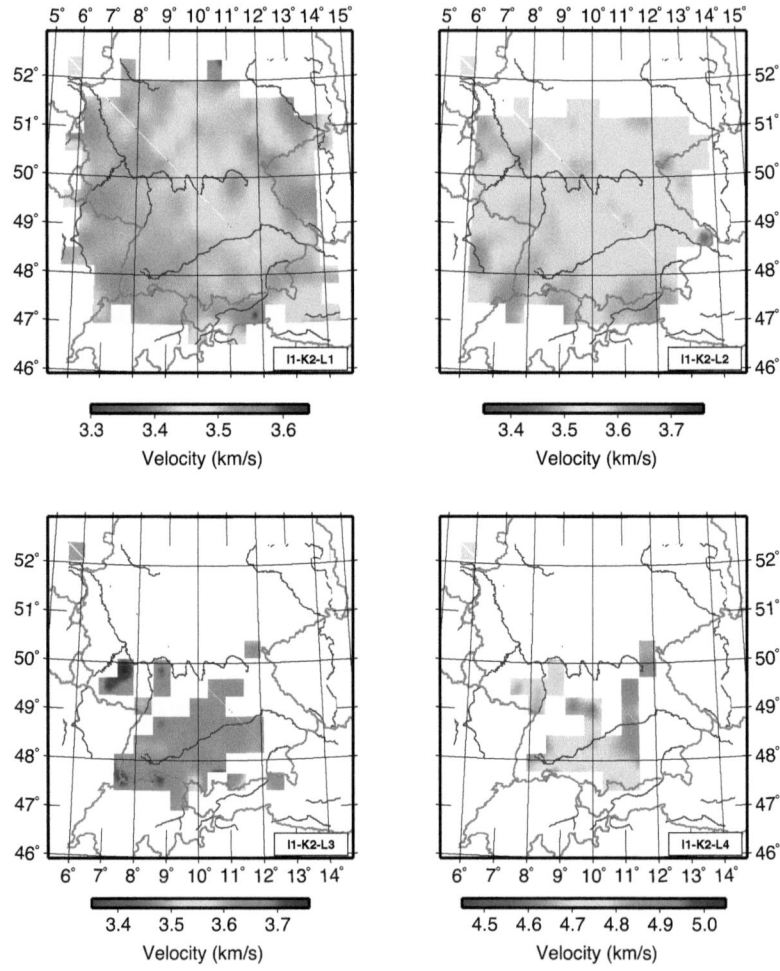

(a) $V_S$ model, isotropic, ifix = 1

## 6.8. CALCULATION AND INTERPRETATION OF $V_P/V_S$-RATIOS

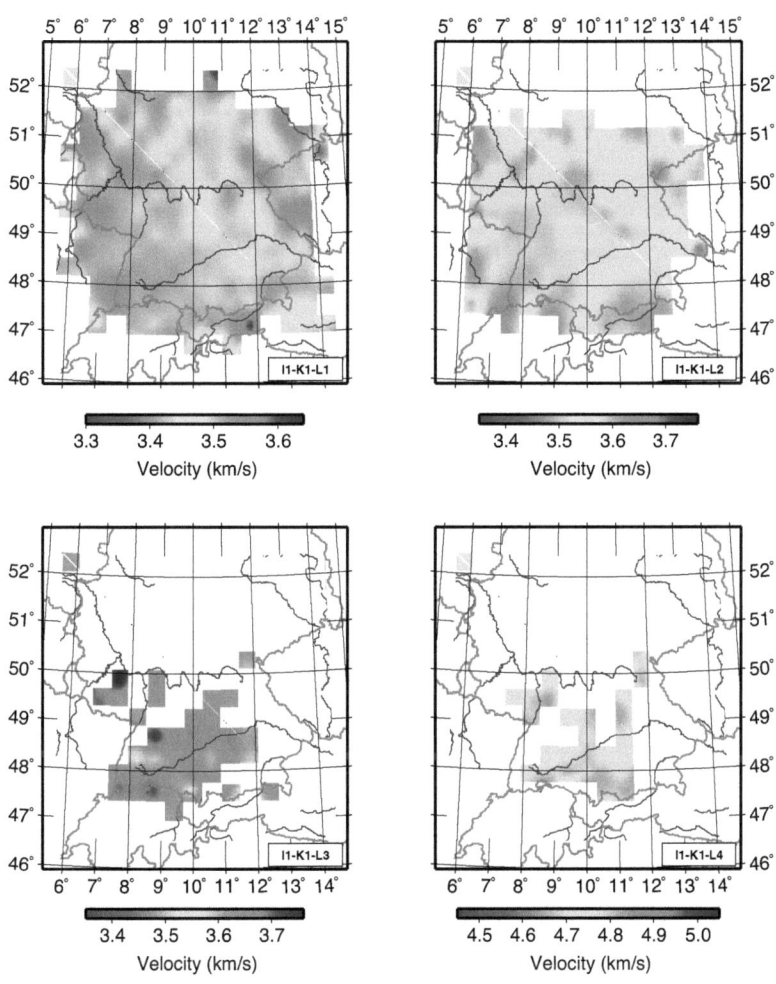

(b) $V_S$ model, anisotropic, ifix = 1

Figure 6.33: The optimal 15×15- $V_S$-models for isotropic and anisotropic inversions with fixed hypocenters. These results will be used together with the optimal $V_P$ wave model to calculate the $V_P/V_S$ ratios.

The topmost layers of the crust partially consist of sediments, so for example limestone or sandstone. The difference between both is the content of quartz, which is higher in the case of sandstone, leading to a lower $V_P/V_S$-ratio of about 1.59 to 1.76, whereas limestones have up to 1.9 (Domenico, 1984). Another effect is the presence of water filling the pores and leading to an increase of the $V_P/V_S$-ratio.

The most abundant mineral in the crust is Plagioclase (31 - 41 %, $V_P/V_S \approx 1.83$) and Orthoclase (7 - 21%, $V_P/V_S \approx 1.93$), also called K-feldspar, followed by quartz (12 - 24 %, $V_P/V_S \approx 1.48$) and hydrous minerals such as micas and amphiboles (See Tables B.3 and B.4). With these minerals the average density of the crust is about $2700\,kg/m^3$. As there are enough differences in the P-wave velocities as well as in the $V_P/V_S$-ratios of the more abundant crustal minerals, these two seismic parameters provide a good mineralogical discriminant. The Crust itself can be divided into two different layers, the Upper and Lower Crust, which are separated by the so called Conrad discontinuity at a depth of about $20\,km$.

The Upper Crust mainly contains Granodiorite or Tonalite and the Lower Crust probably Diorite, Garnet granulite and Amphibolite, which can be separated by their different mineral composition, for example low or high percentage of Quartz, resulting again in high and low $V_P/V_S$-ratios. (Anderson, 2007).

For a detailed list of average crustal abundance, density and seismic velocities of major crustal minerals see Tables 6.7 and B.4 (Anderson, 2007) in appendix B.


- Petrological and geophysical properties of the upper mantle
  The upper mantle, as known so far, has a different mineralogical composition than the crust, mainly consisting of Olivine, and is, seismically, characterized by the sudden increase of the $V_P$ velocity across the Mohorovičić discontinuity (Moho) detected in 1910 by the famous croatian Geophysicist Mohorovičić at a depth range of about $30\,km$ to $50\,km$ underneath continents. This ratio together with the anisotropy is useful to discriminate the mantle mineralogy, so for example to differentiate between Eclogite and Peridotite minerals. The most dominant minerals in the Upper Mantle are Olivine, Orthopyroxene, Clinopyroxene and Garnet and the dominant rock types Harzburgites, Lherzolites, Pyroxenites and Eclogites (Anderson, 1984). Olivine is highly **anisotropic**, having compressional velocities of 9.89, 8.43 and $7.72\,km/s$ along the principal crystallographic axes and a $V_P/V_S$-ratio of about 1.73. Orthopyroxene has velocities ranging from 6.92 to $8.25\,km/s$, also depending on the crystal axis and a $V_P/V_S$-ratio of 1.67. Clinopyroxene has a $V_P/V_S$-ratio of 1.74 and Garnet 1.80. In natural Olivine-rich aggregates (Table B.6), the maximum velocities are about 8.7 and $5.0\,km/s$ for P-waves and S-waves, respectively. With 50% orthopyroxene the velocities are reduced to 8.2 and $4.85\,km/s$, and the composite is nearly isotropic. Eclogites are also nearly isotropic. For a given density, eclogites tend to have lower shear velocities than peridotite assemblages. More details are in Anderson (2007). Based on these properties, the $V_P/V_S$-ratio, together with the anisotropy, is useful to discriminate the mantle mineralogy, so, for example, to differentiate between Eclogite and Peridotite minerals. Probably, here, the different $V_P/V_S$-ratios have to be correlated with the different structural elements than with different mineral and rock composition, which, indeed is correlated in some way whit these anomalies. So, for example, magmatic intrusions have a different composition than the older rock.


For more details, a complete list of the properties of upper-mantle rocks (Christensen and Lundquist, 1982; Manghnani and Ramananotoandro, 1974), Table B.6) and densities and elastic-wave velocities (After Manghnani and Ramananotoandro (1974); Clark (1966); V. Babuska (1972); Jordan (1979)), Table B.5 is provided again in Appendix B.

## 6.8.2  15 × 15 $V_P/V_S$-models

Figure 6.34 illustrates the 15×15 $V_P/V_S$-models computed as discussed. The values of the $V_P$-$V_S$-ratio vary between 1.65 and 1.75 in the center of the model area, but oscillate more widely at the model borders, as the resolution is strongly decreased there.

The anisotropic model (with the corrections of the $P_n$- and $S_n$-phases included, Figure 6.34(b)) and the isotropic model (Figure 6.34(a)) are comparable in the first layer where the $V_P/V_S$-contrasts are more pronounced for the former, but start to become somewhat different in layer two, especially, in the poorer illuminated northeast model area, and more so, in the third and the fourth, the upper mantle layer. The area resolved rapidly decreases in layer three and further in layer four.

For the anisotropic $V_P/V_S$-model one may associate the high $V_P/V_S$-ratios in the first layer - which are more consistent than for the isotropic model - with the mountains along the German-Czech border, namely, the Thüringer Wald, the Böhmer Wald and the Erzgebirge.

According to the depth range of the second layer between 10 and 20 $km$, the $V_P/V_S$-ratios may be interpreted in terms of the mineral composition of the possible crustal rocks, the latter having a non-negligible, though not always unique effect on the former as indicated in Table 6.7 (Anderson, 2007), where the $V_P/V_S$-ratios for various minerals are listed. Based on these literature results, the red areas of layer two may be indicative of rocks with a high content of quartz ($V_P/V_S = 1.48$), whereas the blue areas probably hint of a higher content of Plagioclase ($V_P/V_S = 1.83$) and Orthoclase ($V_P/V_S = 1.93$), also called K-feldspar. Also, the volcanic Vogelsberg area with its basaltic composition is reflected by the deep blue spot north of the Rhine-Main-river with a higher $V_P/V_S$-ratio of about 1.73. Overall, the distribution of the high and low $V_P/V_S$-ratios is different for the anisotropic and isotropic inversion cases.

Since the inverted velocity values in layer three are not very reliable and more ambiguous - as proved in previous sections - they are not discussed further here. On the other hand, for the upper mantle layer four the variations of the $V_P/V_S$- ratios follow more the northern boarder of the alps for the anisotropic than for the isotropic model case. For the model area covering northern Switzerland the $V_P/V_S$-ratios found here are between 1.67 to 1.72 which appears to be in line with the results of Lombardi et al. (2008) who got ratios of about 1.72 under the Black-Forest and of about 1.74 along the Danube river from a receiver-function analysis of seismic phases from teleseismic events. According to Table 6.8 (Anderson, 2007), the reddish area along the river Danube in the anisotropic model may reflect Peridodits with a higher content of Orthopyroxen.

The results of the present section clearly show the necessity of applying the elliptical anisotropic correction for the $P_n$-phases and, to a lesser extent, the $S_n$- phases [13]. With these anisotropic $P_n$- and $S_n$-corrections the seismic features obtained here for the crust and upper mantle underneath Germany, as quantified by, namely, the $V_P/V_S$ ratios, appear to be better interpretable in terms of the natural tectonic features, as well as providing some interesting petrological constraints on the lithospheric composition.

## 6.9  Geological and petrological interpretation of the 3D seismic tomographic models

For an easier description of the structural differences by the refinement along the increasing discretization, the plots for the anisotropic inversions with hypocenters free for the three different bloc discretization and one isotropic inversion result for the 35 × 35 bloc model are all together in Figure 6.35.

The finer bloc discretization makes it more and more difficult to get a correlation between the real

---

[13] As discussed in the Data Chapter, at this stage of the investigation, there is no clear evidence from the present data for strong seismic upper mantle $S_n$-anisotropy underneath Germany

Table 6.7: Average crustal abundance, density and seismic velocities of major crustal minerals. After Anderson (2007)

| Mineral | Volume % | $\rho$ $(g/cm^3)$ | $V_P$ $(km/s)$ | $V_S$ $(km/s)$ | $V_P/V_S$ |
|---|---|---|---|---|---|
| Quartz | 12 | 2.65 | 6.05 | 4.09 | 1.479 |
| K-feldspar | 12 | 2.57 | 5.88 | 3.05 | 1.928 |
| Plagioclase | 39 | 2.64 | 6.30 | 3.44 | 1.831 |
| Micas | 5 | 2.8 | 5.6 | 2.9 | 1.931 |
| Amphiboles | 5 | 3.2 | 7.0 | 3.8 | 1.842 |
| Pyroxene | 11 | 3.3 | 7.8 | 4.6 | 1.696 |
| Olivine | 3 | 3.3 | 8.4 | 4.9 | 1.714 |

Table 6.8: Anisotropy of upper-mantle rocks. From Anderson (2007)

| Mineralogy | Direction | $V_P$ | $V_S - 1$ | $V_S - 2$ | $V_P/V_S$ |
|---|---|---|---|---|---|
| | | Peridotites | | | |
| 100 pct. ol | 1 | 8.7 | 5.0 | 4.85 | 1.74 - 1.79 |
| | 2 | 8.4 | 4.95 | 4.70 | 1.70 - 1.79 |
| | 3 | 8.2 | 4.95 | 4.72 | 1.66 - 1.74 |
| 70 pct. ol, | 1 | 8.4 | 4.9 | 4.77 | 1.71 - 1.76 |
| 30 pct. opx | 2 | 8.2 | 4.9 | 4.70 | 1.67 - 1.74 |
| | 3 | 8.1 | 4.9 | 4.72 | 1.65 - 1.72 |
| 100 pct. opx | 1 | 7.8 | 4.75 | 4.65 | 1.64 - 1.68 |
| | 2 | 7.75 | 4.75 | 4.65 | 1.63 - 1.67 |
| | 3 | 7.78 | 4.75 | 4.65 | 1.67 - 1.67 |
| | | Eclogites | | | |
| 51 pct. ga, | 1 | 8.476 | | 4.70 | 1.80 |
| 23 pct. cpx, | 2 | 8.429 | | 4.65 | 1.81 |
| 24 pct. opx | 3 | 8.375 | | 4.71 | 1.78 |
| 47 pct. ga, | 1 | 8.582 | | 4.91 | 1.75 |
| 45 pct. cpx | 2 | 8.379 | | 4.87 | 1.72 |
| | 3 | 8.30 | | 4.79 | 1.73 |
| 46 pct. ga, | 1 | 8.31 | | 4.77 | 1.74 |
| 37 pct. cpx | 2 | 8.27 | | 4.77 | 1.73 |
| | 3 | 8.11 | | 4.72 | 1.72 |

Manghnani and Ramananotoandro (1974), Christensen and Lundquist (1982).

## 6.9. GEOLOGICAL AND PETROLOGICAL INTERPRETATION, 3D MODELS

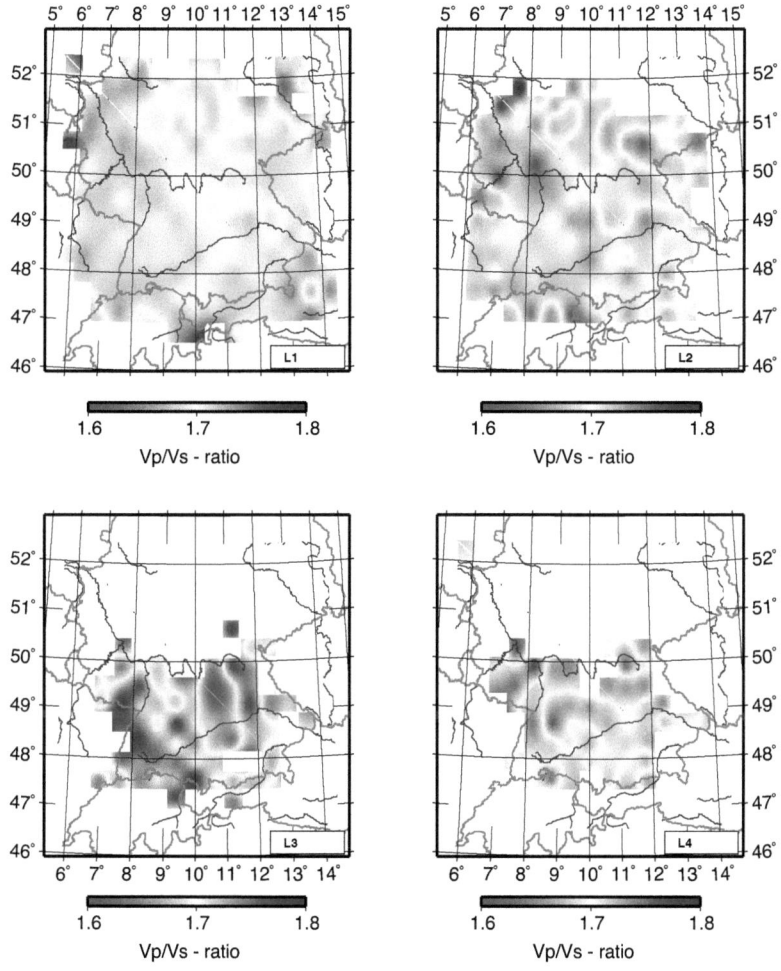

(a) isotropic $V_P/V_S$, ifix $= 1$

184                    CHAPTER 6. SSH-INVERSIONS FOR 3D VELOCITY MODELS

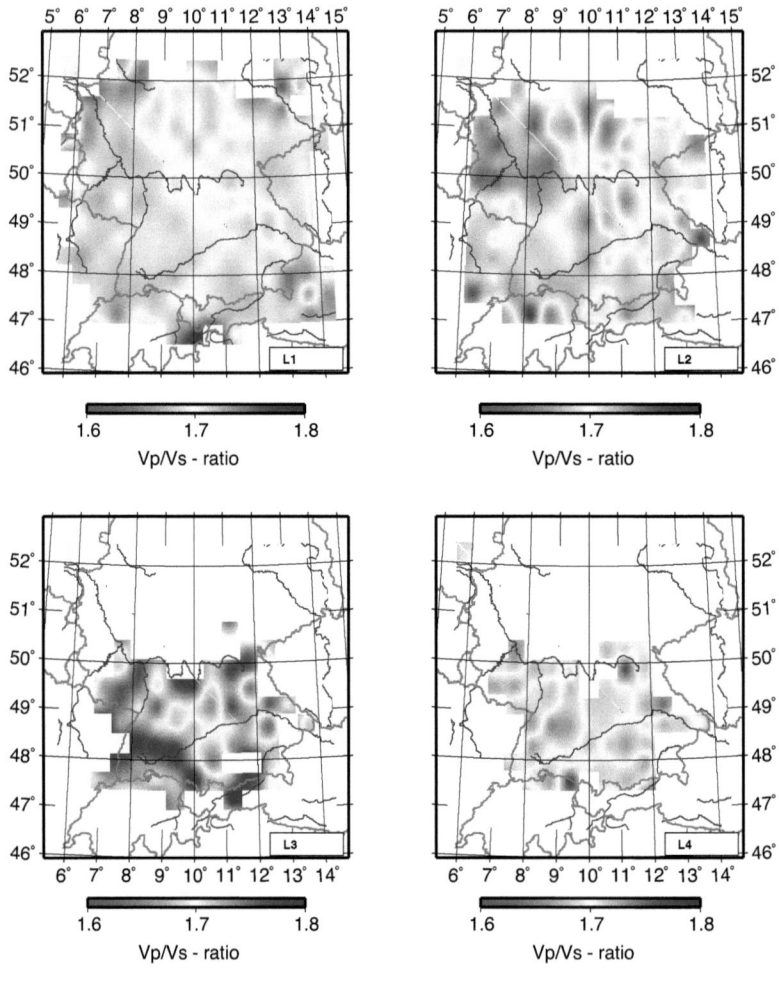

(b) anisotropic $V_P/V_S$, ifix = 1

Figure 6.34: $15 \times 15$- $V_P/V_S$-models for isotropic and anisotropic inversions with fixed hypocenters.

## 6.9. GEOLOGICAL AND PETROLOGICAL INTERPRETATION, 3D MODELS

and the computed structure. In Figure 6.35, the isotropic plot for the $35 \times 35$ models is added to show the complications which occur using that fine bloc discretization, where the anisotropic information is somewhat compensated by it. Nevertheless, the main structures are still better resolved in the anisotropic case, compared to the lower bloc discretization. In the following paragraphs the details for each layer and the distinct bloc discretization are summarized and interpreted in terms of the geology, tectonics and petrology of the crust and upper mantle beneath Germany.

- Layer one

  The blueish area in the middle east of the study area starting at the source of the River Main and following the border between Czechia and Germany rather good follows the Erzgebirge in both, the isotropic and anisotropic case, but with higher $P_n$-velocities in the latter one being more representative for the metamorphic rocks with $P_n$ velocities usually greater than $6\,km/s$. The southern area follows the shape of the Tepla-Barrandium and the Böhmerwald along the southern german-czech boarder, which consists of volcanic and granitoidic rock, explaining the higher velocities.

  The reddish region in the south of the model still has nearly the same shape in the case for the $15 \times 15$, $25 \times 25$ and $35 \times 35$ blocs, except that at $48.5°$ and $11.5°$ a blue positive velocity anomaly occurs, where in the case of the $15 \times 15$ blocs only a blueish homogeneous region appears, penetrating the reddish area. The blue area at $50.5°$ and $9°$ may be correlated with the ancient Vogelsberg volcano, where intrusive basaltic rock leads to a higher velocity. The blue high velocity anomalies are to be correlated with the volcanic basaltic rock west of it, what coincides perfectly with the tectonic map of Germany (Berthelsen, 1992). The Rhine Graben area in between the Black forest and the Vosges is more pronounced in the $25 \times 25$ and $35 \times 35$ models than in the $15 \times 15$ one, where only a slight reddish area occurs indicating the sediments along the Rhine Graben. The region of the former meteoric impact just can be localized at the correct position in the case of the $35 \times 35$ model whereas due to the interpolation in the lower bloc discretization the position is located north or even fuzzy.

  In case of the isotropic inversion, the results for the structure are different in such a way that the anomalies are not localized at the correct position. So, the anomaly at the Nördlinger Ries is more spread. The Rhinegraben Proper is not as clear as in the anisotropic models, which can be seen by the comparison of the $35 \times 35$ models, where the anisotropic correction just has its influence in this layer.

- Layer two

  Again, the major structure is the same in all three cases for the different bloc discretization. In the case of the $25 \times 25$ bloc discretization, in the Molasse region at $48°$ Latitude and $11°$ Longitude, there occurs a high velocity anomaly. But this might also be an artifact because of the lack of rays going through there indicated by the white area there. The positive velocity anomaly at $48.5°$ Latitude and $10°$ Longitude is better to identify in the case of the $35 \times 35$ model than in the $25 \times 25$ model.

  In general, the anomalies in this layer are the same between the isotropic and anisotropic case but with different shape. Again, the anisotropic results are more equal between the cases with hypocenters fixed and the anisotropic case with hypocenters free.

- Layer three

  The reconstruction in this layer only in the lower left triangle is possible in all cases. The blue area along the Rhine river is still present as in the case for the $15 \times 15$ bloc discretization, but with more distinct anomalies.

- Layer four

  This layer is the one where most of the differences occur due to the dominant influence of the anisotropic correction. The blueish area along the river Danube is visible in all three models, but with increasing differentiation for the increasing bloc discretization. At 50.5° and 9.5°, a blue area like in the first layer occurs, which might be correlated again with the Vogelsberg volcanic rock.

The anomalies, that occur in the $S$-wave models, separately described only for the $15 \times 15$ bloc models, correlate in some way with the anomalies in the $P$-wave models. So, for example, the anomaly for the "Nördlinger Ries" has its blue correspondent, but with wider extent, which probably is due to the lower ray density leading to a smearing of the anomalies. Another possibility is, that these anomalies are correlated to the "Steinheimer Becken", which is another part of the former meteoritic impact. Also the blue area in the "Vogelsberg" region and the related volcanic area in the west are represented in the $S$-wave model. But over all, the $S$-wave models are not as representative as the $P$-wave models for the anisotropic correction due to the small amount of about 200 $S_n$-phases that leads to nearly the same velocity structure.

Unfortunately, the number of available $S_n$-phases is not as huge as the one for $P_n$-phases, so the influence of correcting them is of minor influence onto the different velocities in the first three layers. Only in the fourth layer, some small differences occur between the isotropic and anisotropic case.

## 6.9. GEOLOGICAL AND PETROLOGICAL INTERPRETATION, 3D MODELS

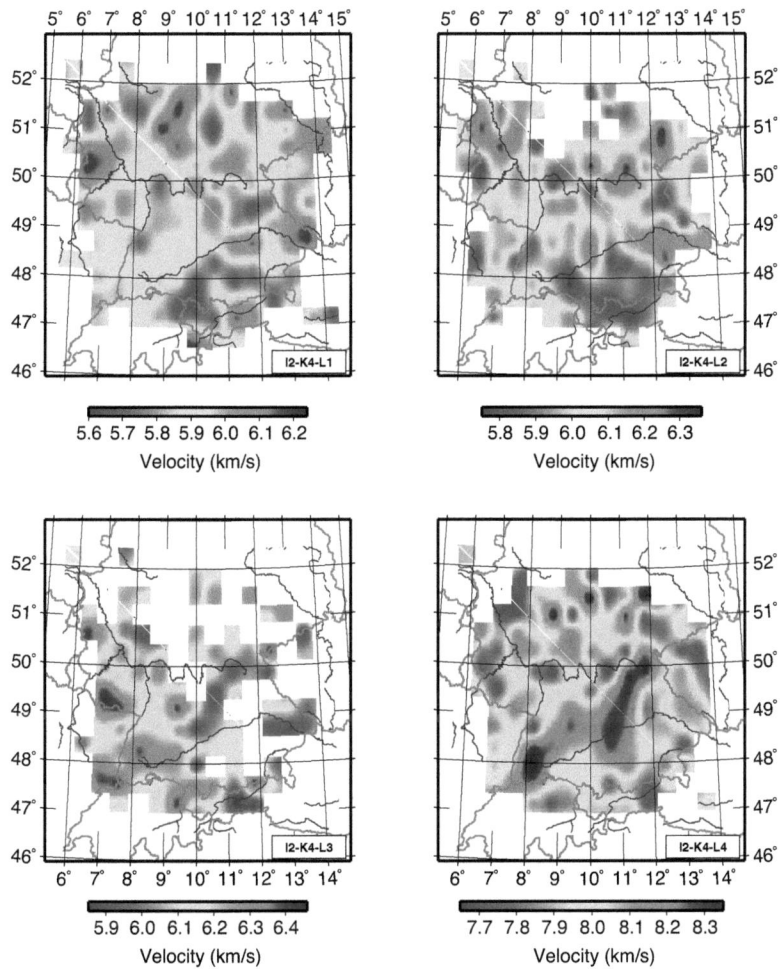

(a) 15 x 15, Gap = 180, ani

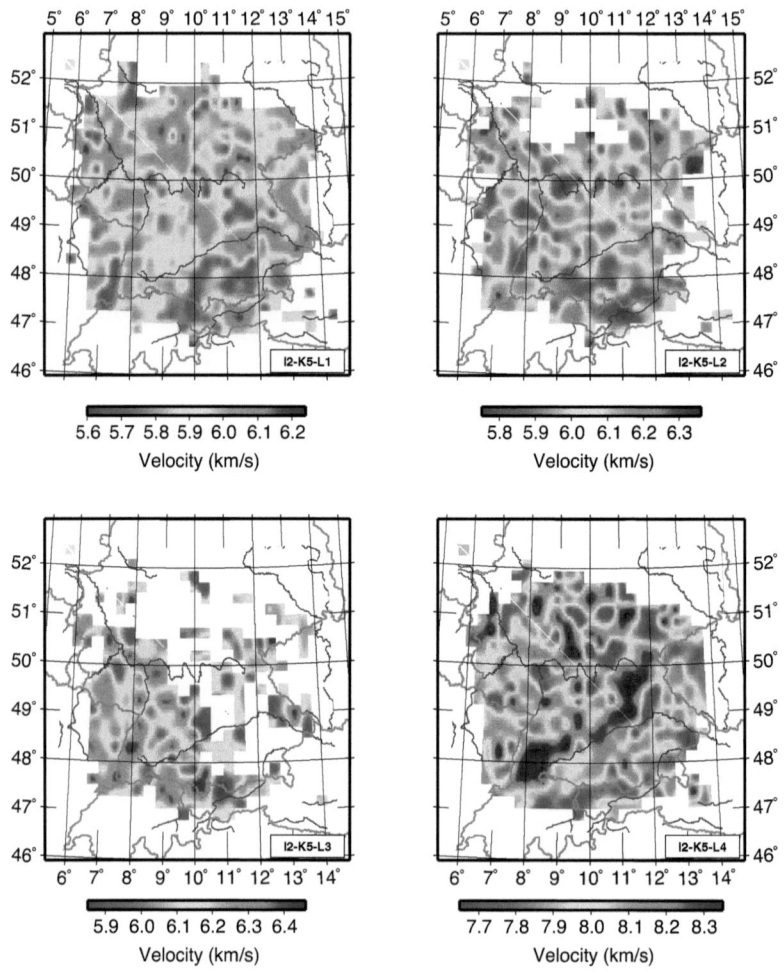

(b) 25 x 25, Gap = 180, ani

## 6.9. GEOLOGICAL AND PETROLOGICAL INTERPRETATION, 3D MODELS 189

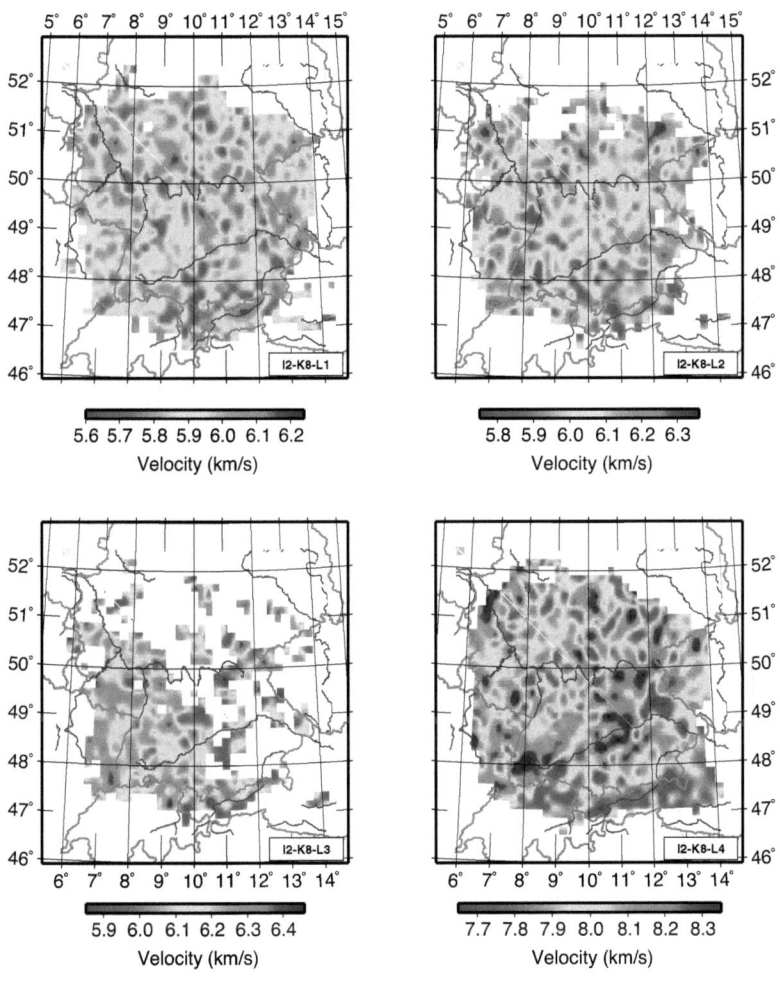

(c) 35 x 35, Gap = 180, iso

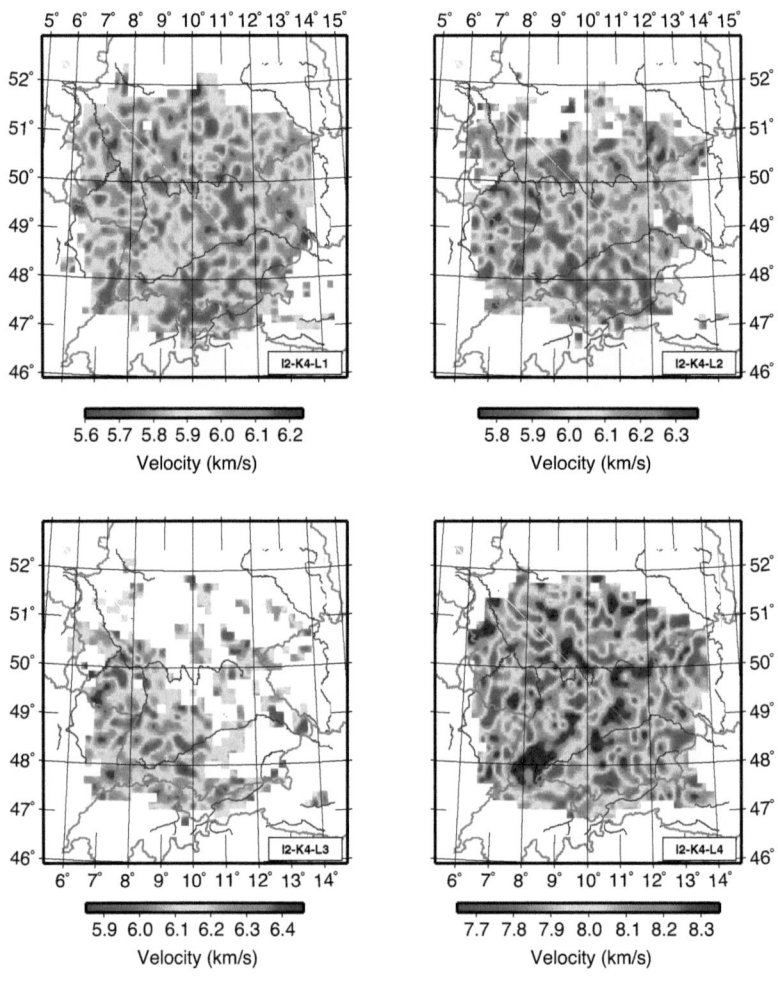

(d) 35 x 35, Gap = 180, ani

Figure 6.35: The comparison for different bloc discretizations, all with GAP = 180°. Figure 6.35(a) anisotropic inversion with 15 × 15 blocs, Figure 6.35(b) anisotropic inversion with 25 × 25 blocs. Isotropic inversion in Figure 6.35(c) and anisotropic inversion in Figure 6.35(d) for 35 × 35 blocs.

# Chapter 7

# Impact of SSH inverted seismic structure on the relocations of the hypocenters

## 7.1 General approach

One of the major questions in the present thesis is that of the influence of the SSH inverted 1D- or 3D-seismic structures in the previous chapters on the relocation of the hypocenters, i.e. their shifts compared to their original locations. As described in Chapter 4, the latter have been calculated here by standard earthquake location procedures[1], employing a standard 1D vertically inhomogeneous velocity model for the crust and upper mantle beneath Germany. Another question of interest is how the elliptical anisotropic correction for the $P_n$- and $S_n$-phases (the anisotropic inversions) affects these hypocentral relocations when compared with those of the isotropic inversions. In other words, since the anisotropic inversion models are seismically as well as statistically (based on the TSS or the RMS which quantify the fit of the model to the data) more trustworthy than the isotropic models, the question is then also what is the bias in the hypocenter determinations, when the anisotropic correction is not included into the SSH-inversion or into the standard hypocenter location procedures of the BGR? In any case, the anisotropic inversion should, necessarily, lead to improvements of the hypocenter relocations[2].

In an attempt to answer these questions, several tests will be carried out in this chapter:

Firstly, full SSH-inversions for seismic structure and hypocenters are done (IFIX=0) and the hypocentral foci determined simultaneously with the new optimal vertically stratified 1D velocity model of Section 7.2 are compared with the original locations. Furthermore, the 1D hypocentral results will then be compared with those obtained simultaneously with the 3D-seismic inversion models of Section 7.4 to see whether the latter provide further improvements of the hypocenter relocations.

Secondly, after the optimal 1D velocity model has been determined, a pure hypocenter relocation with the SSH-program is carried out, whereby the velocity model is kept fixed (IFIX = 2). At the same time the influence of the older MKS-2004 1D velocity model proposed for the crust and upper mantle beneath Germany during recent years (see Chapter 5) on the hypocentral relocations will be investigated. Thirdly, we can use the former computed models (as there are SKKS-2001, MKS-2004 or MKS-2007) to compare the results with the optimal model and to decide, how much is the influence

---

[1] The most widely used earthquake location code is HYPOINVERSE, HYPO2000 http://www.faldersons.net/Software/Hypoinverse/Hypoinverse.html or LOCSAT, which assumes a vertically inhomogeneous seismic velocity model

[2] True hypocenter localizations can only be tested by controlled source data which, in the present dataset are, unfortunately, only available for a few near-surface (0 to 5 $km$ depth range) sources, most of these being mining seismic events in the Saarland region

of the different models on the hypocenter localization, which is on the one hand a possibility to test for the range of the hypocenter localization and on the other hand to test for some further influence of the anisotropic correction along the different models.

Finally, to test the stability of the SSH-relocations with respect to uncertainties in the observed data, additional statistical and synthetic tests are implemented where the recorded arrivaltimes as well as the initial hypocenters are randomly perturbed and their final relocations analyzed. To this regard, emphasis is again put on the investigation of the beneficial effects of the anisotropic $P_n$-traveltime correction.

## 7.2 Hypocentral relocations with the optimal 1D velocity model

In this section are analyzed (1) the *a posteriori* with the optimal 1D model MKS-2007 (Table 7.1 and Figure 7.1 of Chapter 5) relocated hypocenters and (2) together with the former 1D MKS-2004 model as starting model simultaneously relocated original earthquakes. This is done for both the anisotropic and isotropic model cases, so four different relocation sets will be obtained. The most important question here is whether the anisotropic SSH inversion models result in more precise determination of the seismic foci than the isotropic ones, given that the original traveltimes have been affected by the $P_n$-anisotropy. For example, one would expect that for the "slower" isotropic model an originally high $P_n$-velocity along the fast axis of the anisotropy ellipsoid has to be compensated by a shorter length of this traveltime path for the $P_n$-wave, compensating the earlier observed arrivaltime. The only way to do this is to shift the hypocenter to greater depth, because the epicentral parameters are fixed by an even distribution of stations around the focus of the event. On the other hand, in the direction of the slow axis, the observed $P_n$-traveltime is longer and will lead to an upward shift of the focus to lengthen the ray path. In summary, one may expect the possible hypocentral depth range to be larger for the isotropic than for the anisotropic inversion case. Moreover, this effect will mainly occur for events with an uneven and limited station distribution, so that in such a situation - which, unfortunately, is common in practice - not paying attention to the presence of the $P_n$ anisotropy, will lead to more biased hypocentral depth relocations than those obtained with a full anisotropic inversion model.

In the following the tests are done for three different GAP criteria (230°, 180° and 120°), where in the figures only the case with GAP = 180° is shown, the other ones are listed in Table 7.2. Figure 7.2 shows the epicentral shifts (relative to the original epicenters) of the *a posteriori* with the optimal model MKS-2007 seismic events for both the anisotropic and the isotropic SSH-inversion cases with the fixed velocity model (IFIX=2). One may note that the epicentral shifts are spread somewhat less for the former than for the latter model case. Concerning the variation of depth Z, it is less in the anisotropic case (Figure 7.2(d)) where the cumulation around 6 $km$ is more narrow in contrast to the isotropic one (Figure 7.2(c)). Additional statistical details for the relocation results of these two model cases are listed in Table 7.2.

Figure 7.3 shows the epicentral and hypocentral shifts for two subsequent nonlinear iteration steps of anisotropically SSH-computed relocations. One notes that for most of the events[3] the largest differences of up to 6 $km$ are in EW-direction, whereas the ones in NS-direction are less than 2 $km$. Only a slight overall shift can be recognized expressing the rather good stability of the inversion procedure.

The histograms of Figure 7.4 illustrate the frequency distribution of the three coordinate shifts separately. Overall, the shape of the histograms is about the same for the isotropic and anisotropic model inversions for the epicenters. For the hypocentral depths, the range is more narrow in the anisotropic than in the isotropic case (Figure 7.4(f)), so about 10 events are shifted upwards by 10 $km$ for the former and 20 for the latter case. The number of events shifted to greater depths also is less resulting in a higher peak at 6 $km$ in the anisotropic case. In fact, this can also be recognized in more detail

---

[3]Due to the limit of 10 $km$ in the polar plots, some events are not plotted here, though they are in the histograms in Figure 7.4.

## 7.2. HYPOCENTRAL RELOCATIONS, OPTIMAL 1D MODEL

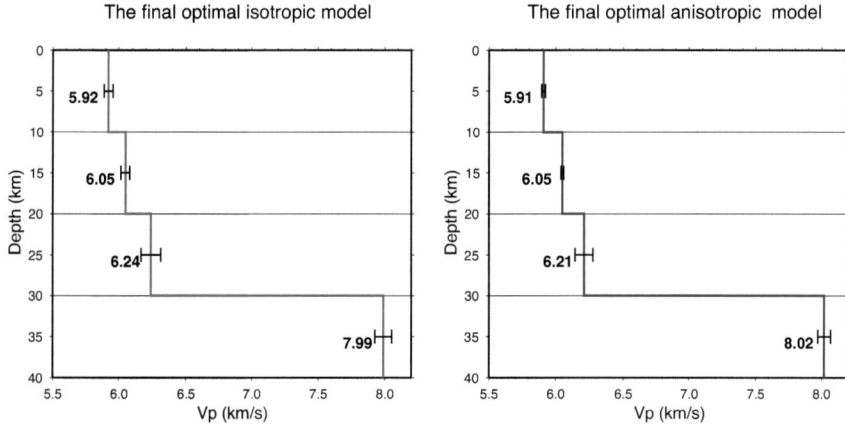

Figure 7.1: The final optimal isotropic and anisotropic 1D velocity models (computed with hypocenters free, i.e. full inversion). Whereas for the upper two layers nearly the same velocities are obtained for the two models, differences occur in layers three and four. The statistical uncertainty of the layer velocities is indicated by the horizontal error bars. Numerical velocity values are listed in Table 7.1.

from Figure 7.5 where the focal depths of the original events have been divided in $5\,km$-classes and the distribution of the shifts for each class is shown.

The absolute differences for the hypocentral depths for the two SSH-inversion cases, $(z_{ani} - z_{iso})$, are shown in Figure 7.6, after the first nonlinear iteration in Figure 7.6(a) and after the second and last nonlinear iteration in Figure 7.6(b). After the first iteration, the differences are not more than $\pm 5\,km$, according to the limit of the maximally allowed depth-shift in the SSH program (set here to $5\,km$ to avoid unrealistic solutions after the first inversion). As this limit is allowed again for the following iteration, a maximally possible depth shift of $\pm 20\,km$ may arise in the fourth and last iteration. From the visual inspection of the various depth histograms in Figure 7.6 one gets the impression that the average relocated hypocentral depths are larger for the anisotropic than for the isotropic inversion case. This also can be grasped from the individual x-, y- and z-histograms of the hypocentral shifts in Figure 7.4 where, although most of the relocated depths of the events lie within the same range for the isotropic and anisotropic inversions, a good number of them are relocated deeper by the anisotropic than by the isotropic inversion.

Finally, the previously discussed *a posteriori* relocations of the optimal models MKS-2007-iso and MKS-2007-ani are compared with the relocations of the full (simultaneous) inversions of the 1D MKS-2004 model. First of all the absolute epicentral shifts obtained with this full SSH-model with respect to the original events are depicted in Figure 7.7. Figure 7.8 then shows the epicentral differences between the MKS-2004 and MKS-2007 relocations. One notices a rather narrow spread of the epicenters and hypocenters between the two model relocations (Figure 7.8). This is an important result as it shows that the *a posteriori* relocation of the hypocenters with a previously determined optimal model is a viable alternative to a full SSH-inversion, i.e., at least for the 1D-model case, the computationally demanding full SSH-inversion might be decoupled into successive and simpler steps of a separate velocity- and hypocenter determination.

Regarding the question raised at the beginning of this Chapter as to whether the relocations of the

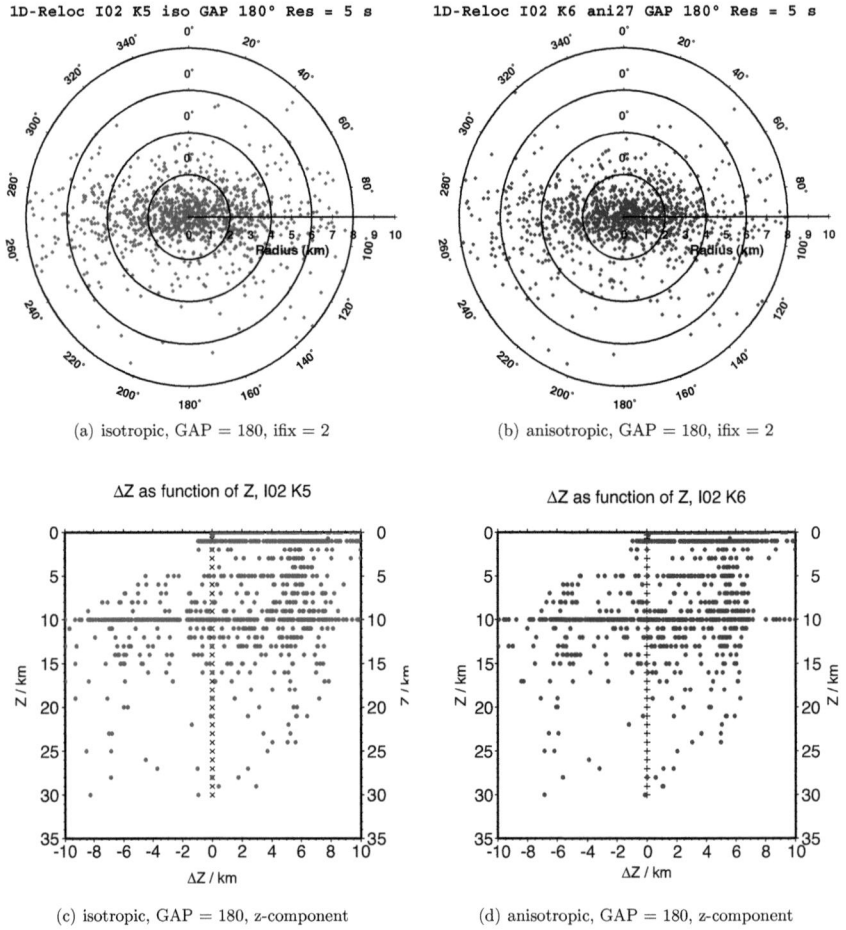

Figure 7.2: The optimal results for the optimal 1D input model MKS-2007 and GAP = 180°, velocity model fixed, IFIX = 2.

## 7.2. HYPOCENTRAL RELOCATIONS, OPTIMAL 1D MODEL

Table 7.1: Final optimal isotropic and anisotropic 1D velocity models (computed with hypocenters free, i.e. full inversion) together with standard errors for the layer velocities (taken from Table 5.5). For comparison, the results for the velocity models after the full SSH-inversion are in the last two lines.

| anisotropic | V1 km/s | V2 km/s | V3 km/s | V4 km/s | $\sigma_{V1}$ km/s | $\sigma_{V2}$ km/s | $\sigma_{V3}$ km/s | $\sigma_{V4}$ km/s | av.RMS s | RMS-hypo s |
|---|---|---|---|---|---|---|---|---|---|---|
| Average velocity models for relocation | | | | | | | | | | |
| no  | 5.92 | 6.05 | 6.24 | 7.99 | 0.033 | 0.033 | 0.075 | 0.064 | 0.9810 | 1.1461 |
| yes | 5.91 | 6.05 | 6.21 | 8.02 | 0.014 | 0.010 | 0.066 | 0.048 | 0.9471 | 1.1461 |
| Velocity models as result of the full SSH-inversion | | | | | | | | | | |
| no  | 5.92 | 6.05 | 6.26 | 8.00 | - | - | - | - | 0.9676 | 1.1461 |
| yes | 5.92 | 6.06 | 6.26 | 8.03 | - | - | - | - | 0.9244 | 1.1461 |

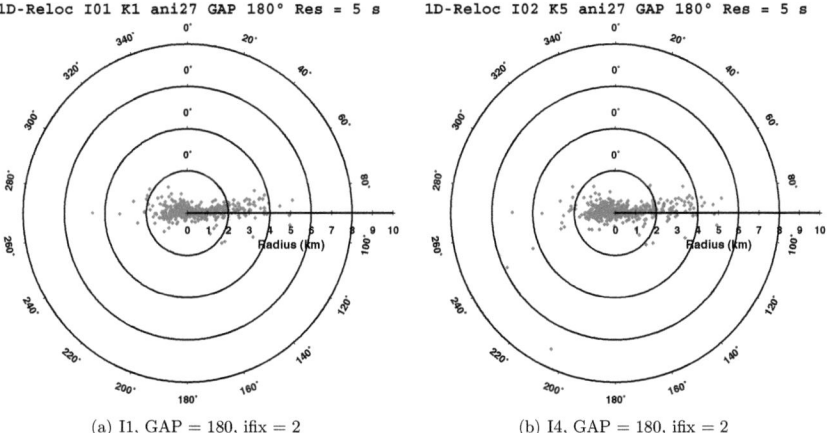

(a) I1, GAP = 180, ifix = 2    (b) I4, GAP = 180, ifix = 2

Figure 7.3: Differences in the epicenter relocations obtained simultaneously (IFIX = 2) for the first (I = 1) and the second (I = 2) nonlinear iteration step. Left panel: results after the first iteration; right panel: results after the second and last iteration. The radius axis for the shift is up to 10 km.

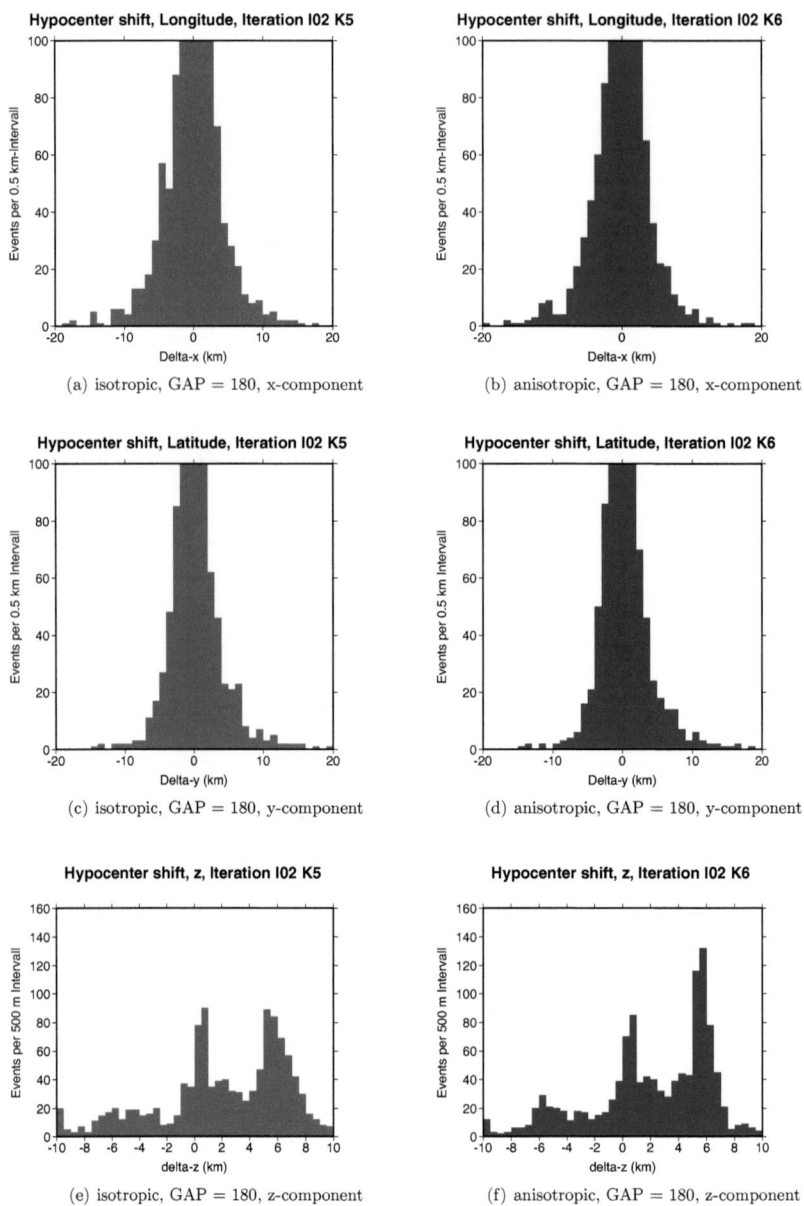

Figure 7.4: Histograms of the x-, y- and z-shifts of the hypocenters for the isotropic (left columns) and the anisotropic (right columns) inversion with the fixed velocity model (IFIX = 2)

## 7.2. HYPOCENTRAL RELOCATIONS, OPTIMAL 1D MODEL

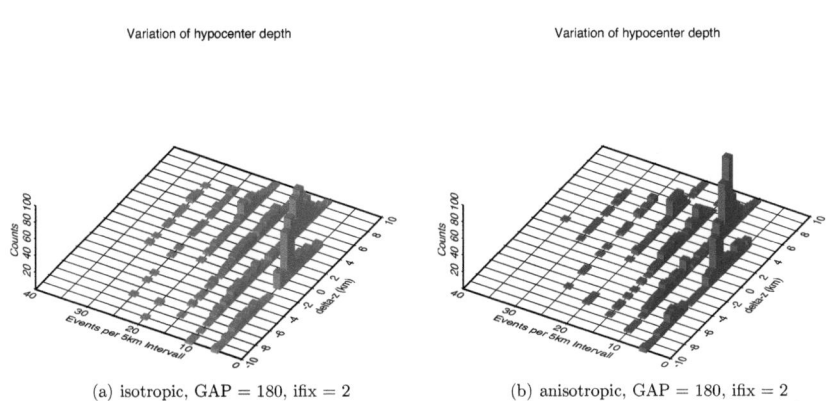

Figure 7.5: Relocations of the hypocentral depths for the isotropic (red) and anisotropic (blue) inversion cases, fixed velocity model (IFIX = 2).

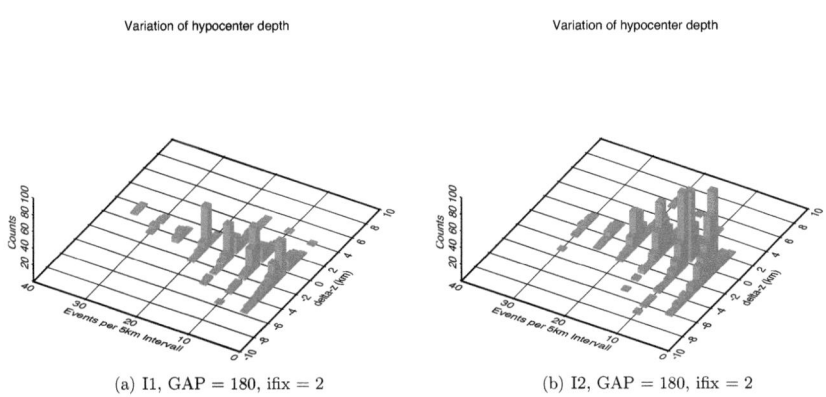

Figure 7.6: Differences of the hypocentral depth relocations with a fixed velocity model (IFIX = 2) between the anisotropic and isotropic inversion cases. Left panel: After the first iteration; right panel: After the second and last iteration.

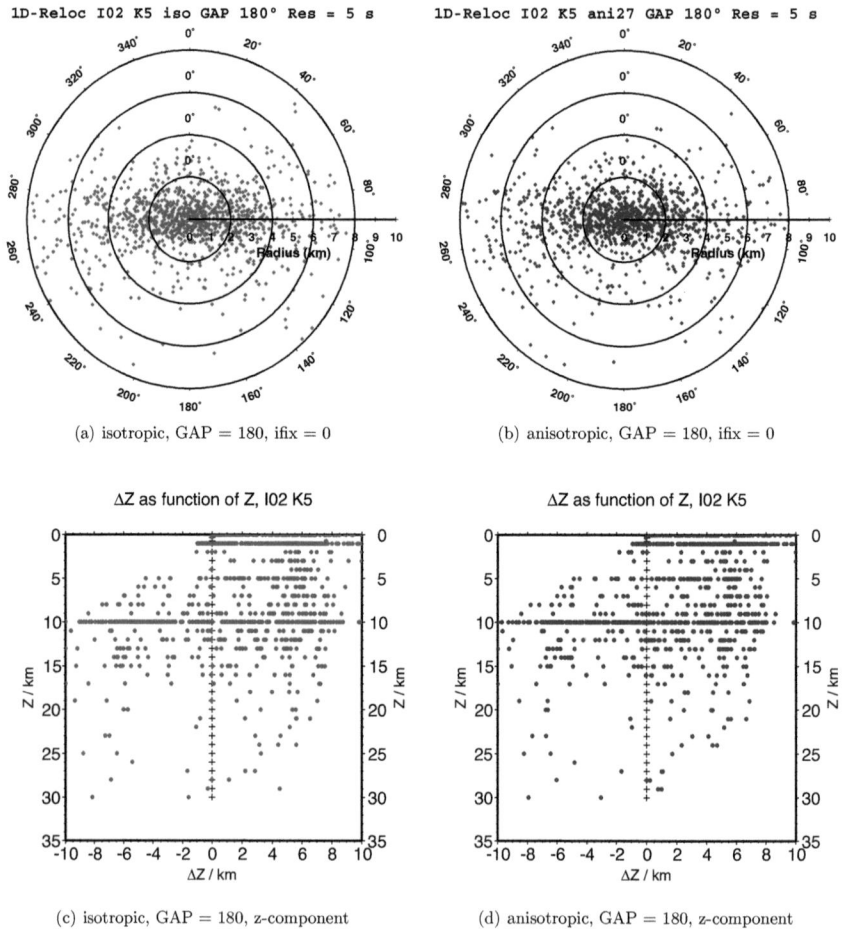

Figure 7.7: Simultaneously with the initial 1D-model MKS-2004 relocated hypocenters (full SSH-inversion) using GAP = 180°. The radius axis for the shift is up to $10\,km$.

## 7.2. HYPOCENTRAL RELOCATIONS, OPTIMAL 1D MODEL

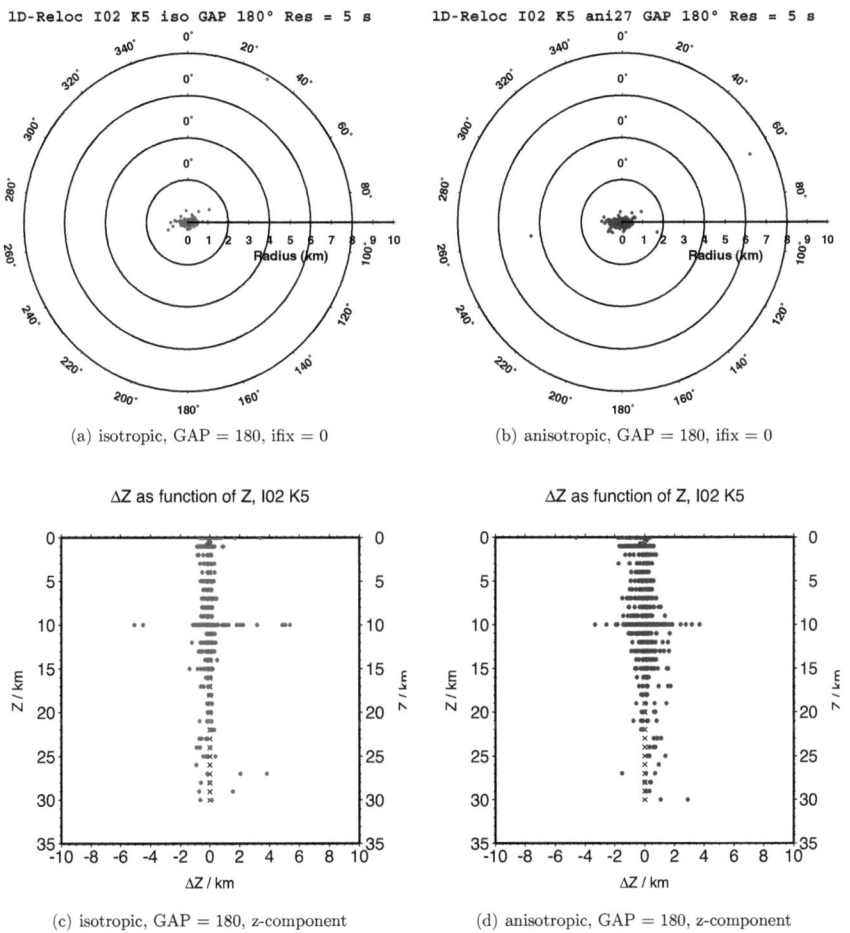

Figure 7.8: Hypocenter relocation differences between the results of a complete inversion (hypocenters and velocity, IFIX=0) with the 1D input model MKS-2004 and those obtained *a posteriori* with the optimal model MKS-2007 (IFIX=2). The radius axis for the shift is up to 10 $km$.

Table 7.2: Statistical results for the optimal isotropic and anisotropic SSH inversions for three different gaps of 230°, 180° and 120°. Δ-x, Δ-y and Δ-z denote the average coordinate shifts between the relocated and the original hypocenter. Imp-R is the improvement in percent for the RMS relative to the isotropic result for GAP = 230°. Imp-T is the improvement between TSS-Hypo and TSS-SSH for the distinct result.

| ani | GAP<br>deg | I | RMS<br>s | Imp-R<br>% | TSS-SSH<br>$s^2$ | TSS-hypo<br>$s^2$ | Imp-T<br>% | Δ-x<br>km | Δ-y<br>km | Δ-z<br>km |
|---|---|---|---|---|---|---|---|---|---|---|
| full inv (IFIX = 0), MKS-2004 | | | | | | | | | | |
| no  | 230 | 4 | 1.0367 | 0.0  | 28622 | 38626 | 26 | 5.72 | 2.70 | 5.30 |
| yes | 230 | 4 | 0.9867 | 4.8  | 25929 | 38626 | 33 | 6.45 | 3.05 | 6.27 |
| no  | 180 | 4 | 0.9676 | 6.7  | 20345 | 28545 | 20 | 5.45 | 2.61 | 5.67 |
| yes | 180 | 4 | 0.9244 | 11.0 | 18570 | 28546 | 35 | 5.33 | 2.61 | 5.34 |
| no  | 120 | 4 | 0.9989 | 3.6  | 11778 | 14264 | 17 | 4.81 | 2.37 | 5.18 |
| yes | 120 | 4 | 0.9355 | 10.0 | 10331 | 14264 | 28 | 4.81 | 2.39 | 5.41 |
| only reloc (IFX = 2), MKS-2007 | | | | | | | | | | |
| no  | 230 | 2 | 1.0369 | 0.0 | 28633 | 38626 | 26 | 4.91 | 2.43 | 5.04 |
| yes | 230 | 3 | 1.0001 | 3.5 | 26635 | 38626 | 31 | 5.96 | 2.81 | 4.77 |
| no  | 180 | 3 | 0.9743 | 6.0 | 20628 | 28546 | 28 | 5.41 | 2.60 | 4.99 |
| yes | 180 | 2 | 0.9385 | 9.5 | 19140 | 28546 | 33 | 4.99 | 2.30 | 4.29 |
| no  | 120 | 3 | 0.9846 | 5.0 | 11444 | 14264 | 20 | 4.91 | 2.43 | 5.04 |
| yes | 120 | 2 | 0.9443 | 8.9 | 10526 | 14264 | 26 | 4.81 | 2.41 | 4.87 |

anisotropic SSH-inversions are more trustworthy than those of the isotropic inversions, the previous figures do not provide a clear answer. Although there are obviously differences in the relocations for the two cases, without a statistical analysis of the individual hypocentral relocation errors - which is computationally not feasible in a comprehensive manner for the fully coupled SSH inversion case - it is practically not possible to decide for one or another hypocenter relocation. The only "objective" information (in data space) are the final residuals of the fitted nonlinear SSH-models to the observed traveltime data as quantified by the TSS or RMS (listed in Table 7.2) and based on this criterion the fully-coupled anisotropic SSH-model (velocities and hypocenters) must be better than the isotropic one. Indeed, Table 7.2 indicates that whereas the isotropic SSH inversion results in a datafit that is about 25% better than the original BGR-data set (with the HYPO-located events), the anisotropic inversion provides an additional improvement (reduction of the RMS) of more than 10%. However, it will be interesting to see how the datafit is improved further when the hypocenters are relocated simultaneously with the 3D velocity model.

It is well agreed upon that for high quality tomography with regional traveltimes, the GAP-criterion should at least be set to 180° or, even better, to 120° for a particular event to have a relatively homogeneous ray coverage and so ensuring a good hypocentral relocation. Using such a small gap would also reduce the impact of bias effects due to the presence of azimuthal anisotropy in the routine isotropic hypocenter determinations (where anisotropy is not taken into account), as anisotropically affected traveltimes in one direction might be offset by the corresponding traveltimes in the opposite direction. For larger gaps, however, this is not the case anymore, so ignoring anisotropy in such a situation would lead to larger mislocations. This is indicated by the stronger variations of the hypocentral relocations in the anisotropic inversion model which, theoretically, should provide more reliable hypocenter determinations.

On the other hand, the statistical results for the various values of the GAP as listed in Table 7.2, while consistently showing a better datafit for the anisotropic than for the isotropic models, do not provide clear support for this argument, i.e. the relative anisotropic improvements for the large gaps are not

Table 7.3: Standard deviations $\sigma_x$, $\sigma_y$ and $\sigma_z$ for the individual hypocentral shifts and the average RMS for three formerly used 1D vertically stratified four layer velocity models.

| model | ani | $\sigma_x$ $km^2$ | $\sigma_y$ $km^2$ | $\sigma_z$ $km^2$ | Av-event-RMS $s$ | RMS $s$ | I | K | RMS-hypo $s$ |
|---|---|---|---|---|---|---|---|---|---|
| MKS 2004 | no  | 5.68 | 2.61 | 6.47 | 0.8148 | 0.9520 | 4 | 6 | 1.1461 |
|          | yes | 5.52 | 2.61 | 6.38 | 0.7906 | 0.9257 | 4 | 6 | 1.1461 |
| SKKS 2001| no  | 6.63 | 2.51 | 6.50 | 0.8227 | 0.9609 | 4 | 6 | 1.1461 |
|          | yes | 5.62 | 2.53 | 6.44 | 0.8085 | 0.9375 | 4 | 8 | 1.1461 |
| MKS 2007 | no  | 5.28 | 2.39 | 5.00 | 0.8087 | 0.9743 | 4 | 7 | 1.1461 |
|          | yes | 5.31 | 2.60 | 5.22 | 0.7894 | 0.9244 | 3 | 5 | 1.1461 |

better than for the small ones. This issue will be analyzed further in Section 7.5.2 where numerical tests of the sensitivity of the relocations to various initial parameters will be carried out.

## 7.3 Relocations with former 1D velocity models

In this section, the effects of two formerly used 1D velocity models SKKS-2001 and MKS-2004 onto the hypocenter relocations are investigated and compared with those obtained with the final optimal 1D model MKS-2007 above. The result is shown in Figure 7.9, where the differences (along the three coordinate directions) of the hypocentral relocations of the SKKS-2001 and MKS-2004 models relative to the preliminary optimal 1D model MKS-2007 are plotted for both the isotropic (red colors) and anisotropic model (blue colors) variants. From these panels one may observe that there occur particular large shifts for the depths and these are shifted more downwards for the anisotropic inversion cases by values between 0.5 to 2 $km$, but which is in the overall relocation accuracy proposed by the errors given in the initial data-set.

Further, quantitative details are listed in Table 7.3, where one notes that the RMS successively increases when going from the old models SKKS-2001 and MKS-2004 to the new model MKS-2007, and for each model the anisotropic inversion case always has a lower RMS, too. Clearly, based on the objective criterion of the best datafit, the new regional 1D-velocity model MKS-2007 discussed in the previous section is the optimal one. Also the standard variations, which are computed in respect to the original hypocenters in the data-set are minimal for the new model MKS2007, especially for the $x$- and $z$-components.

## 7.4 Hypocenter relocations with the optimal 3D velocity models

The same analysis as in the previous section will now be applied to the hypocentral relocations obtained with the various optimal 3D velocity models discussed in detail in Chapter 6, "SSH-inversions for 3D velocity models" (cf. Figures 6.25, 6.26 or 6.35). Emphasis will be again on the comparison of the anisotropic and isotropic inversion cases to check the influence of the anisotropic $P_n$-traveltime correction.

Figure 7.10 and 7.11 show the epicentral shifts (relative to the original epicenters) and the hypocentral shifts of the simultaneously relocated seismic events with the optimal 3D models (see Figures 6.26, 6.30 and 6.32) for the $15 \times 15$, the $25 \times 25$ and the $35 \times 35$ bloc models, respectively, for both the anisotropic and the isotropic SSH-inversion cases. One notices that for all three bloc-models, most of the relocated epicenters are within a radial distance of $\pm 4\,km$ around the original epicentral position. Basically it is hard to see major differences between the anisotropic and isotropic inversion cases, though it appears that the epicentral shifts are somewhat more limited (the points are more concentrated around the origin of the polar plot) for the anisotropic than for the isotropic inversion model, particularly for the

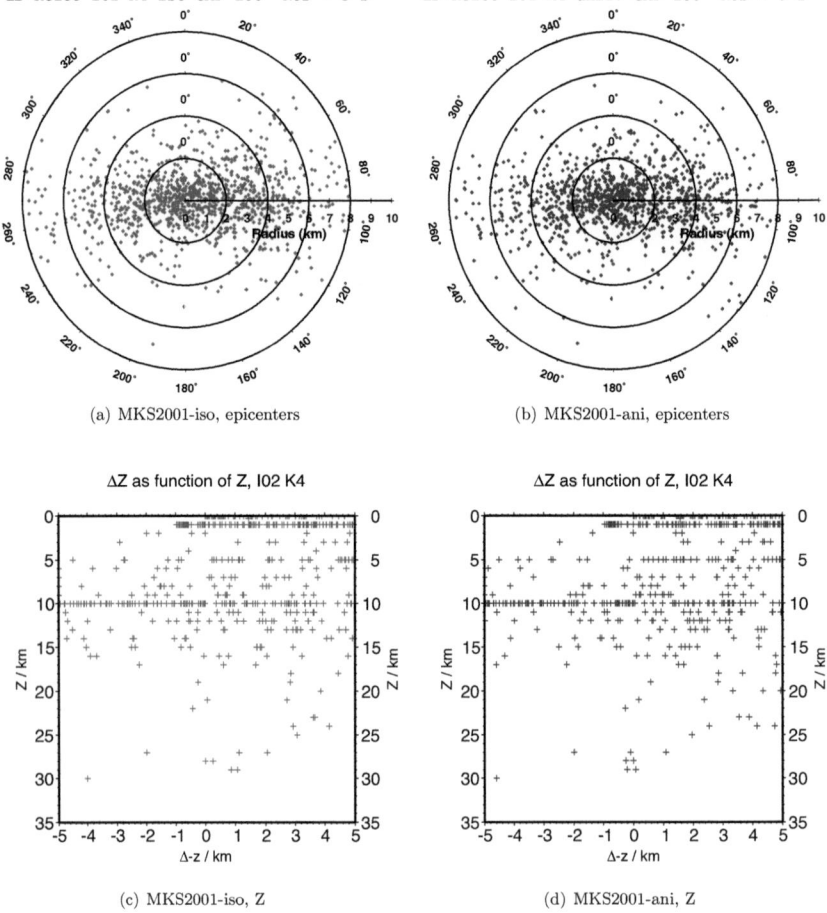

(a) MKS2001-iso, epicenters

(b) MKS2001-ani, epicenters

(c) MKS2001-iso, Z

(d) MKS2001-ani, Z

## 7.4. HYPOCENTER RELOCATIONS, OPTIMAL 3D VELOCITY MODELS

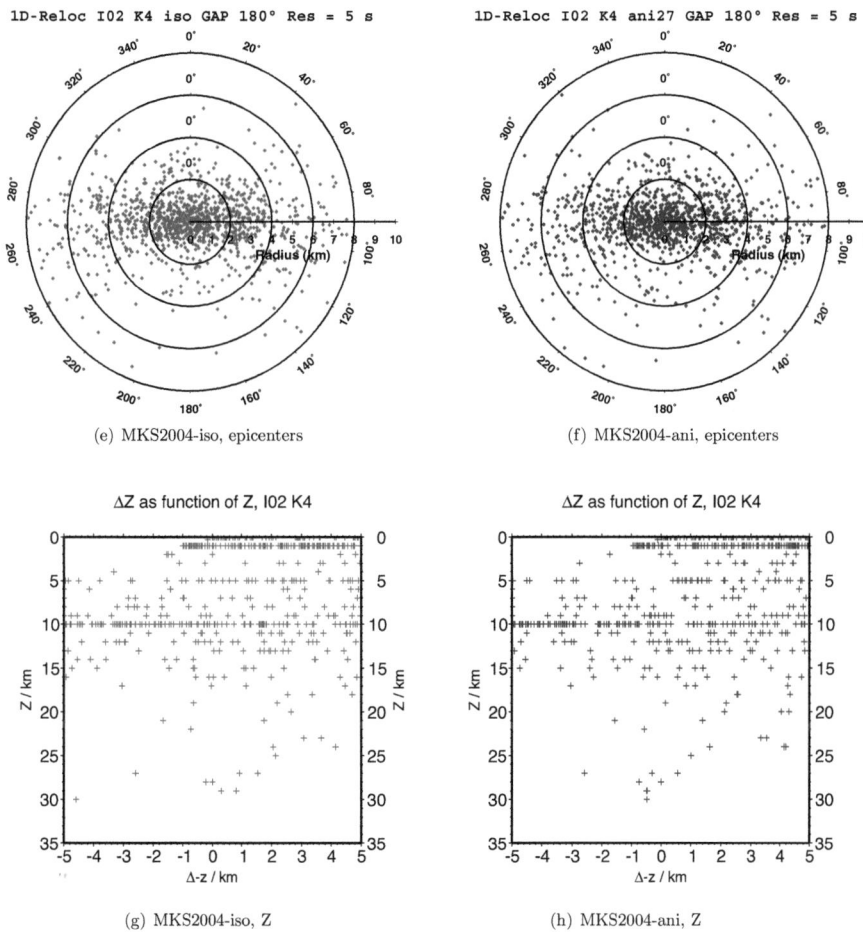

Figure 7.9: Differences (along the three coordinate directions) of the hypocentral relocations of the SKKS-2001 and MKS-2004 models relative to the optimal 1D models MKS-2007. The radius axis for the shift is up to $10\,km$.

Table 7.4: Average differences between the original and the newly relocated epicenters for the various isotropic and anisotropic 1D- and 3D-velocity models (simultaneous inversion) discussed. $\bar{x}$ and $\bar{y}$ are the average horizontal components of the epicentral shifting vector, $\bar{r}$ the average magnitude of the shift, $\bar{\phi}$ the average shift direction and RMS measures the datafit of the corresponding model. Imp-R is the total improvement for the RMS, relative to the isotropic result and $\Delta_x$, $\Delta_y$ and $\Delta_z$ the changes for hypocenters relative to the original ones.

| model | I | K | $\bar{x}$ km | $\bar{y}$ km | $\bar{\phi}$ deg | $\bar{r}$ km | RMS s | Imp-R % | $\Delta_x$ km | $\Delta_y$ km | $\Delta_z$ km |
|---|---|---|---|---|---|---|---|---|---|---|---|
| 1D iso-opt | 4 | 6 | -2.76 | -10.54 | 211 | 3.47 | 0.9743 | 0 | 5.28 | 2.39 | 5.00 |
| ani-opt | 2 | 6 | -1.14 | -9.54 | 261 | 3.46 | 0.9385 | 3.7 | 5.31 | 2.60 | 5.22 |
| 3D 15x15 iso | 4 | 7 | 3.11 | -9.05 | 137 | 3.84 | 0.7474 | 23 | 4.85 | 2.13 | 5.02 |
| ani | 4 | 6 | -0.07 | -9.54 | 281 | 3.69 | 0.7279 | 25 | 4.38 | 1.93 | 4.44 |
| 3D 25x25 iso | 4 | 7 | -0.17 | -8.74 | 279 | 3.10 | 0.7630 | 22 | 3.93 | 1.70 | 4.47 |
| ani | 4 | 6 | -1.35 | -9.26 | 264 | 3.44 | 0.7130 | 27 | 4.24 | 1.87 | 4.48 |
| 3D 35x35 iso | 4 | 5 | -0.10 | -8.81 | 281 | 2.74 | 0.7439 | 24 | 3.94 | 1.73 | 4.52 |
| ani | 4 | 7 | -0.90 | -8.71 | 264 | 2.78 | 0.7008 | 28 | 3.97 | 1.76 | 4.68 |

finer bloc discretizations. Additional quantitative results from these SSH-inversion cases are summarized in Table 7.4 from which it can be seen, among others, that the datafit gets better, i.e. the RMS decreases with increasing complexity of the 3D velocity model. Also, in successive order the datafit improves when going from an isotropic 1D-velocity model to an anisotropic one, improves further for an isotropic 3D velocity model and even more for an anisotropic 3D model.

For a better evaluation of the quality of the hypocentral relocations obtained above for the distinct 3D-bloc models the former are compared with the *a posteriori* relocations from the optimal 1D-velocity model MKS-2007 (isotropic and anisotropic either) discussed in the previous section. The epicentral and hypocentral differences are shown in Figures 7.12 and 7.13, respectively. Obviously, the variations for the different models, isotropic as well as anisotropic, are only minor and no clear picture emerges as to whether the hypocenter relocations with the 3D-velocity models are systematically different from those obtained with the optimal 1D-velocity model MKS-2007. One may infer from the figures that for the anisotropic model inversions the variations of the epicenters and hypocenters are slightly less than for the isotropic ones, that depends on the use of an optimal isotropic and another, anisotropic computed model.

For the hypocentral depths (Figure 7.13) also no clear picture arises for the three bloc discretizations, neither for the isotropic nor for the anisotropic model cases, but the relocation shifts of the epicenters and hypocenters are slightly larger for the isotropic than for the anisotropic inversion case.

To better quantify the different results obtained so far some extra quantities are computed and summarized in Table 7.4. Thus $\bar{x}$, $\bar{y}$ are the average horizontal components of the shifting vector for all epicenters from which an average shifting vector $\bar{r}$ with average magnitude $\bar{r}$ and average angle $\bar{\phi}$ can be calculated. Basically one would expect that the length of this vector gets smaller with increasing model complexity, as more degrees of freedom for the adjustment of the unknown parameters (hypocenters and velocities) are available. The values in the table indicate that this holds for the 1D- and the $15 \times 15$ velocity bloc models, while for the $25 \times 25$ bloc models the $\bar{r}$- values are less for the isotropic case and nearly the same as for the $35 \times 35$ bloc models. The average radius $\bar{r}$ ranges between 3.83 $km$ and 2.74 $km$, indicating the average shift.

For the 1D-velocity models the average shift is in the direction $\bar{\phi}$ of about 236° for the isotropic and 201° in the anisotropic case. The situation becomes completely different for the $15 \times 15$ bloc model, where $\bar{\phi}$ is 137° in the isotropic and nearly opposite (281°) in the anisotropic case. For the other 3D-models, the direction is about 279° and 281° for the isotropic- and 264° for both anisotropic

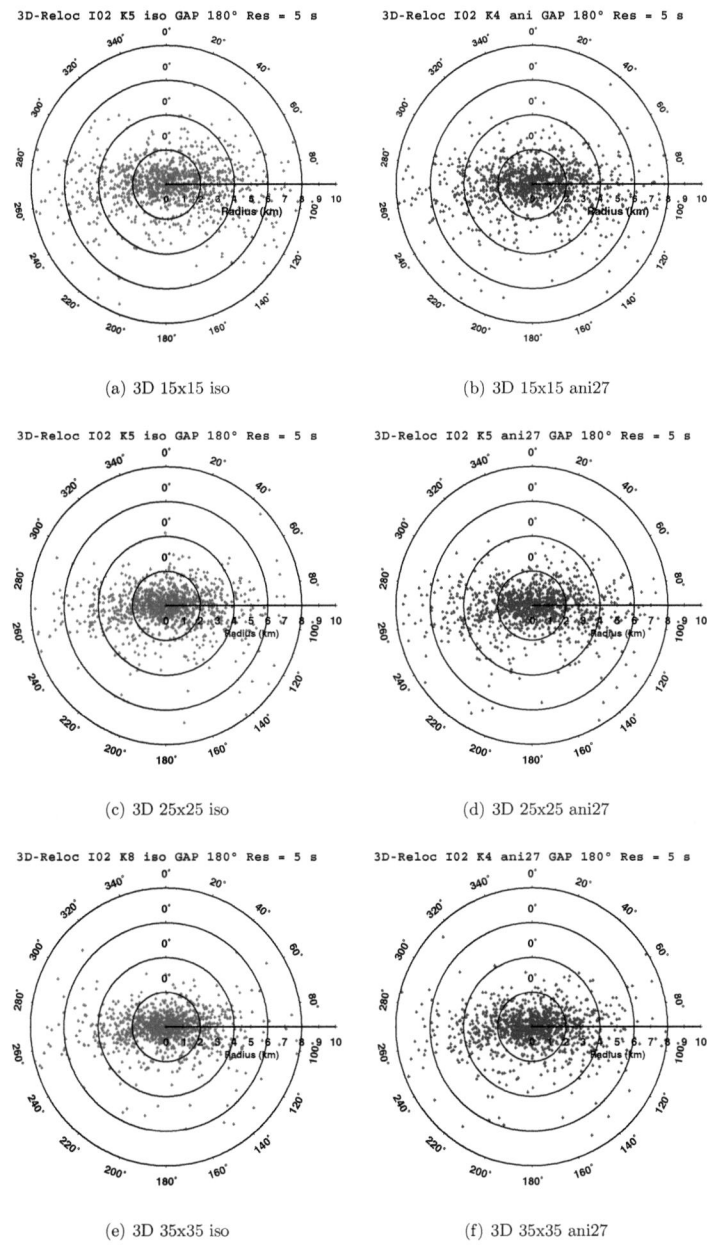

Figure 7.10: Epicentral shifts of the original events after full simultaneous inversions (IFIX = 0) with the optimal 3D velocity models without (left panels) and with anisotropic $P_n$-corrections (right panels), for the $15 \times 15$ (top row), $25 \times 25$ (middle row) and $35 \times 35$ bloc model (bottom row). The radius axis for the shift is up to $10\,km$.

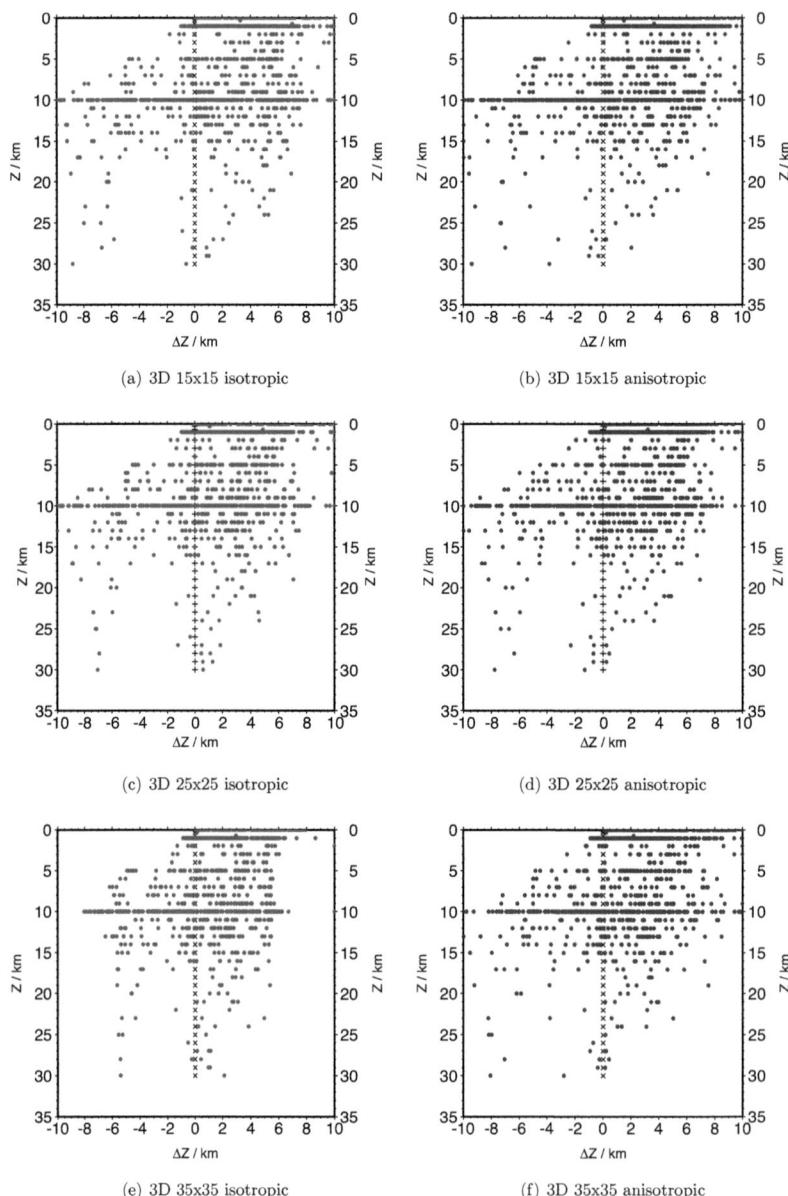

Figure 7.11: The distribution of the hypocentral depths shifts $\Delta Z$ over depth $Z$ for the three different bloc discretizations, simultaneous inversion.

## 7.4. HYPOCENTER RELOCATIONS, OPTIMAL 3D VELOCITY MODELS

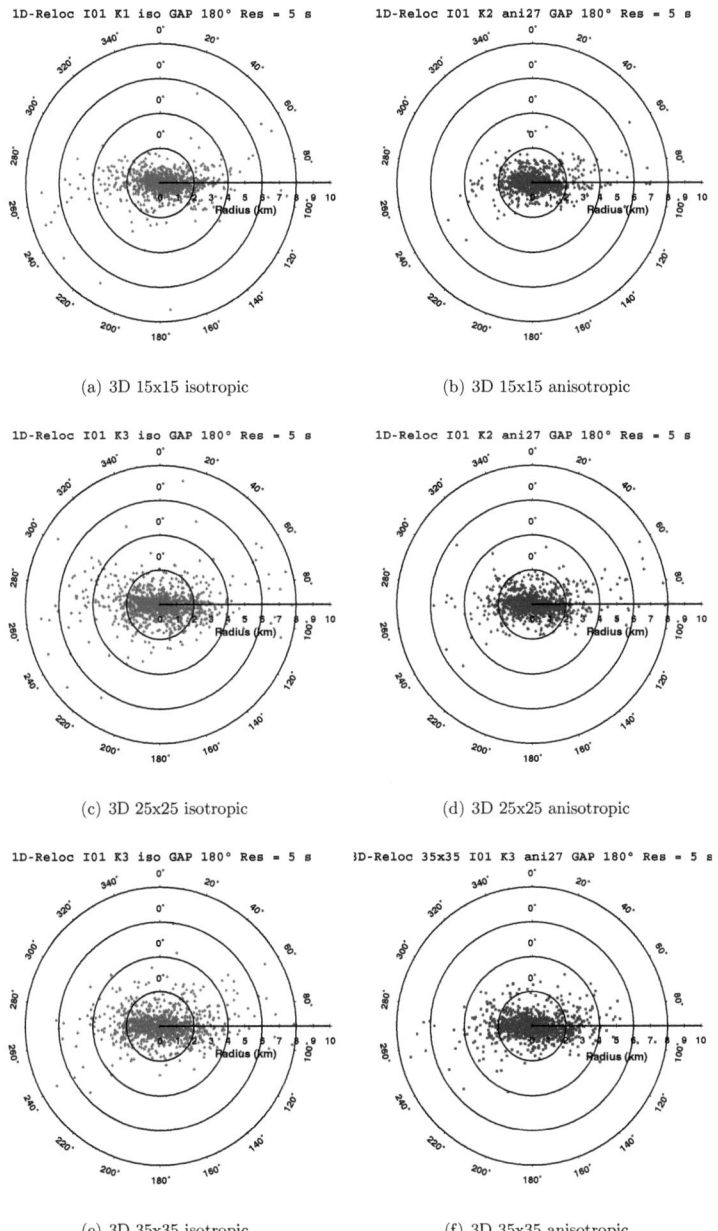

Figure 7.12: Relocation differences between the epicenters of the previous figure (relocated simultaneously (IFIX=0) with the various optimal 3D velocity models) and those relocated *a posteriori* (IFIX=2) in the previous section with the optimal MKS-2007 model.

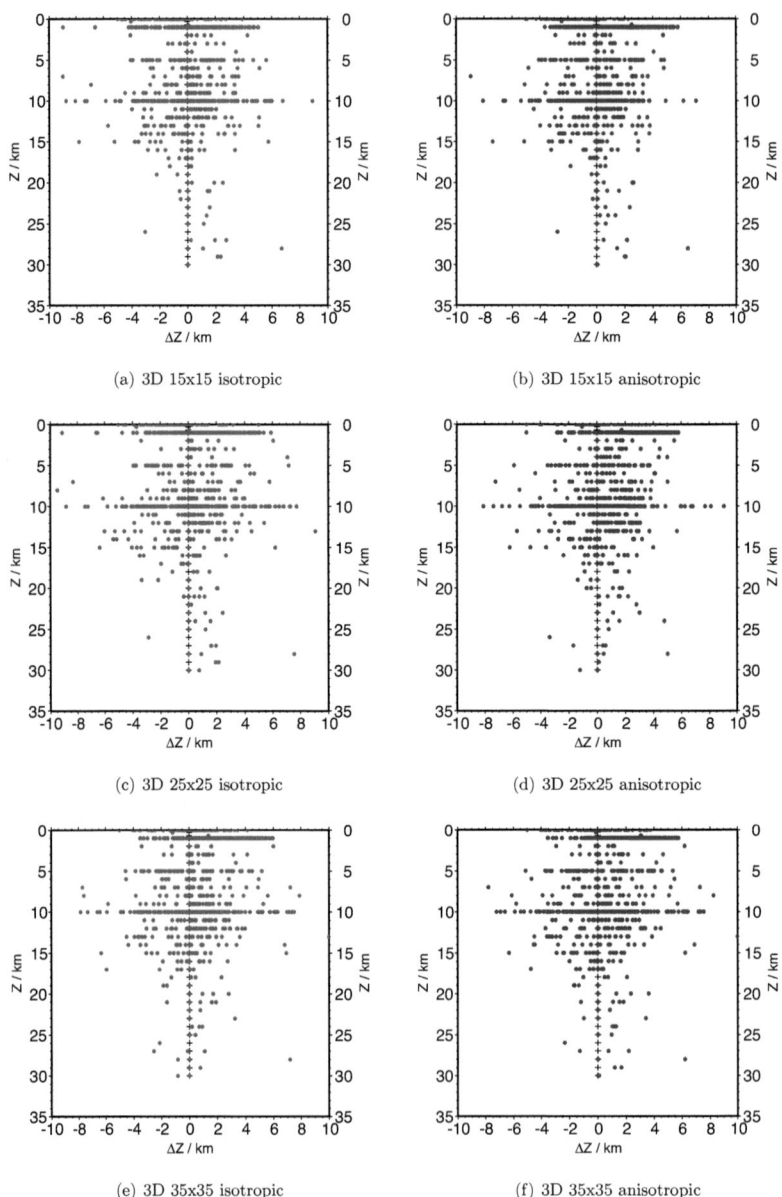

Figure 7.13: The distribution of the hypocentral depths for the three different bloc discretizations.

## 7.5. SENSITIVITY TESTS

cases. The consistency of the direction $\overline{\phi}$ of the shift vectors in Table 7.4 for all anisotropic inversion models, unlike those of the isotropic ones, may be seen as an indication of the stabilization of the SSH-inversion through the incorporation of the anisotropic $P_n$-traveltime correction.

It should be noted that the differences between the isotropically and anisotropically relocated hypocenters (as quantified by the $\Delta$ values) are per se not an indication to decide in favor of one of the two models, except if one assumes that the original epicenters are the best localized ones. Thus the $\Delta$ values are larger for the isotropic 3D $15 \times 15$ bloc model but smaller for the isotropic $25 \times 25$ and $35 \times 35$ bloc models than for their anisotropic counterparts. The only "objective" criterion to judge the quality of the relocations is the value RMS, which stands for the goodness of the datafit. The conclusion now is, that the anisotropic $P_n$-correction is most effective for the 1D velocity model and the $15 \times 15$ bloc model. For the models with higher bloc discretizations, the effectiveness of the anisotropic $P_n$-correction is only minor since its traveltime effect may be absorbed by the increased number of velocity parameters in the 3D models, leading to a slightly different isotropic velocity structure, especially in the fourth (Upper Mantle) layer, than one would get from a true anisotropic velocity model.

## 7.5 Sensitivity tests of the hypocenter relocations with synthetic data

In this section various sensitivity tests are performed to analyze the reliability of the hypocentral relocations under various conditions, namely, (1) the influence of isotropic and anisotropic inversions of an anisotropically generated data-set, (2) the influence of the initial hypocenter position and (3) the influence of errors in the arrivaltime measurements.

### 7.5.1 Influence of anisotropically generated traveltime data

Here a kind of plausibility test is performed to analyze the influence of synthetically generated anisotropic arrivaltime data on the relocations of the hypocenter, when the former are inverted isotropically, i.e. the anisotropic $P_n$-traveltime correction is not taken into account. Indeed, as discussed in detail in Chapter 4, one must assume that the original data has been affected by the upper mantle $P_n$-anisotropy, so foregoing this effect in a routine hypocenter determination with a 1D-standard velocity model must lead to some bias in the relocations, the amount of which will be analyzed in the subsequent paragraphs.

To generate the synthetic dataset, the original distribution of the hypocenters and recording stations as indicated in the original phase list for a particular event is selected. Then, using the SSH-ray tracing subroutine, theoretical arrivaltimes are generated for the optimal 1D-velocity model MKS-2007, with the elliptical anisotropic $P_n$-phase correction included. This synthetic dataset is then used in the subsequent hypocenter determinations of the original events.

The relocations of the original epicenters and hypocenters determined in this way for both the isotropic and the anisotropic inversion are shown in Figure 7.14. From the upper two panels of Figure 7.14, showing the epicentral shifts, one can clearly recognize the advantageous effect of the anisotropic hypocenter inversion (relocation), as the blue dots are concentrated closer around the origin of the polarplot than the red ones. Moreover, whereas the elongated form of the relocated epicenter cloud for the isotropic inversion still hints of a systematic relocation bias in this case, for the anisotropic inversion the cloud is circular and more concentrated, indicating clearly the importance of the anisotropic $P_n$-traveltime correction for reliable epicentral relocations. This situation is even more clearly pronounced for the hypocentral depth relocations shown in the lower two panels of Figure 7.14, as the anisotropic inversion model is concentrating the hypocenters stronger toward the center line than the isotropic model. The variation decreases with increasing depth, further showing the stabilizing effect of the anisotropic corrected $P_n$-phases together with steeper $P_g$ rays. This interesting phenomenon is

Table 7.5: Results for the isotropic and anisotropic inversions of a synthetically generated dataset with $P_n$-anisotropy included. The $\sigma$ values are computed for the difference between the new and old hypocenter results.

| ani | GAP deg | TSS $s^2$ | unbias. $s$ | aver.RMS $s$ | Av-Event-RMS $s$ | $\sigma_x$ km | $\sigma_y$ km | $\sigma_z$ km |
|---|---|---|---|---|---|---|---|---|
| no | 180 | 2205 | 0.3553 | 0.3148 | 0.1994 | 4.72366 | 1.07236 | 2.12253 |
| yes | 180 | 855 | 0.2213 | 0.1961 | 0.0856 | 2.08134 | 1.02070 | 1.32624 |

further corroborated by the quantitative results listed in Table 7.5. Thus, the values of the variances show that the deviations from the initial, original hypocenters are smaller for the anisotropic than for the isotropic inversion (relocation).

### 7.5.2 Sensitivity test with randomly shifted hypocenters

To further test the stability of the SSH-inversions and the maximally possible accuracy of the hypocentral relocations, the initial hypocenter values are randomly shifted in the $x$-, $y$- and $z$-direction by amounts drawn from a gaussian distribution with given standard deviations (That can be chosen separately for each coordinate direction, here it is $10\,km$ for the x- and y-component and $2\,km$ for the z-component). During the relocation inversions, the relocated hypocenters shall be shifted back near toward their original positions that have been computed before simultaneously with the optimal 1D velocity model in Chapter 5. The remaining difference between the position of the relocated and the original hypocenter will give an indication of the range of uncertainty of the hypocenter relocation, id est a measure of its error.

In Figure 7.15, the results of this relocation test are shown. One observes from the various panels of this figure that the epicentral as well as the hypocentral shifts are somewhat smaller for the anisotropic than for the isotropic relocation model, as the relocations are aligned in a more narrow band around the zero line for the latter than for the former case. From these panels and Table 7.6 where the most important statistical results of the relocations are summarized one may also deduce that with the present distribution of earthquakes and recording stations, the accuracy of the hypocentral relocations is about $\pm 2\,km$, though some vertical (z)-shifts show extremes of more than $6\,km$.

Table 7.6: Statistical results of the relocations obtained simultaneously with the optimal 1D velocity model and the *a posteriori* determined relocations after initially randomly shifting the hypocenters. *mean* denotes the average differences along the three coordinate directions between the new and original hypocenters; $\sigma$, the corresponding standard deviation; RMS$_{new}$, the RMS of the data-fit of the final relocation model; and RMS$_{org}$, the RMS of the original optimal 1D SSH-model.

| | iso | | | ani | | |
|---|---|---|---|---|---|---|
| | x | y | z | x | y | z |
| $mean$/km | 0.2014 | -0.0585 | -0.2638 | 0.1852 | -0.0298 | -0.1848 |
| $\sigma$/km | 1.8623 | 1.9250 | 1.9306 | 1.8313 | 1.8962 | 2.1181 |
| $RMS_{new}$/s | | 0.7992 | | | 0.7808 | |
| $RMS_{org}$/s | | 0.8166 | | | 0.7999 | |

## 7.5. SENSITIVITY TESTS

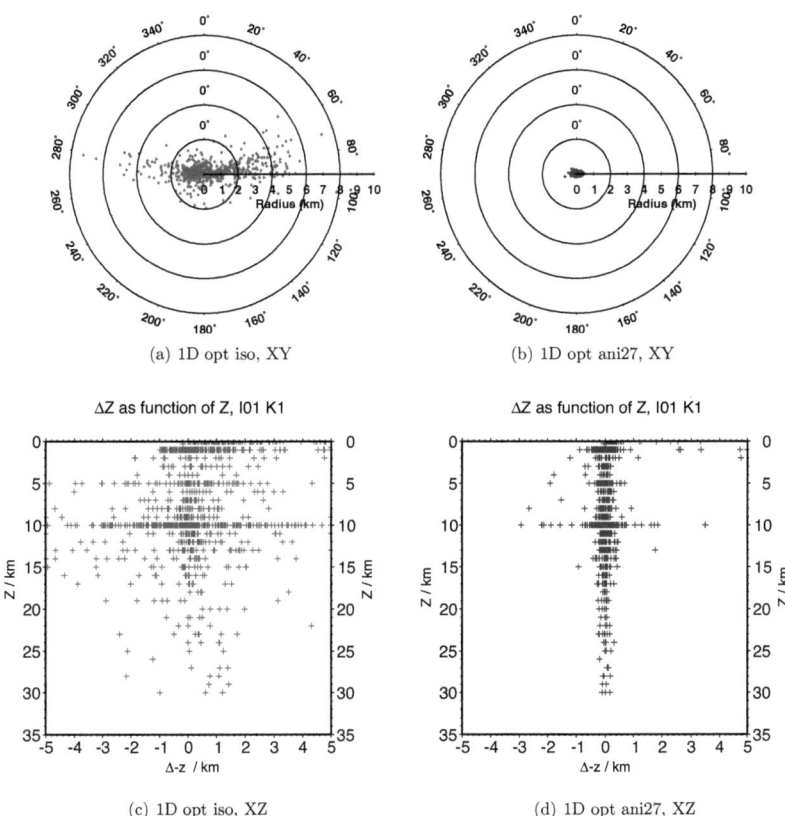

(a) 1D opt iso, XY

(b) 1D opt ani27, XY

(c) 1D opt iso, XZ

(d) 1D opt ani27, XZ

Figure 7.14: Comparison of isotropically and anisotropically relocated hypocenters for a synthetic data-set with anisotropic generated traveltimes for $P_n$-phases. Red dots mark the epicentral differences between the initial and the isotropic relocated hypocenters, blue dots between the initial and anisotropic one.

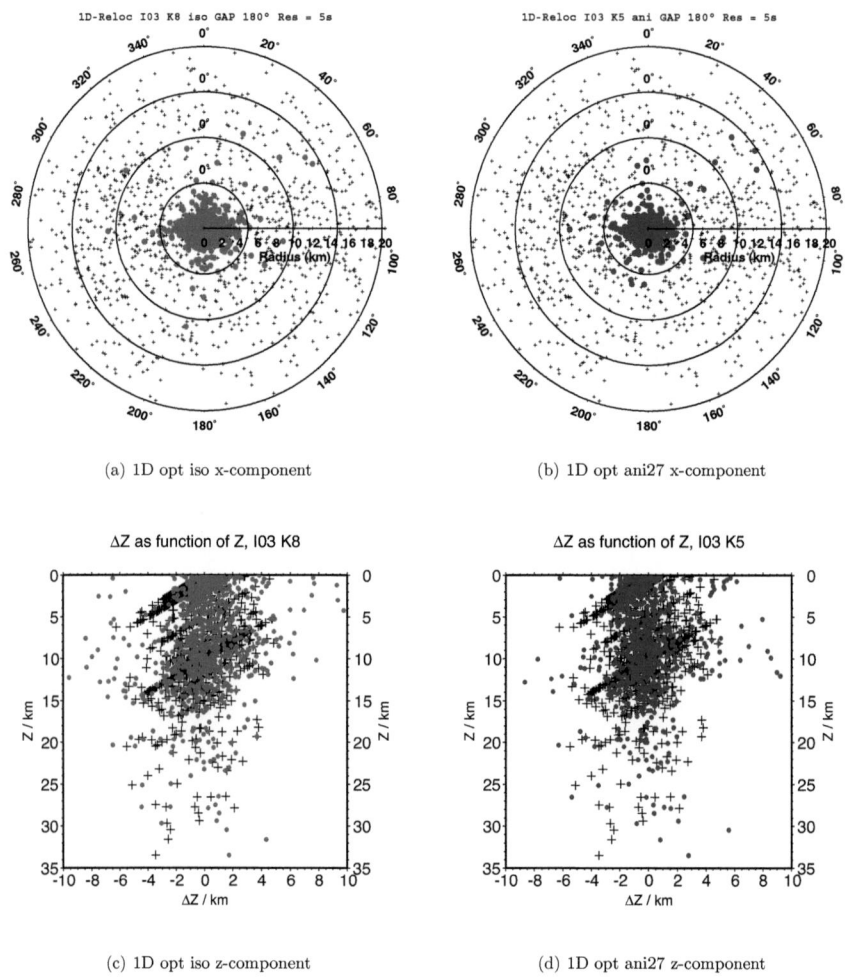

Figure 7.15: Comparison between relocations obtained simultaneously with the optimal 1D velocity model and the *a posteriori* determined relocations after initially randomly shifting the hypocenters (black crosses) for isotropic (left panels) and anisotropic (right panels) inversion cases. The scale of the radius is up to 20 km.

## 7.5.3 Sensitivity tests with randomly perturbed arrivaltimes

The following tests are to investigate the influence of the arrivaltime uncertainties on the hypocentral relocations. To that avail the observations (arrivaltimes) as available in the original dataset are randomly perturbed by time-errors that have been drawn from a Gaussian distribution with standard deviations $\sigma$ values. The latter have been chosen here as $0.1$, $0.2\,s$ and $0.3\,s$ whereby the lower value represents about the size of arrivaltime picking errors in many seismograms, and the higher values add up other errors. The new datasets created in this way are then used to relocate the hypocenters whereby, again, inversions without and with anisotropic $Pn$-traveltime corrections are performed.

Likewise to the previous figures, the results of the various model runs for the hypocentral relocations are shown in Figures 7.16, 7.17 and 7.18. In Tables 7.7 and 7.8 the most salient statistical results of these relocation tests are listed. Based on these figures and tables, two major aspects are noteworthy. Firstly, the statistical results for the residuals in Table 7.7, namely, TSS and RMS clearly indicate that the anisotropic relocations are again better than the isotropic ones and this holds for all values of $\sigma$. So, even if one has inaccurate data, the influence of the anisotropic traveltime distortion is still dominant and must be taken into account through the appropriate $P_n$-traveltime correction in the inversion process. Secondly, the uncertainty range of the hypocentral relocations which is indicated by the values of the standard deviation $\sigma_x$, $\sigma_y$ and $\sigma_z$ in the three coordinate directions, respectively, in Table 7.8 can be deduced. One notes from the table and the figures that the hypocentral uncertainty range increases, expectedly, with the size of the *a prior* imposed arrivaltime error. However, even for the largest value of the latter, the former is not larger than $1.5\,km$ in the x-direction, not more than $1\,km$ in the y-directions and less than $2.5\,km$ for the depth z.

Table 7.7: Statistical results for the datafit for the various models of pure hypocenter relocations with randomly disturbed arrivaltimes of $\sigma_t = 0.1\,s$, $0.2\,s$ and $0.3\,s$. The results are assessed by TSS or RMS. RMS-unbiased is the RMS corrected by the number of unknown model parameters. Av.-event-RMS gives the data-fit for the event localization. TSS-hypo and RMS-hypo are from the original used data.

| ani | $\sigma_t$ | K | TSS | RMS-unbiased | RMS | av.-event-RMS | TSS-hypo | RMS-hypo |
|-----|------------|---|------|--------------|-------|---------------|----------|----------|
|     | $s$        |   | $s^2$ | $s$         | $s$   | $s$           | $s^2$    | $s$      |
| no  | 0.1        | 1 | 246  | 0.1186       | 0.1049 | 0.1231       | 243      | 0.1043   |
| yes |            | 1 | 246  | 0.1185       | 0.1049 | 0.1231       | 243      | 0.1043   |
| no  | 0.2        | 1 | 1134 | 0.2547       | 0.2254 | 0.2412       | 922      | 0.2032   |
| yes |            | 1 | 1082 | 0.2487       | 0.2202 | 0.2361       | 922      | 0.2032   |
| no  | 0.3        | 1 | 2065 | 0.3437       | 0.3042 | 0.3578       | 2049     | 0.3030   |
| yes |            | 1 | 2151 | 0.3507       | 0.3105 | 0.3413       | 2049     | 0.3030   |

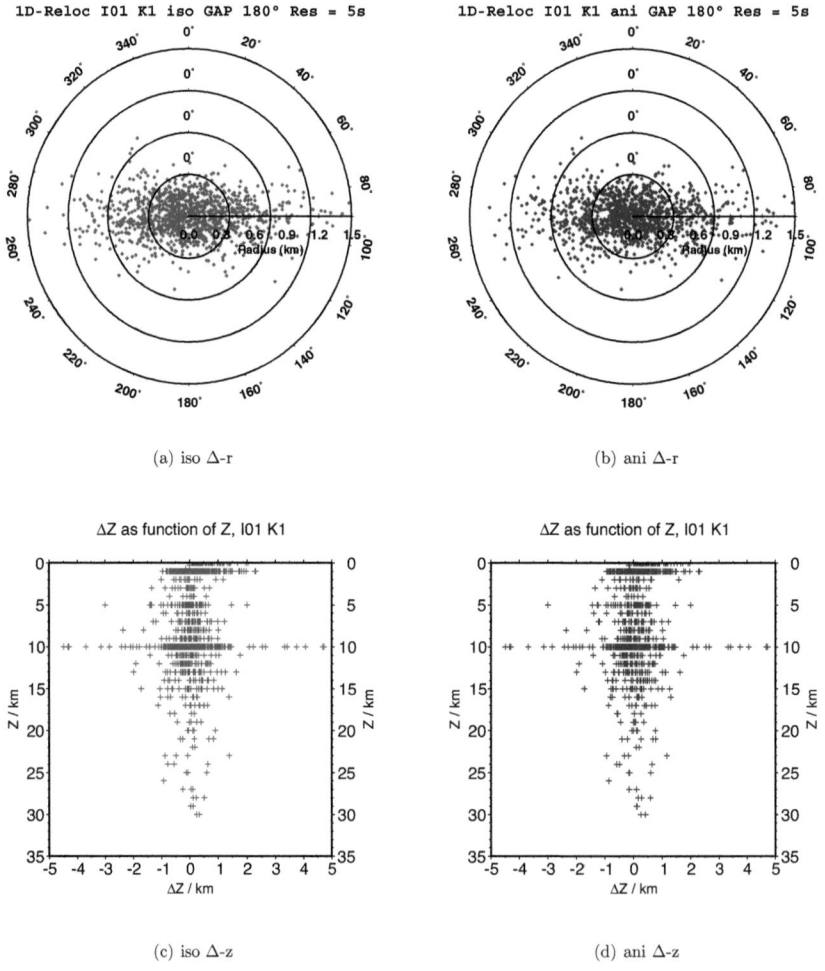

Figure 7.16: Comparison between isotropically and anisotropically relocated hypocenters with randomly perturbed arrivaltimes of $\sigma = 0.1\,s$.

## 7.5. SENSITIVITY TESTS

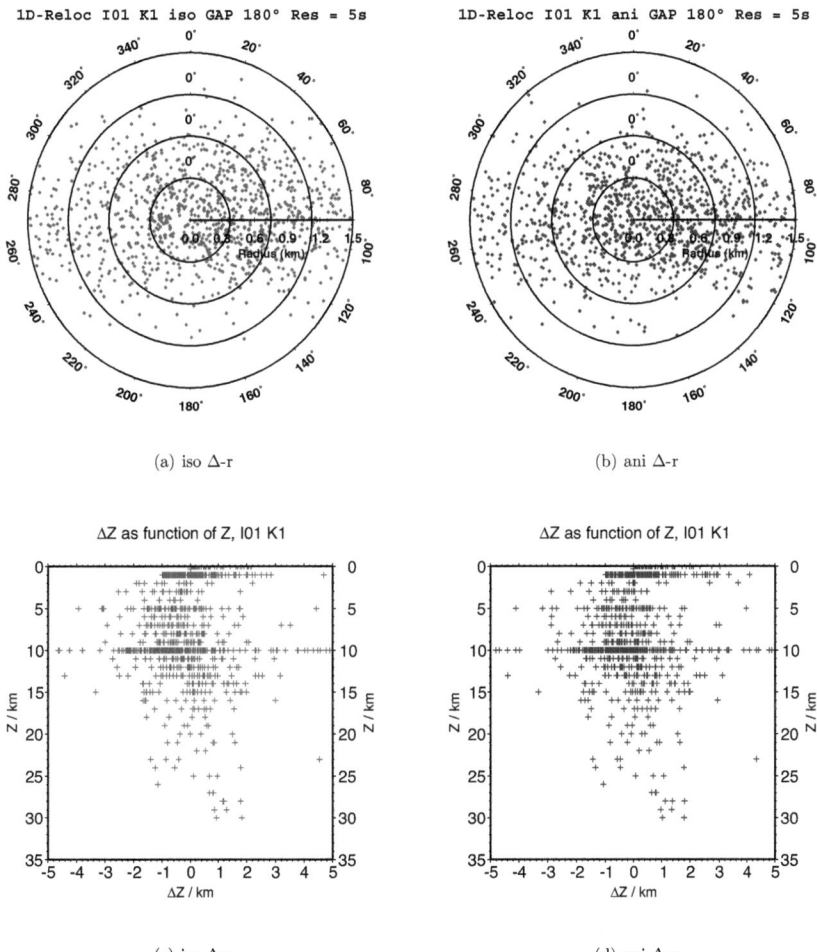

Figure 7.17: Similar to Figure 7.16, but with arrivaltime errors of $\sigma = 0.2\,s$.

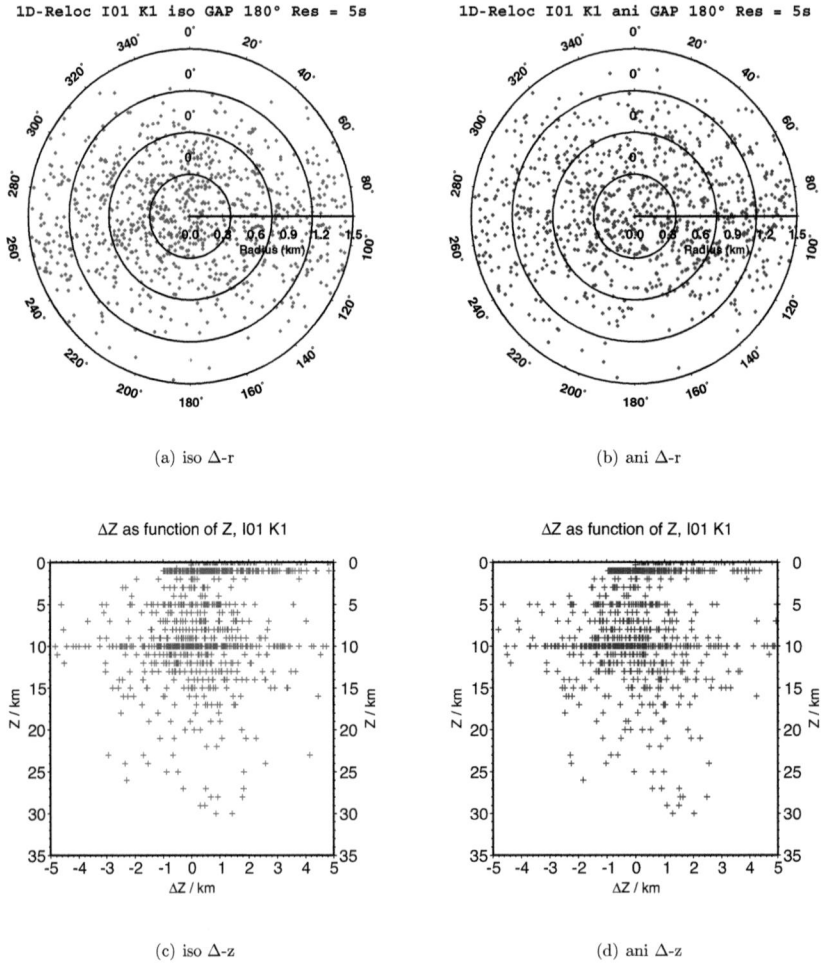

(a) iso Δ-r

(b) ani Δ-r

(c) iso Δ-z

(d) ani Δ-z

Figure 7.18: Similar to Figure 7.16, but with arrivaltime errors of $\sigma = 0.3\,s$.

Table 7.8: Statistical results for the various models of pure hypocenter relocations with randomly disturbed arrivaltimes of $\sigma_t = 0.1\,s$, $0.2\,s$ and $0.3\,s$. The mean values $\overline{x}$, $\overline{y}$ and $\overline{z}$ denote the average differences between the original and the relocated hypocenters in the three coordinate directions, respectively, and $\sigma_x$, $\sigma_y$ and $\sigma_z$ the corresponding standard deviations.

| ani | $\sigma_t$ | $\overline{x}$ | $\overline{y}$ | $\overline{z}$ | $\sigma_x$ | $\sigma_y$ | $\sigma_z$ |
|---|---|---|---|---|---|---|---|
|     | s   | km      | km     | km     | s      | s      | s      |
| no  | 0.1 | -0.0142 | 0.0002 | 0.3247 | 0.4364 | 0.1697 | 1.7154 |
| yes |     | -0.0148 | 0.0004 | 0.3235 | 0.4398 | 0.1697 | 1.7157 |
| no  | 0.2 | 0.1578  | 0.0306 | 0.3393 | 1.0178 | 0.3907 | 2.4367 |
| yes |     | 0.1131  | 0.0231 | 0.3699 | 0.9559 | 0.3665 | 2.3164 |
| no  | 0.3 | 0.1424  | 0.0295 | 0.3906 | 1.3970 | 0.5265 | 2.6401 |
| yes |     | 0.1005  | 0.0224 | 0.4537 | 1.3476 | 0.5088 | 2.5890 |

## 7.6 Summary of the major results of the hypocentral relocations

The encompassing SSH-analysis of the hypocentral relocations carried out for various model conditions and test scenarios allows us to clarify the following points:

**Relocations with the optimal 1D velocity models**

For the two optimal 1D velocity models, namely the isotropic and the anisotropic one, the hypocentral relocations are statistically more reliable when the anisotropic correction for the $P_n$-phases is included (see Table 7.2). All model variants show improvements between 3.5% to 10% for RMS relativ to the isotropic case with GAP = 230° and between 17% and 35% for the differences between $TSS_{hypo}$ and $TSS_{SSH}$, whereby these improvements are consistently better for the anisotropic than for the isotropic model cases (see Table 7.2). Also, as one would expect, the RMS-improvements are larger as the $GAP$ is decreased, as more reliable data is selected for the inversion process.

Concerning the average shifts of the hypocenters for the isotropic and anisotropic model cases ($\Delta_x$, $\Delta_y$ and $\Delta_z$ in Table 7.2), for the anisotropic case these shifts are bigger for $GAP = 230°$, but smaller for $GAP = 180°$ and nearly the same for $GAP = 120°$. This behavior is due to the fact that **poorly** determined hypocenters are more and more eliminated when reducing the gap. The relatively large shifts for $GAP = 230°$ for the anisotropic model are most likely due to the fact that the originally (with such a large gap) **poorly** determined hypocenters are now better relocated when applying the anisotropic correction. This explains also the better datafit (measured by TSS or RMS) with an improvement of 33%, compared with only 28% for the isotropic model. The mean differences between the isotropically and anisotropically computed positions ($\Delta_{iso}$ - $\Delta_{ani} = \delta$) of the hypocenters in the three coordinate directions are $\delta_x = -0.13$, $\delta_y = 2.43$ and $\delta_z = 0.27$. The situation reverts itself when the gap is reduced to $GAP = 180°$, where the differences become $\delta_x = 0.42\,km$, $\delta_y = 0.3\,km$ and $\delta_z = 0.7\,km$ and more so, for $GAP = 120°$, where it is $\delta_x = 0.1\,km$, $\delta_y = 0.09\,km$ and $\delta_z = 0.17\,km$. Thus, when smaller gaps are used, i.e. only high-quality and well-observed earthquakes are considered, the differences between the isotropic and anisotropic inversions become smaller, so that these hypocenters may as well be relocated as by standard location procedures. Concerning the depth variations, they are overall also less for the anisotropic than for the isotropic inversion case. The statistical measure of the RMS confirms the relocation results, as the datafit is consistently better for the former than for the latter.

Comparing the pure (*a posteriori*) determined hypocenter relocations with those obtained from the simultaneous inversion with the optimal 1D velocity models, the results with respect to both the datafit and the hypocentral shifts (differences of only about $1\,km$) are nearly the same. This interesting effect holds for both isotropic and anisotropic models and may explain the usability of the optimal 1D model for a pure hypocenter relocation, thus avoiding the computationally intensive full (coupled) SSH inversion for both velocity structure and hypocenters.

## Relocations with the optimal 3D velocity models

The comparisons of the hypocentral inversion results with three 3D velocity models of increasing bloc-resolutions among each other and with those of the optimal 1D models above show an increasing datafit in all four cases when the anisotropic $P_n$-traveltime correction is applied, whereby the largest improvement difference of about 30% occurs between the isotropic 1D- and 3D velocity models. The anisotropic corrected 3D-models lead to a further $4-9\%$ improvement of the datafit. As for the hypocentral shifts $\Delta_x$, $\Delta_y$ and $\Delta_z$, their variances (or standard deviations) are different for the isotropic and anisotropic inversion cases.

There are also differences between the average hypocentral deviations for isotropic and anisotropic relocated hypocenters ($\delta = \Delta_{iso} - \Delta_{ani}$). Thus, negative values indicate bigger variations for the anisotropic inversion case. For the $15 \times 15$ bloc model, the values are $\delta_x = -0.47$, $\delta_y = -0.2$ and $\delta_z = -0.58\,km$, representing a bigger deviation in the anisotropic case. With finer bloc-discretizations, these deviations become smaller ($0.31$, $0.17$, $0.01\,km$) for the $25 \times 25$ bloc model and are nearly the same for the $35 \times 35$ bloc model ($0.03$, $0.03$, $0.16\,km$). This leads to the conclusion that the anisotropic correction is most effective for lower 3D-bloc discretizations and the least for the 1D velocity models.

Overall the epicentral (radial) shift variation is about $2\,km$ whereas the hypocentral (depth) shift goes up to $\pm 5\,km$. Another conclusion is that the differences between these hypocenter shifts $\Delta_x, \Delta_y$ and $\Delta_z$ as well as the datafit between the isotropic and anisotropic inversions are somewhat smaller for the $35 \times 35$ bloc models. This could be due to the fact that the anisotropic effects are somewhat compensated by velocity perturbations in the increased number of blocs of layer four.

A more detailed picture arises from the analysis of the average shift vector of the epicenters as specified by its magnitude $\overline{r}$ and angle $\overline{\phi}$. Thus, whereas the absolute values of $\overline{r}$ do not provide a hint in favor of a particular model solution (isotropic or anisotropic one), the values for the average angle $\overline{\phi}$ are nearly the same for the various anisotropic inversion models, but differ quite a bit for their isotropic counterparts. This can be taken as evidence that the epicentral relocations are more consistent, i.e. more stable for the anisotropic than for the isotropic inversion models.

## Sensitivity tests with synthetic anisotropic generated data

The various sensitivity tests with synthetic anisotropically generated data, indicate clearly the influence and the importance of the inclusion of the $P_n$-anisotropy for reliable hypocentral relocations. Thus, whereas the isotropic inversion of the anisotropic traveltime data still hints of a systematic hypocenter mislocation, with values of $4.7\,km$, $1.1\,km$ and $2\,km$ for $\sigma_x$, $\sigma_y$ and $\sigma_z$, respectively, for the anisotropic inversion the original hypocenters are better relocated and the shifts reduced to $2\,km$, $1.0\,km$ and $1.3\,km$. Also the datafit, as measured by the RMS, is reduced from $0.3148s$ for the isotropic to $0.1961s$ for the anisotropic inversion case.

The sensitivity tests with randomly disturbed initial hypocenters show again the importance of the anisotropic $P_n$-traveltime correction but also the overall stability of the hypocenter determination process. Thus, even by initially shifting the original hypocenters by $\Delta X$, $\Delta Y = 10\,km$ and $\Delta Z = 2\,km$, the relocated hypocenters are shifted back toward their original positions with a relocation misfit of only a few $km$ (see 7.6) for the anisotropic inversion model.

Using input arrivaltime data with random errors to mimic observation errors, the uncertainty range of the hypocentral relocations has been assessed further. Thus, although the hypocentral uncertainty range as well as the RMS increases somewhat with the size of the *a priori* imposed arrivaltime errors ($\sigma$-t $= 0.1\,s$, $0.2\,s$ and $0.3\,s$), the mislocations are no more than $1.5\,km$, $1\,km$ and $2.5\,km$ for the $x$-, $y$-, and $z$-direction. The hypocentral shifts are less for the anisotropic than for the isotropic models (similar to the hypocentral results of the optimal 1D-velocity models) for the cases with $\sigma = 0.2\,s$ and $\sigma = 0.3\,s$, unlike the relocations with the model case with $\sigma = 0.1\,s$ where the anisotropic shifts are larger than the isotropic ones. Also, the optimal solutions are obtained in the same iteration step with the same damping parameter, so that they are readily compared.

# Chapter 8

# Summary

In the present thesis the existence of seismic anisotropy ($P_n$-anisotropy) in the upper mantle underneath Germany and adjacent areas and its possible influence on the computation of the seismic velocity structure and the relocation of regional earthquakes in the study region is investigated by means of a 3D tomographic SSH (Simultaneous Structure and Hypocenters) inversion technique (Koch, 1985a). Until recently, upper mantle anisotropy has been neglected in the interpretation of the seismic signals, i.e. the seismic velocity has been computed without considering the former, and hypocenters have been located assuming mostly simple isotropic 1D velocity models. Such an isotropic approach may be warranted, when only direct $P_g$- or $S_g$-crustal phases are observed, i.e. close to the epicenter. However, for regional events, $P_n$- or $S_n$-phases are usually detected at distances larger than $100\,km$, so these will be influenced by the assumed upper mantle anisotropy which, in turn, will bias the structural as well as the hypocentral determinations.

Based on this premise, the present work is innovative in two fundamental ways: not only the 3D seismic structure of the crust and upper mantle underneath Germany is investigated using regional traveltime data, but the hitherto well-accepted $P_n$ wave anisotropy of the upper mantle across the region is also included in the SSH-inversions. In addition to provide new information on the laterally inhomogeneous crustal and anisotropic upper mantle velocity structure across Germany, the effects of the latter on the hypocentral locations of the regional earthquakes are also investigated. The results of this analysis form another important product of the present thesis and should help regional seismic networks to better locate seismic events throughout Germany in the future.

For the present tomographic study a dataset that consists of 10028 seismic events recorded between 1975 and 2003 with 46550 $P_g$-, 12804 $P_n$-, 50309 $S_g$- and 3903 $S_n$-phases at 320 seismic recording stations across Germany and adjacent areas is available. With such a huge dataset, a good and rather homogeneous coverage of the whole model area by seismic rays can be achieved allowing the application of high-resolution SSH-tomography. As discussed further down, various quality checks of the data are to be performed to select an optimal subset of seismic events and arrivaltimes for input into the SSH-procedure that turned out to be not a trivial, but rather time-consuming task of the thesis work. A major effort has been put into the statistical analysis of the traveltime residuals, and it could be shown that the $P_n$-phases, unlike the $P_g$-phases, show consistent sinusoidal azimuthal variations, providing another clear evidence for upper mantle anisotropy across the region.

After selection of an appropriate dataset, both isotropic and anisotropic tomographic SSH-inversions are carried out to find the optimal average 1D vertically stratified velocity model usable, for example, in regional standard hypocenter relocation, as well as for an optimal starting model[1] for the later SSH inversions for the 3D structure of the crust and upper mantle. Thereby SSH-inversions with fixed hypocenters, i.e. inversion for velocity structure alone and for both, hypocenters and structure are compared to test the influence of the seismic structure onto the hypocentral locations. For the

---
[1]Selecting the optimal 1D-velocity model as a starting model for subsequent 3D velocity inversions ensures that the starting solution is as close as possible to the local minimum of the nonlinear objective function

full SSH-inversions, the resulting velocity models appear to become petrologically more reasonable (Chapter 5).

Simultaneously, the hypocentral shifts relative to the originally reported hypocentral locations, namely the depths are analyzed. To this regard, the use of the $P_n$-phases is of major importance for the isotropic, but more so for the anisotropic inversion models. In fact, the anisotropic elliptical correction for the $P_n$-phases results in a further improvement of the fit of the model (structure and hypocenters) to the observed data, corroborating the evidence for anisotropy in the upper mantle beneath Germany.

The emphasis of the present thesis is then devoted to the determination of 3D tomographic seismic models underneath Germany, as well as with and without upper mantle $P_n$-anisotropy included. Again, the anisotropic 3D models fit the observed data better and show seismic features that are more consistent with known geological and tectonic properties of the region than the isotropic models. Finally, the effects of the 3D models on the hypocentral locations are investigated in detail.

## 8.1 Theory

For seismological studies, seismic waves, propagating through an elastic medium, are used to detect the properties of the latter. In general, the seismic wave is described by the general wave equation 2.4, where the elasticity tensor $c_{ijkl}$ theoretically has 81 components, but which, because of several compatibility conditions, goes down to 21 for the general 3D anisotropic medium. For the regular isotropic medium which is the subject of most seismic studies the elasticity tensor reduces eventually to only two, the Lamé constants $\lambda$ and $\mu$, so the wave equation can be split in two equations, describing $P$- and $S$-waves, respectively.

To apply seismic ray theory in tomographic earth studies, one uses the high-frequency approximation of the wave equation which eventually leads to the Eikonal equation for the traveltime $T(x, y, z)$ (2.8). As discussed in detail in Chapter 2, for practical tomographic applications, this high frequency approximation is sufficiently precise if some minimal criteria with respect to the desired geometrical resolution are fulfilled. The major ingredient of seismic ray theory is that the propagation of a seismic wave is described by a ray whose path through an medium (earth) from source to station is such that its traveltime is a minimum (Fermat's principle). Practically the ray-tracing used in the present SSH-method is based on an iterative shooting method as explained in Koch (1985b) for direct phases. In Koch (1993a) the ray-tracing has been extended to incorporate refracted and reflected phases as well.

To extend the tomographic regional study to an anisotropic upper mantle with a horizontally oriented axis of the anisotropy ellipse, a particular kind of "weak anisotropy" (Silver, 1996) is assumed. Thereby, an "elliptical correction", defined by a velocity contrast between the maximum and minimum value of the anisotropy ellipse and the azimuthal direction of its major (fast) axis is applied to the average isotropic velocity and the ensuing traveltime effect is calculated. For a vertically stratified 1D velocity model, the whole upper mantle layer velocity is corrected, whereas for a 3D laterally heterogeneous model, the correction is applied to each velocity bloc separately.

Once, the forward problem is solved, i.e. new traveltime residuals are calculated, correction values for the velocity structure and the hypocentral parameters are computed through application of the SSH-inversion method, described, for example in Koch (1985a, 1993a). In the SSH-technique, the seismic inverse problem is solved through application of a nonlinear, iterative least squares method to the linearized and discretized version of the nonlinear integral equation, relating the local seismic velocity along the ray with the integral traveltime. Eventually this results, for practical purposes, in an overdetermined, linear system of equations in each iteration step for an update $\Delta x$ of the solution vector $x_i$, which is solved by a damped least squares method (which in the present SSH technique is based on the classical normal equations). In this method, also called the Ridge regression or the

Levenberg-Marquardt method, a series of trial-and-error solutions are computed using different values of the damping (ridge) parameter on the diagonal of the system matrix. Among these solutions the optimal one is found, i.e. the one which results in the best fit of the model to the traveltime data after new application of the ray-tracing procedure with the updated seismic model. In the next iteration, a new system matrix of the Fréchet derivatives, i.e. the partial derivatives of the arrivaltimes with respect to the model parameters is set up and the system of equations with the new right hand side of the arrivaltime residuals is solved again for a set of damping parameters. Basically the whole nonlinear least squares problems consists in an outer loop of nonlinear and an inner loop of linear iterations. Further details of the inversion procedure as they relate to evaluation of the important statistical properties of the inverted models are discussed in the applications sections.

## 8.2 Anisotropy

In the Section Anisotropy, the origins and different kinds of seismic anisotropy are discussed. The upper mantle consists of about 70 % olivin and of pyroxene for most of the remaining part. The mineral olivine, in particular, exhibits strong anisotropic behavior along its different crystal axes. Overall, one may assume that for a heterogeneous distribution of this crystal, no anisotropic effect is visible. However, in the earth, a special effect takes place, whereby the orientation of these crystals is affected by the movement of the overlaying crust, due to large-scale tectonic stress. This leads to strain which causes minerals and their crystals to orient themselves along that direction. By this process - called also lattice-preferred orientation (LPO) - seismic anisotropy in the upper mantle may occur that follows the finite strain ellipsoid (FSE), with the fast a-axis of olivine aligned with the longest axis of the finite strain ellipsoid. In fact, LPO in olivine minerals is most likely the dominant origin of anisotropy observed at the crust mantle boundary and in the first kilometers of the upper mantle in many regions in the world, including central Europe, as indicated first by the eminent studies of Bamford (1973); Fuchs (1975), and corroborated more more recently by Enderle et al. (1996a); Babuška and Plomerová (2000).

The first signs of $P_n$-anisotropy in regional traveltimes using a subset of the present dataset have been detected by Schlittenhardt (1999). These traveltimes have then been analyzed by Song et al. (2001b) by means with a 1D anisotropic time-term method and, subsequently, by Song and Koch (2002) with an extended 2D-technique and the form and the direction of the anisotropy in the upper mantle underneath Germany been determined. The results of these regional traveltime studies are consistent with those of the earlier studies mentioned, namely, a large-scale $P_n$-anisotropy ellipse across the region, with its fast axis oriented in an azimuthal direction of about 27° NE.

## 8.3 Initial data analysis

The initial data analysis carried out in Chapter 4, Data, first shows the effects of upper mantle seismic anisotropy on the azimuthal variation of the $P_n$-wave residuals, namely a sinusoidal undulation with a minima at azimuths of about 30° and 210°, and maxima shifted by, expectedly, 90°, representing delayed or advanced arrivaltimes, respectively. For the $P_g$-phases, no such anisotropic variations can be detected, whereas for the $S_n$-phases no clear picture emerges. This might be due to the fact that the seismic records do not distinguish between $S_H$- and $S_V$-phases, both of which may be affected differently by upper mantle anisotropy. Notwithstanding, the later inversion models appear to indicate the possibility of some $S_n$-anisotropy, so the final jury to this regard is still out.

Another objective of the initial data analysis has been the analysis of various data selection criteria for later use in the SSH-inversions.

- **Criterion RES**
  The first investigated criterion has been the maximally allowed traveltime residual for a phase

record to be included in the inversion process (criterion RES). The analysis indicates that RES = ±5 s is a good choice, as this value is a good compromise to eliminate mispicks, i.e. wrongly assigned and falsely detected phases while still keep important information about traveltime variations caused by anisotropy. Overall, the RES-limitation is not as restrictive to the amount of usable data as the GAP- or NOBS criterion below, as it rejects fewer records, while still being effective in the elimination of the ambiguous ones.

- **Criterion NOBS**
  The tests for the NOBS-criterion, the number of observations per event, shows that NOBS = 8 is a good value, being large enough for eliminating most of the earthquakes with poor initial relocations, while still providing at least 4 degrees of freedom (beyond the four hypocentral parameters) to be used for the seismic velocity determination. Setting NOBS to a higher value will not significantly reduce the spread of the data, but will cut down the number of seismic events and seismic rays too much, resulting in poorer geometrical resolution of the tomographic model, as indicated visually by the ray-coverage and, more quantitatively, by the resolution matrix.

- **Criterion GAP**
  The criterion GAP describes the maximally allowed azimuthal gap between pairs of event-station rays for a particular earthquake. For large gaps the observing stations are lying within a small azimuthal cone from the seismic source which, mathematically, leads to a high degree of collinearity and ill-conditioning of the hypocentral system matrix, ergo to an ill-posed inverse problem. This effect will be particularly strong for earthquakes located at the borders of the study area, as they are "seen" by stations only within a limited aperture.

  The detailed analysis of the GAP criterion shows that significant shifts of the hypocenter, especially of the epicenter, will occur for gaps larger than 180°, when an isotropic velocity model is used for anisotropically influenced $P_n$-traveltimes, i.e. it is wrongly assumed that there is no upper mantle anisotropy, as rays will then be missing from the opposite azimuthal direction to compensate the anisotropically-induced traveltime bias. In contrast, when the anisotropic correction is included, events with gaps much higher than 180° can still be relocated reasonably well. Another observation of this preliminary analysis has been that the strongest reduction of the TSS or the RMS occurs for a gap of 120°, as then only high quality seismic events are incorporated in the SSH-inversion. On the other hand, for 3D SSH-inversions with more than 15 × 15 blocs in a model layer, the restrictive gap of 120° reduces the ray coverage for the individual velocity blocs too much. Thus, for the very fine 3D tomographic seismic models, a gap of 180° has been used in the inversion calculations.

- **Criterion Anisotropy level**
  Another important parameter for the subsequent SSH inversions has been the proper choice of the anisotropy level, i.e. the velocity contrast of the anisotropy ellipse. For this, a range of levels was evaluated. Based on the best fit of the anisotropic arrivaltimes to the observed ones, the optimal anisotropy level could be fixed between ±2.5 % and ±3.0 %.

Taking all things together, RES ≤ 5 s, GAP ≤ 180° and NOBS ≤ 8 were chosen as base values of these selection criterion to extract an adequate data set in the tomographic SSH inversion computations in the following chapters. Notwithstanding, these base values were adjusted slightly to study their final effects on the statistical quality of the inversion models obtained.

## 8.4 Analysis and interpretation of 1D-vertically inhomogeneous seismic velocity models

Using the optimal input parameters from the initial data analysis, numerous isotropic as well as anisotropic SSH-inversions for 1D-seismic velocity models are performed. At first, the optimal value of the azimuth of the fast $P_n$-velocity axis of the anisotropy ellipse was identified by testing values in the range from 25° to 35° and assessing the minimal RMS. Differences in the direction of the fast axis are obtained, depending on whether the hypocenters are fixed or free (full SSH-inversion) in the SSH-inversions. Thus, the first category delivers a fast direction close to 35° and the second one at about 27°. As the last value is also more in line with the results of other seismic anisotropy studies across Germany (Fuchs, 1975; Enderle et al., 1996b; Song et al., 2001a), one must conclude that there is a strong effect of anisotropy onto the hypocentral locations, which only a full SSH-inversion can quantify.

In a next step, the effects of the average Moho depth, which is known to vary over several $km$ across Germany with a remarkable uprise to about 25 $km$ in the southern Rhinegraben (Mechie, 2005) and a deepening towards the Alps (Song et al., 2004), has been investigated. By varying the Moho depth in the SSH-inversions, the optimal average depth (minimizing the RMS) is found to be about 30 $km$ across Germany. While there is certainly a biasing traveltime effect of a sloping Moho onto both, the velocity structure and the hypocenters (cf. Koch and Kalata (1992)), this cannot be quantified further at this stage of the investigation, given to the present limitations of the SSH-program. However, there is some evidence that at least in the present study crustal thickness variations may not be too detrimental as the good azimuthal ray coverage across Germany may partly lead to a compensation of $P_n$-wave traveltime residuals which may arise from a sloping Moho.

To test the sensitivity of the model space, i.e. final velocity model and hypocenters onto different initial (starting) velocity models, low- and high-1D-velocity models are set up, with the limits for the layer velocities being chosen as 5.00, 5.50, 6.00 and 7.50 $km/s$ with a final RMS at 1.3429 $s$ for the low velocity model, and 6.00, 6.50, 7.00, 8.50 $km/s$ with a final RMS of 1.2989 $s$ for the high velocity model, after inversion with anisotropic $P_n$-traveltime correction. The computations for these extreme velocity variations result in the narrow range at 5.83, 5.93, 6.02 and 7.87 $km/s$ and 5.96, 6.12, 6.26 and 8.08 $km/s$ for the low- and high-velocity model, respectively. The two final velocity distributions show that the initial starting model has to be chosen carefully within a more narrow range, to get a reasonable result for the velocity structure, as it can be determined by other seismic investigations. On the other hand, the final RMS values show a clear superiority of the high-velocity starting model.

To further investigate the reasonableness of the somewhat artificial velocity discontinuities of the four-layer model, a finer discretized seven-layer input model, each of which with a thickness of 5 $km$, was tested. Both SSH-inversion variants, i.e. "hypocenters fixed" or "hypocenters free", exhibit a velocity jump to 7.97 $km/s$ at the crust-upper mantle boundary. Also, for both isotropic and anisotropic SSH-inversion cases with "hypocenters fixed", the velocity in the lower crustal layer six shows a petrologically unrealistic high value. In the second set of inversions with "hypocenters free", the velocity depth distribution is smoother for the crust, with basically no velocity discontinuity at a depth of 25 $km$, layer six. This means that there is no hint of a Conrad discontinuity in the region.

Based on the various SSH-inversions of Chapter 5, the final optimal 1D velocity models, isotropic as well as anisotropic ones, are fixed to those listed in Table 8.1, whereby each of them is an average of optimal 1D-velocity models obtained from several computations with different starting models. Compared to the earlier computed models SKKS2001, MKS2004 and MKS2005, the new model MKS2007 has a slightly increased velocity of 6.21 $km/s$ - compared to 6.15 $km/s$ in the old model - in layer three, whereas the other layer velocities are nearly the same. The difference between the two latter models is due to the introduction of the parameter $\Delta_{cut-off}$ (chosen in the present application to be at 100 $km$) that effectively eliminates wrongly assigned $P_n$-phases below that distance. Considering the standard deviations $\sigma$ for the distinct layer velocities, the MKS2007 velocity model is in the range of the other

Table 8.1: The final optimal 1D velocity model MKS2007 (full inversion, i.e. hypocenters free), with and without $P_n$-anisotropy included. The standard errors for the layer velocities are taken from Table 5.5.

| ani | V1 $km/s$ | V2 $km/s$ | V3 $km/s$ | V4 $km/s$ | RMS $s$ | TSS $s^2$ | $\sigma_{V1}$ $km/s$ | $\sigma_{V2}$ $km/s$ | $\sigma_{V3}$ $km/s$ | $\sigma_{V4}$ $km/s$ |
|---|---|---|---|---|---|---|---|---|---|---|
| no | 5.92 | 6.05 | 6.24 | 7.99 | 0.9743 | 20627 | 0.033 | 0.033 | 0.075 | 0.064 |
| yes | 5.91 | 6.05 | 6.21 | 8.02 | 0.9385 | 19140 | 0.014 | 0.010 | 0.066 | 0.048 |

models, except for the first layer. Moreover, as the values of $\sigma$ are somewhat lower for the anisotropic than for the isotropic inversion model, one must conclude that the anisotropic $P_n$-traveltime-correction is indeed indispensable for a reliable determination of an average crustal velocity model underneath Germany and, as will be seen later, for the relocations of the hypocenters as well.

A further corroboration of the advantage of the anisotropic velocity model comes from the fact that the isotropic inversion results in a relatively low $P_n$-velocity of $7.99\,km/s$ which is petrologically less reasonable than the corresponding average anisotropic $P_n$-velocity of $8.02\,km/s$. This lower isotropic $P_n$-velocity is compensated by a small increase of the velocity in layer three in the lower crust. This is another hint for the better reliability of the anisotropic corrected model, as there exists no remarkable discontinuity in the crust, as proven before by the seven-layer model.

Concerning the main solution criteria, TSS and RMS, which basically are the only objective measures to select the optimal inversion model, they successively decrease for the different 1D-model variants (isotropic, anisotropic, hypocenters free, fixed and various gaps, see Table 5.7). Different initial starting models (MKS2001 and MKS2004) lead to slightly different final velocity models with slightly different RMS values. However, as these changes are minor, it appears to be legitimate to choose the final optimal velocity model as an average of these variants. As for the RMS values obtained with the isotropic and the anisotropic inversion models, their improvements with respect to the RMS from the original HYPO-relocations as reported in the original dataset are about 30 % and 35 %, respectively.

Finally, the vertically inhomogeneous $V_P/V_S$-ratio of the crust and upper mantle has been investigated. To that avail, separate, full inversions (hypocenters free) for 1D-S-wave models have been performed. However, unlike the laterally inhomogeneous $V_P/V_S$-ratios to be discussed later, no significant deviations of the $V_P/V_S$-ratio from its standard value $\sqrt{3}$ has been found neither for the isotropic nor for the anisotropic inversion model.

## 8.5 Analysis and interpretation of 3D laterally heterogeneous seismic velocity models

The optimal 1D isotropic and anisotropic vertically inhomogeneous velocity models found above have been used as the starting model in the SSH-inversions for 3D laterally heterogeneous velocity models. During the course of this part of the study, the 3D models have been increasingly refined, starting with a lateral discretization into $15 \times 15$ blocs (lateral extension of about $40 \times 40\,km$ per bloc) and finally going up to $35 \times 35$ blocs, (lateral extension of about $16 \times 16\,km$). For each of the models, the inversion solutions for the isotropic, as well as the anisotropic case (i.e. incorporating the anisotropic velocity correction for the $P_n$-phases) are examined and the sensitivity of the solution to the data is estimated by means of various tests for resolution, covariance and other trade-off characteristics of the data- and the model-space (see Chapter 4).

## 8.5. ANALYSIS AND INTERPRETATION OF 3D VELOCITY MODELS

### 8.5.1 Synthetic tests

Before proceeding with the real 3D velocity inversions for the crustal and lithospheric seismic structure underneath Germany, some synthetic tests with randomly distributed anomalies and checkerboard tests are done to evaluate the reconstruction (resolution) capabilities of the traveltime data with regard to various model variants, used in the subsequent real 3D inversions. From these initial synthetic inversion tests the need for the elliptical anisotropic $P_n$-velocity correction in the real subsequent 3D tomographic study beneath Germany has also been recognized, as the anisotropic inversion test models are consistently better resolved than the isotropic ones.

The traveltime data for the synthetic tests are generated with an initially homogeneous 3D velocity model, but with some of the blocs perturbed by some velocity anomaly. These synthetic tests clearly show the power of the dataset to reconstruct the anomalies along with different levels of noise and are able to identify differences between isotropic and anisotropic inversions. Thus, isotropic inversions of, initially, anisotropically generated traveltimes deliver wrong reconstructions, whereby the original velocity anomalies are either split into several ones or shifted to different positions. Also the hypocenters are not well relocated, as their positions are shifted by some few $km's$ in the three coordinate directions. This holds especially for the hypocentral depths which are more biased for the isotropic than for the anisotropic inversion models.

The most salient results of these synthetic tests with respect to various issues can be summarized as follows:

- **Choice of the optimal damping parameter**
  One of the most controversial issue in geophysical inverse theory concerns the optimal choice of the damping (ridge) parameter $\lambda$ which is normally necessary to stabilize the inversion process and to "scan" the solution range for the optimal inversion model. The procedure used here to select the optimal model is that of minimal RMS of the nonlinear misfit function of the traveltime residuals. But comparing the reconstructed model of the smallest $\lambda$ with that of the minimal RMS, the situation is not always clear, as the reconstruction of the anomalies appears to be best for the lowest $\lambda$, whereas the minimal RMS would indicate a much higher $\lambda$, as found also from the "trade-off" analysis, discussed below. As $\lambda$ is increased, the resolution of the synthetic anomalies decreases until some of them disappear. These, somewhat confusing results suggest that, in some nonlinear tomographic inversion cases, the concept of minimal RMS (the objective criterion in the data space) to select the optimal $\lambda_{opt}$ does not hold and must be complemented by a subjective criterion (information) in the model space. Also the behavior of the velocity solutions for the fourth (upper mantle) layer is quiet different as strong undulations occur here for small $\lambda$ which fade away when $\lambda$ is increased to its "optimal" value for minimal RMS. This suggests that for the most reliable inversion model the damping values $\lambda$ may have to be chosen separately for each layer or even for each bloc which means, technically, that a non-unit matrix must be added onto the diagonals of the normal equation system to damp the solution parameters individually.

- **Trade-offs**
  Various trade-off curves to select the optimal damping parameter, i.e. the optimal models, have been investigated, whereby the Tikhonov regularization (TR-method) turns out to be the most powerful. In the TR-method trade-off curves between the norm of the data fit and the norm of the model parameters (velocities and hypocenters) are drawn as a function of the damping (regularization) parameter $\lambda$ and one attempts to find a good compromise - usually defined by the knee-point of the curve - between the two. The results gathered here for the optimal $\lambda_{opt}$ are well within the range of the optimal $\lambda_{opt}$ found in the previous tests discussed above. For the $15 \times 15$ bloc synthetic test, in particular, the $\lambda$ at the knee-point of the trade-off curve defines also the mimimal norm solution of the nonlinear objective function, i.e. the minimal RMS.

The superiority of the anisotropic over the isotropic inversion of the synthetic (anisotropic) model data is also indicated by the difference in the optimal ridge parameter $\lambda_{opt}$ found in the two cases. Thus, $\lambda_{opt}$ is not only lower for the anisotropic than for the isotropic inversion, i.e. the anisotropic solution needs to be stabilized less than the isotropic one, but the RMS as well, i.e. the data-fit is better for the former than for the latter. Moreover, as both the velocity structure and hypocenter relocations obtained by the isotropic inversion are completely false, a trade-off analysis in this case appears to be a senseless undertaking.

- **Checkerboard tests**
  The results of the checkerboard tests where synthetically generated, anisotropically corrected arrivaltimes for a checkerboard-like structure are inverted isotropically as well as anisotropically, also show the necessity of the anisotropic correction for $P_n$-phases. Whereas for all three bloc-discretizations analyzed (15 × 15, 25 × 25 and 35 × 35 blocs in each layer), the anisotropic inversions deliver fairly good results, the results for the isotropic inversion fail just in the case with 15 × 15 blocs and get worse for the higher bloc discretizations. This again shows the necessity of the anisotropic correction for $P_n$-phases to properly resolve the 3D seismic structure across the region.

  The influence of the checkerboard velocity solution onto the hypocentral relocations, i.e. their shifts relative to their original positions, has also been investigated in detail. For all three model discretizations, the average hypocentral shifts, i.e. the hypocentral mislocations, increase with increasing statistical noise $\sigma$ added to the theoretical arrivaltimes. Thus with $\sigma = 0.3\,s$, the average hypocentral shifts are $1.5\,km$ for the x-, $0.3\,km$ for y- and $5\,km$ for the z-component. The checkerboard tests also show that, surprisingly, the velocity anomalies are somewhat better resolved for SSH-inversions with the hypocenters fixed, as when the hypocenters are allowed to move as in the full inversion. Thus, there is indeed some feedback effect of the hypocentral locations onto the seismic structure.

- **Resolution**
  As the figures in Section 6.4 show, the overall geometrical resolution of the various tomographic 3D models discussed is quiet good, even for the very finely discretized 35 × 35 bloc model, although its resolution becomes a bit patchy in some parts of the model area. In agreement with the theory of resolution embedded in linear inverse theory (see Chapter 2), the geometrical resolution of the velocity solution depends on the choice of the damping parameter $\lambda$. Thus, for small $\lambda$ the resolution is nearly perfect for most of the model area, with values for the diagonal elements of the resolution matrix $R$ being close to one. The figures show, expectedly, that although the resolution values decrease as $\lambda$ is increased to stabilize the inverse solution, this deterioration is only gradual and, more importantly, acts more or less homogeneously over most of the model volume, particularly for the blocs of the upper crustal layers which are well covered by seismic rays.

## 8.5.2 Structural interpretation of the 3D seismic models

Once the theoretical capabilities of the present regional data-set to infer the laterally heterogeneous seismic velocity structure in the crust and upper mantle beneath Germany through SSH-inversion have been shown (whereby the importance of the anisotropic $P_n$-phase correction became clear), "real" 3D P-wave, S-wave as well as 3D -$V_P/V_S$- seismic tomography models for the region have been determined.

- **3D P-wave models**
  The relative comparison of the three 3D P-wave velocity models of increasing bloc-resolutions (15 × 15, 25 × 25 and 35 × 35 blocs in each of the four layers) as well as with the optimal 1D velocity model above shows (1) an increasing fit of the computed to the observed arrivaltimes with increasing complexity of the model - with an improvement of about 30% for the isotropic

## 8.5. ANALYSIS AND INTERPRETATION OF 3D VELOCITY MODELS

3D-model over the isotropic 1D model - and (2) a consistent further improvement of the data fit between 4 to 9% for all anisotropic, $P_n$-velocity corrected models relative to the corresponding isotropic (no anisotropic correction) ones.

Among the most salient features of the different 3D P-wave models ($15 \times 15$, $25 \times 25$ and $35 \times 35$ bloc discretization) which can be related to existing tectonic and geological features across the study region, the extended Molasse area, north of the Alps, the Vogelsberg volcano and the upper Rhinegraben are noteworthy. Also an anomalous structure near the Nördlinger Ries can be detected and the structure in the east mostly follows the mountain there. The influence of the anisotropic $P_n$-phase correction on the reconstruction of the seismic structure can be recognized by comparing the isotropic and anisotropic SSH-inversion variants, particularly, for the full SSH-inversion ("hypocenters free"). Expectedly, the most prominent differences occur in the upper mantle layer four where the anisotropic $P_n$-correction takes place. Thus, the fast (blue) area extending from the southwest to the northeast section of layer four is somewhat broader and of slightly different shape for the isotropic (Figure 6.25(c)) than for the anisotropic inversion case (Figure 6.25(b)). Also the anomalous structure in layer four between 50° and 51° Latitude and 8° and 10° Longitude, indicated by the blue area (high velocity), is more dominant and another one occurs representing the volcanic intrusions west of the Vogelsberg. For the isotropic inversion case only a red area (low velocity) shows up with opposite values for the velocity anomalies. Some of the seismic anomalies inferred here appear to be related to the known and expected local geological features as illustrated, for example, in the tectonic map of Germany of Berthelsen (1992).

- **3D S-wave models**
  For the pure S-wave SSH-inversions only $15 \times 15$ $V_S$-bloc models have been computed, as the limited number of usable $S$-phases does not allow finer lateral resolutions of the model, particularly for the upper mantle layer four which is only sparsely covered by $S_n$-rays. This has as consequence that there may only be a rather small influence of a possibly existing $S$-wave anisotropy, so that isotropic and anisotropic $S$-wave inversions basically lead to the same 3D S-wave model structure for the crust (see Figure 6.6.1.5). Furthermore, the existence of $S$-wave anisotropy is difficult to investigate due to the fact that mixing of $S_V$- and $S_H$-components may occur which cannot be investigated here. Nevertheless, the overall distribution of the $V_S$-model is different from that of the corresponding $15 \times 15$ $V_P$-model, which hints of laterally varying $V_P/V_S$-ratios, as discussed in the following paragraph.

- **3D-$V_P/V_S$-ratios**
  From the $15 \times 15$ $V_P$- and $V_S$-models, laterally varying $V_P/V_S$-ratios for the crust and upper mantle underneath Germany have been calculated. Overall, the anisotropically (for $P_n$ and $S_n$) corrected $V_P/V_S$-model shows nearly the same $V_P/V_S$-ratio structure for the anisotropic and isotropic case in the first layer. Starting at layer two, the isotropically and anisotropically computed $V_P/V_S$-ratios clearly indicate major differences. The most remarkable result shows up in the upper mantle layer four where in the model area covering northern Switzerland the $V_P/V_S$-ratio's range between 1.67 to 1.72. This appears to be more in line with the results of Lombardi et al. (2008) who get ratios of about 1.72 under the Black-Forest and of about 1.74 along the Danube river from a receiver-function analysis of teleseismic phases.

It should be mentioned that a this stage of the thesis work the results of the $S$-wave tomography above, as well as those of the $V_P/V_S$-ratio's can only be considered as preliminary. The $S$-wave inversion approach used here assumes that the hypocenters are fixed to the originally reported locations. Thus, both the $P$-wave and the $S$-wave inversions have been performed with "hypocenters fixed". This simplified procedure should be replaced by a direct inversion for the $V_P/V_S$-ratio which will insure a more precise calculation of the latter, thus avoiding the uncertainties in the seismic structures inverted when the hypocenters are fixed. However, this is not possible yet due to the poor amount of usable

$S_n$-phases.

Nevertheless, the results obtained in here for the 3D $P$- and $S$-wave models as well as for the $V_P/V_S$-ratio's clearly show the necessity of applying the elliptical anisotropic correction for the $P_n$-phases and, to a lesser extent, the $S_n$-phases[2]. With these anisotropic $P_n$- and $S_n$-corrections, the seismic features obtained here for the crust and upper mantle underneath Germany appear to be better interpretable in terms of the natural tectonic features, as well as providing some interesting petrological constraints on the lithospheric composition.

## 8.6 Effects of the seismic structure on the hypocentral relocations

A main objective of the present thesis has been the understanding of the influence of the SSH-inverted 1D- or 3D-seismic structure on the relocations of the original earthquakes and, in particular, the effects of the - now well-accepted - upper mantle $P_n$-anisotropy underneath Germany. Since the seismic events used here have been located by standard earthquake location procedures with a 1D vertically inhomogeneous velocity model for the crust and upper mantle for the study region, the outcome of this relocation analysis should assist seismologists with the evaluation of the reliability of their hypocenter determinations.

- **Hypocenters relocated with the 1D velocity models**
  The analysis of the optimal 1D velocity models show that the differences (shifts) between the original and relocated hypocenters are usually less for the anisotropic than for the isotropic model, except for the models where a large gap of 230° has been specified. This is because an earthquake observed with such a large gap of the ray coverage, it is usually poorly located, due to the azimuthal unbalance of the arrivaltimes which means that anisotropic distortions may become particularly strong. Relocating such an event with an isotropic model, i.e. neglecting the $P_n$-anisotropy, will lead to a stronger mislocation than if the existing anisotropy is considered. The superiority of the anisotropic model for large gaps is also indicated by the fact that its datafit is about 5% better than that of the isotropic model for cases "hypocenters free" and about 3.5% for cases "hypocenters fixed". For smaller gaps, i.e. 180° and 120°, the datafit undergoes a further increase of up to 11%, respective 9.5%. Over the whole gap range, the inversions with anisotropically corrected $P_n$-phase consistently result in a better datafit of 3.3% to 6.4% when compared with the corresponding isotropic case.

  Comparing the pure (a posteriori) determined hypocenter relocations with those obtained from the simultaneous inversion with the optimal 1D velocity models, the results with respect to both the datafit and the hypocentral shifts (differences of only about 1 $km$) are nearly the same. This interesting effect holds for both isotropic and anisotropic models and may explain the usability of the optimal 1D model for a pure hypocenter relocation, thus avoiding the computationally intensive full (coupled) SSH inversion for both velocity structure and hypocenters.

- **Hypocenters relocated with former 1D velocity models**
  The relocations of the hypocenters obtained with the final optimal 1D-velocity model have also been compared with those of two former 1D velocity models, MKS-2004 and SKKS-2001, that have been in use over the years of the present study. These two former models provide slightly different data-fits and the overall hypocentral relocations differences compared to the optimal 1D model above are about 1.5 $km's$. On the other hand, the hypocentral differences between the isotropic and anisotropic variants are only minor, though always less for the anisotropic case.

- **Hypocenters relocated with the 3D velocity models**
  The inter-comparison of the hypocentral relocation with the three 3D velocity models of increasing bloc-resolutions and with the optimal 1D velocity models shows in all four cases an

---
[2] As discussed in the data chapter, at this stage of the investigation, there is no clear evidence from the present data for strong seismic upper mantle $S_n$-anisotropy underneath Germany.

increasing fit of the observed arrivaltimes when the anisotropic $P_n$-velocity correction is applied. Expectedly, the largest improvement occurs when going from the 1D- to the 3D-models where, in the case of the isotropic inversion, the former is then about 30%. The anisotropically inverted 3D-models lead to an additional improvement of 2 to 5%, relative to the related isotropic case.

The epicentral (radial) shifts are about $2\,km$, whereas the depth-shifts are up to $\pm 5\,km$, depending on the depth of the original event. Overall are these shifts somewhat smaller and appear to be more consistent - which hints of a better stability of the solution - for the anisotropic than for the isotropic inversion cases, particularly, for the coarser block models. The differences between the average hypocentral shifts resulting from the isotropic and anisotropic SSH-inversions have also been calculated. With finer bloc-discretization, these become smaller and are nearly the same for the 35 bloc model. This hints of a fundamental phenomenon of seismic tomography, that a sufficiently finely discretized lateral heterogeneous isotropic velocity model is also able to fit anisotropic traveltimes to a satisfactory degree (W. Zürn, private communication). These results indicate, on the other hand, that the anisotropic $P_n$-velocity correction is rather effective for coarse 3D-models, but most effective for the 1D-models.

- **Sensitivity tests of the hypocenter relocations**
  Various sensitivity tests with synthetic traveltime data are performed to analyze the reliability of the hypocentral relocations when (1) the anisotropy is not considered, (2) the initial hypocenter positions are varied, and (3) errors in the arrivaltime measurements are included.

The various sensitivity tests with synthetic anisotropically generated data, indicate clearly the influence and the importance of the inclusion of the $P_n$-anisotropy for reliable hypocentral relocations. Thus, whereas the isotropic inversion of the anisotropic traveltime data still hints of systematic hypocentral mislocations (with values of $4.7\,km$, $1.1\,km$ and $2\,km$ in the three coordinate directions), for the anisotropic inversion the original hypocenters are better relocated (with the named misfit values reduced by a factor of two) and closer to the original ones. Also the datafit, as measured by the RMS, is reduced significantly (by more than 40 %) for the anisotropic relative to the isotropic inversion case.

Using input arrivaltime data with random errors to mimic observation errors, the uncertainty range of the hypocentral relocations has been assessed further. Thus, although the hypocentral uncertainty range as well as the RMS increases somewhat with the size of the *a priori* imposed arrivaltime errors, the mislocations are not more than $1.5\,km$, $1\,km$ and $2.5\,km$ for the three co-ordinate directions. Similar to the the hypocentral relocations of the optimal 1D-velocity model above, the hypocentral shifts obtained with two noisy models ($\sigma = 0.2\,s$ and $\sigma = 0.3\,s$) are less for the anisotropic than for the isotropic inversion cases.

The sensitivity tests with randomly disturbed initial hypocenters show again the importance of the anisotropic $P_n$-velocity correction, but also the overall stability of the hypocenter determination process. Thus, even by initially shifting the original hypocenters by $\Delta X$, $\Delta Y = 10\,km$ and $\Delta Z = 2\,km$, the relocated hypocenters are shifted back toward their original positions with a relocation misfit of only a few $km$ (see 7.6) for the anisotropic inversion model.

These results clearly indicate that for future regional earthquake relocations across Germany the use of a 3D crustal und upper mantle seismic velocity model that takes the upper mantle $P_n$-anisotropy into account is highly recommended. In fact, in the Appendix this recommendation has been further elaborated and is supported by the results of the analysis of the relocation of the Waldkirch event of Dec. 5, 2004, using the methods and findings discussed throughout the present thesis.

## 8.7 Concluding remarks

A major task of this thesis has been the introduction of an elliptical velocity correction for the $P_n$-phases to explain the azimuthal variations of up to $\pm 5\,s$ in the arrivaltime residuals which occur when an isotropic standard velocity model is employed. The azimuthal direction of the fast anisotropy axis found here through numerous trial-and error tests has been set at about 27° which is in good agreement with other previous seismic studies. With this anisotropic $P_n$-phase correction, the observed traveltimes can be correlated with physical, mineralogical and tectonic properties of the crust and upper mantle beneath Germany. To further test the reliability of the correction, numerous SSH-inversions were done with different starting models, different input selection parameters and various other parameters that control the proper functioning of the inversion code.

The different SSH-computations can be separated into two major families, one with a full simultaneous inversion for seismic structure and hypocenters, and another one where the hypocenters are fixed to their original position, as reported in the original dataset and determined by standard location methods with a 1D vertically inhomogeneous velocity model. Both sets of SSH-inversions are done with and without elliptical anisotropic correction for the $P_n$-phases so that, overall, four variants of inversion models are being investigated throughout the thesis.

The SSH inversions with a 1D vertically stratified velocity model, with "hypocenters free" (full inversion) and anisotropic $P_n$-phase correction included, result in optimal velocity models that vary more gradually with depth than those of the SSH-inversions where this correction has been omitted (isotropic cases). In fact, for the isotropic inversions a velocity discontinuity between layer two and three arises, instead, which could be interpreted as a Conrad discontinuity. However, since the latter has not been observed in the numerous other regional seismic studies (mostly refraction seismics), it must be considered an artifact of the wrong model approach here, i.e. isotropic upper mantle. Also, the $P_n$-velocity in the upper mantle layer is also somewhat higher for the anisotropic than for the isotropic inversion and more in line with those of the other seismic studies mentioned.

The results of the SSH inversions with a 3D laterally heterogeneous velocity model depend slightly on which of the four different SSH-inversion variants (see above) are used. Whereas the inversions with "hypocenters fixed" and no anisotropy-correction deliver a more consistent picture for the seismic crustal and upper mantle structure that appear to be more in line with the known existing tectonic and geological features underneath Germany, the one with the anisotropic correction show major discrepancies. This gives reason to believe that the original hypocenters have inconsistent locations resulting out of the use of an isotropic velocity model. Correcting the $P_n$-phases for anisotropy will give a better data-fit, but on the other hand parts of the traveltime residuals will be projected into the seismic structure. For both inversion cases, the much of traveltime residuals are projected into lower crustal layer three which is insufficiently reconstructed due to a lack of seismic events in this part of the crust, resulting in anomalous high velocity variations in both cases.

Turning to the family of SSH-inversion with "hypocenters free", the isotropic inversions cases will result in both a wrong hypocentral relocation and a wrong seismic structure. This bias will be significantly corrected once the anisotropic $P_n$-phase correction has been applied where then the seismic structure is more in line (better than the isotropic case with "hypocenters fixed" above) with the tectonics and geology of the study region. Also, the seismic structure in layer three becomes smoother and more reliable as the $P_n$-phases which alone probe this layer are not isotropically biased anymore.

The final conclusion for the 3D laterally heterogeneous models is that, with the anisotropic $P_n$-phase correction included, the seismic structure inferred as well as the relocations of the seismic events are indeed better, as measured by the only objective criterion, the data fit, quantified by the RMS. In fact, the minimum RMS of the anisotropic SSH-inversions is about 20% lower than that of the isotropic inversions.

## 8.8 Further developments and outlook

The following items, which turned out to be questionable during the course of this thesis, should be considered and investigated further in future studies of this kind:

## 8.8. FURTHER DEVELOPMENTS AND OUTLOOK

- The stabilization of the hypocentral relocations by fixing the origin-times of the events as computed by means of external Wadati-Diagrams, where possible, i.e. where more than eight P- and S-phase observations are available.

- Extending the seismic record data base in order to get a better ray coverage in the northeast and east of Germany where, at the moment, the ray-density is much less than in the southwest of the study region.

- The Moho is assumed to be a plain with an average depth of about $30\,km$ which, of course, is only correct as a first approximation. To include the real Moho undulations, some major changes have to be done in the SSH-program. Eventually, one may wish to directly invert for the lateral Moho depth variations themselves.

- The anisotropic elliptical $P_n$-phase correction is done across the whole study area, i.e. only one large anisotropy ellipse is assumed. However, the 2D time-term regional traveltime analysis of Song et al. (2004), the refraction seismic analysis of Enderle (1998) and that of Kummerow (2004) hint of some rotation of the anisotropy axis across Germany. Thus possible local variations of upper mantle anisotropy ellipse should also be considered in future SSH-studies.

- The possibility of crustal seismic anisotropy which might influence the traveltimes of $P_g$-phases, as found by Vavryčuka et al. (2004) for the Moldanubian in the Czech Republic, has not been considered. Although, so far no visible signs of anisotropy have been observed in the residuals, a more thorough analysis is required. Since crustal anisotropy may have different origins than upper mantle anisotropy, namely, it is most likely caused by partly fluid-filled aligned fractures in the upper crust and laminations in the lower crust (Meissner et al., 2006), although lattice-preferred orientation (LPO) cannot be ruled here (Rabbel et al., 1998), such an extended study will be very elucidative.

- Although the traveltime analysis of the $S$-phases does yet give a clue for azimuthal variations like, unlike the $P$-phases, such a possibility may not yet been ruled out. Anisotropy strongly influences $S$-waves as these are split into two orthogonally polarized waves which travel with different speeds and directions. Although this so-called shear wave splitting has been observed for teleseismic traveltimes crossing steeply the upper mantle underneath Germany, the $S$-phases recorded in the present dataset do not provide information about their polarization, i.e. it is not known to which of the two branches of $S$-waves they belong to.

# Appendix A

# Computational details of the sensitivity studies for the 1D velocity models

## A.1 Results of different 1D velocity model computations

In the following, the results of the different SSH-computations to test the influence of the Moho depth, are listed in detail, where in Chapter 5, only the optimal results are shown. Tables A.1 and A.2 contain the results for the computations without $\Delta_{cut-off}$, separated for two different periods (1975 to 2003 and 1997 to 2003) of registrations.

In Tables A.3 and A.4, the computation is done again for these two different time periods, but now with $\Delta_{cut-off} = 100\,km$.

To test the influence of the thickness of layer three, the computations are done again for the two different time periods, with the results now in Tables A.5 and A.6. This, in fact has a major influence onto the velocity in layer three, which has stronger variations than in the case where the third layer is kept fix.

In Table A.7, the results for different residual limits and two different gap criteria (180° and 120°) and both, the isotropic and anisotropic case are listed, mainly showing the influence of the RES criterion.

Table A.1: Variations of the $V_P$-velocity as a result of varying the Moho depth between 27 and 33 $km$ using the recordings between 1975 - 2003, V-mod-in is SKKS2001 and no reassignment of $P_n$ by $P_g$-phases below a certain $\Delta_{cut-off}$ value.

| Layer | L1 | L2 | L3 | L4 | |
|---|---|---|---|---|---|
| Depth | 0 - 10 $km$ | 10 - 20 $km$ | 20 - 30 $km$ | ≥30 $km$ | |
| Moho x | V1 | V2 | V3 | V4 | RMS |
| $km$ | $km/s$ | $km/s$ | $km/s$ | $km/s$ | $s$ |
| 27 | 5.93 | 6.07 | 5.91 | 7.94 | 1.2866 |
| 28 | 5.92 | 6.06 | 6.05 | 7.96 | 1.2904 |
| 29 | 5.92 | 6.07 | 6.16 | 7.99 | 1.3033 |
| 30 | 5.92 | 6.07 | 6.35 | 8.00 | 1.3132 |
| 31 | 5.92 | 6.06 | 6.33 | 8.00 | 1.3053 |
| 32 | 5.92 | 6.07 | 6.40 | 8.01 | 1.3200 |
| 33 | 5.91 | 6.07 | 6.42 | 8.06 | 1.3279 |

Table A.2: As the previous table, 1997-2003 dataset. The computation for $31\,km$ failed. The optimal depth for the Moho seems to be at $27\,km$.

| Layer<br>Depth | L1<br>0 - 10 km | L2<br>10 - 20 km | L3<br>20 - 30 km | L4<br>$\geq$30 km | |
|---|---|---|---|---|---|
| Moho x<br>km | V1<br>km/s | V2<br>km/s | V3<br>km/s | V4<br>km/s | RMS<br>s |
| 27 | 5.90 | 6.07 | 6.05 | 7.94 | 1.3269 |
| 28 | 5.89 | 6.07 | 6.21 | 7.95 | 1.3362 |
| 29 | 5.89 | 6.07 | 6.25 | 7.97 | 1.3444 |
| 30 | 5.89 | 6.06 | 6.38 | 7.97 | 1.3482 |
| 31 | - | - | - | - | - |
| 32 | 5.89 | 6.07 | 6.29 | 7.99 | 1.3481 |
| 33 | 5.89 | 6.06 | 6.35 | 8.00 | 1.3502 |

Table A.3: Similar to the previous table, but for the 1975-2003 data set, and with $\Delta_{cut-off}$ to reassign $P_n$- to $P_g$-phases below $= 100\,km$.

| Layer<br>Depth | L1<br>0-10 km | L2<br>10-20 km | L3<br>20-30 km | L4<br>$\geq$30 km | |
|---|---|---|---|---|---|
| Moho x<br>km | V1<br>km/s | V2<br>km/s | V3<br>km/s | V4<br>km/s | RMS<br>s |
| 26 | 5.90 | 6.05 | 5.84 | 7.90 | 1.3018 |
| 27 | 5.91 | 6.05 | 5.91 | 7.92 | 1.2889 |
| 28 | 5.91 | 6.05 | 6.03 | 7.94 | 1.2903 |
| 29 | 5.90 | 6.04 | 6.09 | 7.95 | 1.2876 |
| 30 | 5.89 | 6.04 | 6.22 | 7.97 | 1.3026 |
| 31 | 5.89 | 6.03 | 6.31 | 7.98 | 1.2973 |
| 32 | 5.89 | 6.04 | 6.27 | 7.99 | 1.2980 |
| 33 | 5.89 | 6.03 | 6.29 | 8.01 | 1.2978 |

Table A.4: Similar to the previous table, but for the 1997-2003 dataset.

| Layer<br>Depth | L1<br>0-10 km | L2<br>10-20 km | L3<br>20-30 km | L4<br>$\geq$30 km | |
|---|---|---|---|---|---|
| Moho x<br>km | V1<br>km/s | V2<br>km/s | V3<br>km/s | V4<br>km/s | RMS<br>s |
| 27 | 5.92 | 6.07 | 6.03 | 7.94 | 1.3216 |
| 28 | 5.92 | 6.06 | 6.18 | 7.95 | 1.3207 |
| 29 | 5.91 | 6.05 | 6.14 | 7.97 | 1.3168 |
| 30 | 5.91 | 6.04 | 6.24 | 7.98 | 1.3181 |
| 31 | 5.91 | 6.05 | 6.12 | 8.00 | 1.3223 |
| 32 | 5.91 | 6.02 | 6.22 | 8.00 | 1.3216 |
| 33 | 5.91 | 6.01 | 6.10 | 8.01 | 1.3345 |

## A.1. RESULTS OF DIFFERENT 1D VELOCITY MODEL COMPUTATIONS

Table A.5: Effect of changing the L3 and L4 boundaries in parallel, such that the thickness of L3 stays at 10 km. 1975 - 2003 dataset, $\Delta_{cut-off} = 100\,km$. The optimal Moho depth is again at $29\,km$. x is the depth of the Moho.

| Layer<br>Depth | L1<br>0 - 10 km | L2<br>10 - 20 km | L3<br>20 - x km | L4<br>x km | |
|---|---|---|---|---|---|
| Moho x<br>km | V1<br>km/s | V2<br>km/s | V3<br>km/s | V4<br>km/s | RMS<br>s |
| 27 | 5.91 | 6.02 | 6.03 | 7.92 | 1.2902 |
| 28 | 5.91 | 6.03 | 6.04 | 7.93 | 1.2852 |
| 29 | 5.91 | 6.03 | 6.10 | 7.96 | 1.2838 |
| 30 | 5.89 | 6.04 | 6.21 | 7.97 | 1.3053 |
| 31 | 5.90 | 6.04 | 6.26 | 7.98 | 1.2955 |
| 32 | 5.89 | 6.05 | 6.31 | 7.99 | 1.3015 |
| 33 | 5.89 | 6.05 | 6.41 | 8.00 | 1.3083 |

Table A.6: Similar to the previous table, but for the 1997 - 2003 dataset. x is the depth of the Moho.

| Layer<br>Depth | L1<br>0 - 10 km | L2<br>10 - 20 km | L3<br>x - 20 km | L4<br>x km | |
|---|---|---|---|---|---|
| Moho x<br>km | V1<br>km/s | V2<br>km/s | V3<br>km/s | V4<br>km/s | RMS<br>s |
| 27 | 5.93 | 6.03 | 6.02 | 7.94 | 1.2897 |
| 28 | 5.93 | 6.04 | 6.05 | 7.96 | 1.3086 |
| 29 | 5.93 | 6.05 | 6.05 | 7.98 | 1.2824 |
| 29.5 | 5.93 | 6.05 | 6.10 | 7.99 | 1.2846 |
| 30 | 5.92 | 6.05 | 6.16 | 7.99 | 1.2902 |
| 31 | 5.92 | 6.05 | 6.07 | 8.00 | 1.2996 |
| 32 | 5.92 | 6.07 | 6.62 | 8.02 | 1.3150 |
| 33 | 5.91 | 6.07 | 6.48 | 8.03 | 1.3048 |

Table A.7: Statistical results for computations with different residual limits RES and for two different values of the gap, GAP = 180° and GAP = 120°, with and without elliptical anisotropic correction. The value of RES to start or to reject the anisotropic information is marked in red.

| RES $s$ | I | K | TSS $s^2$ | RMS $s$ | RMS$_{av}$ $s$ | TSS$_{hypo}$ $s^2$ | RMS$_{hypo}$ $s$ | V1 $km/s$ | V2 $km/s$ | V3 $km/s$ | V4 $km/s$ |
|---|---|---|---|---|---|---|---|---|---|---|---|
| input MKS2004 | | | 33831.5 | 1.8690 | 1.6134 | 2984.2 | 0.4792 | 5.90 | 6.10 | 6.10 | 8.10 |
| GAP = 180°, without anisotropy | | | | | | | | | | | |
| 01 | 2 | 2 | 24859.1 | 1.6021 | 1.3830 | 2984.2 | 0.4792 | 5.90 | 5.94 | 6.24 | 8.13 |
| 02 | 4 | 6 | 20079.4 | 1.1806 | 1.0379 | 12069.7 | 0.8047 | 5.92 | 6.02 | 6.31 | 8.01 |
| 03 | 3 | 9 | 19183.3 | 1.0929 | 0.9637 | 19725.3 | 0.9772 | 5.91 | 6.06 | 6.34 | 8.03 |
| 04 | 3 | 8 | 19576.9 | 1.0833 | 0.9553 | 25418.7 | 1.0886 | 5.92 | 6.05 | 6.27 | 8.00 |
| 05 | 4 | 5 | 20150.8 | 1.0918 | 0.9632 | 28533.0 | 1.1461 | 5.92 | 6.05 | 6.26 | 8.00 |
| 10 | 3 | 8 | 24203.5 | 1.1852 | 1.0460 | 37537.9 | 1.3027 | 5.92 | 6.07 | 6.27 | 8.00 |
| 15 | 3 | 8 | 25050.6 | 1.2055 | 1.0639 | 38970.1 | 1.3270 | 5.92 | 6.07 | 6.27 | 7.99 |
| 20 | 3 | 8 | 26251.8 | 1.2337 | 1.0888 | 40455.5 | 1.3517 | 5.92 | 6.07 | 6.27 | 7.99 |
| GAP = 180°, with anisotropy | | | | | | | | | | | |
| 01 | 3 | 2 | 25995.2 | 1.6383 | 1.4142 | 2984.2 | 0.4792 | 5.90 | 5.93 | 6.17 | 8.14 |
| 02 | 4 | 7 | 19107.9 | 1.1516 | 1.0125 | 12069.7 | 0.8047 | 5.92 | 6.02 | 6.28 | 8.04 |
| 03 | 3 | 6 | 17013.2 | 1.0292 | 0.9075 | 19725.3 | 0.9772 | 5.92 | 6.04 | 6.20 | 8.03 |
| 04 | 2 | 8 | 18807.8 | 1.0618 | 0.9364 | 25418.7 | 1.0886 | 5.92 | 6.06 | 6.27 | 8.03 |
| 05 | 4 | 7 | 18193.4 | 1.0479 | 0.9244 | 28545.5 | 1.1461 | 5.92 | 6.06 | 6.26 | 8.03 |
| 10 | 4 | 9 | 22141.9 | 1.1336 | 1.0005 | 37537.9 | 1.3027 | 5.92 | 6.05 | 6.14 | 8.01 |
| 15 | 4 | 7 | 22847.7 | 1.1512 | 1.0161 | 38970.1 | 1.3270 | 5.92 | 6.05 | 6.15 | 8.01 |
| 20 | 4 | 6 | 24086.1 | 1.1818 | 1.0430 | 40455.5 | 1.3517 | 5.92 | 6.05 | 6.19 | 8.01 |
| 25 | 4 | 6 | 25444.3 | 1.2142 | 1.0716 | 42456.8 | 1.3842 | 5.92 | 6.05 | 6.19 | 8.01 |
| GAP = 120°, without anisotropy | | | | | | | | | | | |
| 01 | 4 | 8 | 14268.3 | 1.6402 | 1.4293 | 1606.8 | 0.1381 | 5.91 | 5.95 | 6.22 | 8.13 |
| 02 | 4 | 9 | 12303.5 | 1.2344 | 1.0939 | 6666.6 | 0.5939 | 5.93 | 6.03 | 6.36 | 8.04 |
| 03 | 2 | 6 | 11953.7 | 1.1543 | 1.0262 | 10404.0 | 0.7551 | 5.93 | 6.06 | 6.38 | 8.03 |
| 04 | 3 | 7 | 12064.1 | 1.1433 | 1.0160 | 12756.2 | 0.8857 | 5.93 | 6.05 | 6.30 | 8.03 |
| 05 | 2 | 6 | 11359.4 | 1.1035 | 0.9810 | 14264.2 | 1.0042 | 5.95 | 6.06 | 6.32 | 8.02 |
| 10 | 2 | 7 | 14207.1 | 1.2230 | 1.0870 | 18967.1 | 1.1951 | 5.92 | 6.08 | 6.15 | 8.03 |
| 15 | 4 | 5 | 13782.8 | 1.2040 | 1.0701 | 19673.5 | 1.2198 | 5.94 | 6.05 | 6.07 | 7.97 |
| 20 | 4 | 6 | 14289.8 | 1.2256 | 1.0892 | 20600.9 | 1.2497 | 5.94 | 6.06 | 6.07 | 7.98 |
| 25 | 4 | 6 | 14928.1 | 1.2525 | 1.1131 | 21450.1 | 1.2775 | 5.94 | 6.06 | 6.07 | 7.98 |
| GAP = 120°, with anisotropy | | | | | | | | | | | |
| 01 | 4 | 4 | 14220.5 | 1.6374 | 1.4269 | 1606.8 | 0.4795 | 5.91 | 5.96 | 6.32 | 8.11 |
| 02 | 4 | 9 | 11838.7 | 1.2109 | 1.0730 | 6666.6 | 0.8052 | 5.93 | 6.04 | 6.49 | 8.04 |
| 03 | 4 | 5 | 11563.5 | 1.1353 | 1.0093 | 10404.0 | 0.9573 | 5.93 | 6.06 | 6.30 | 8.05 |
| 04 | 4 | 9 | 11332.0 | 1.1080 | 0.9847 | 12756.2 | 1.0448 | 5.93 | 6.06 | 6.27 | 8.05 |
| 05 | 4 | 8 | 10330.6 | 1.0524 | 0.9355 | 14264.2 | 1.0993 | 5.93 | 6.06 | 6.27 | 8.04 |
| 10 | 2 | 8 | 13212.0 | 1.1794 | 1.0483 | 18967.1 | 1.2560 | 5.94 | 6.06 | 6.05 | 7.99 |
| 15 | 4 | 7 | 12708.3 | 1.1561 | 1.0276 | 19673.5 | 1.2785 | 5.94 | 6.06 | 6.12 | 8.00 |
| 20 | 3 | 4 | 13485.8 | 1.1906 | 1.0581 | 20600.9 | 1.3077 | 5.95 | 6.06 | 6.06 | 7.98 |
| 25 | 4 | 7 | 14099.7 | 1.2172 | 1.0818 | 21450.1 | 1.3343 | 5.94 | 6.06 | 6.13 | 8.00 |

## A.2. TRAVELTIME VARIATIONS FOR ANISOTROPY OVER DISTANCES

Table A.8: Differences between isotropic and anisotropic traveltimes for ±5% velocity variation along the Moho boundary.

| $P_n$-Distance | t-iso | t-ani-min | t-ani-max | ±δ-t |
|---|---|---|---|---|
| 50  | 6.25  | 6.09  | 6.41  | 0.16 |
| 100 | 12.50 | 12.19 | 12.81 | 0.31 |
| 150 | 18.75 | 18.48 | 19.22 | 0.47 |
| 200 | 25.00 | 24.37 | 25.63 | 0.63 |
| 250 | 31.25 | 30.47 | 32.03 | 0.78 |
| 300 | 37.50 | 36.56 | 38.44 | 0.94 |
| 350 | 43.75 | 42.66 | 44.84 | 1.09 |

## A.2 Traveltime variations for anisotropy over distances

The differences between the isotropic and anisotropic traveltimes for a hypothetical hypocenter at the surface (see A.1). The calculation is done with a simple two layer velocity model with $6.0\,km/s$ in the crust, depth = $30\,km$ and $8.0\,km/s$ in the upper mantle. Out of this, a critical angle of $sin(\alpha_c)/sin(90°) = 6.0/8.0$ results at $48.59°$. The critical distance $\Delta_c$ out of this is $30\,km \cdot tan(\alpha_c) = 68\,km$.

## A.3 Traveltime variations for dipping Moho over distances

The formula (Lay and Wallace, 1995) with $t_u$ for updipping and $t_d$ for downdipping layers is in A.1. From this equation, the traveltime differences between the isotropic and anisotropic case are calculated as difference by $2\,\Delta h_2 \cdot cos(i_c) \cdot cos(\Theta)/v_1 + x \cdot sin(i_c \pm \Theta)/v_1 - x \cdot sin(i_c)/v_1$ for distance x (see Table A.9).

$$t_u = \frac{2h_2 \cdot cos(i_c) \cdot cos(\Theta)}{v_1} + \frac{x \cdot sin(i_c - \Theta)}{v_1} \quad t_d = \frac{2h_1 \cdot cos(i_c) \cdot cos(\Theta)}{v_1} + \frac{x \cdot sin(i_c + \Theta)}{v_1} \quad (A.1)$$

Table A.9: Differences for dipping Moho relative to epicentral distance and dip $\alpha$. $P_n$-distance is between the Moho entering and leaving ray, $\Delta$-h the difference for the depth.

| $P_n$-Distance | $\alpha = 0.1°$ | | $\alpha = 0.2°$ | |
|---|---|---|---|---|
| km | $\Delta h$ km | $t_{0.1}$ s | $\Delta h$ km | $t_{0.2}$ s |
| 50  | 0.44 | 0.09 | 0.87 | 0.21 |
| 100 | 0.87 | 0.21 | 1.75 | 0.43 |
| 150 | 1.31 | 0.31 | 2.62 | 0.63 |
| 200 | 1.74 | 0.42 | 3.49 | 0.85 |

# Appendix B
# Material properties of various earth rocks and minerals

In the following tables, several seismological relevant material properties are listed which are important to describe and to interpret the influence of anisotropy and of the $V_P/V_S$ ratio analyzed in the present study. This information will be especially useful in Chapter 6 where a detailed study for different lateral block discretizations for both P- and S-wave velocities, as well as the $V_P/V_S$-ratio will be done.

Table B.1: $V_P/V_S$-ratios for some major rocks after Volarovich and Budnikov (1978/79)

| Rock | $V_P/V_S$ | | | | $V_P$ km/s | | | | $V_S$ km/s | | | |
|---|---|---|---|---|---|---|---|---|---|---|---|---|
| kbar | 0 | 0.5 | 1 | 2.0 | 0 | 0.5 | 1 | 2.0 | 0 | 0.5 | 1 | 2.0 |
| Dolomite | 1.788 | 1.763 | 1.781 | 1.821 | 4.47 | 4.62 | 4.72 | 4.88 | 2.50 | 2.62 | 2.65 | 2.68 |
| Limestone | 1.866 | 1.892 | 1.906 | 1.923 | 5.69 | 5.79 | 5.85 | 5.96 | 3.05 | 3.06 | 3.07 | 3.10 |
| Diorite | 1.914 | 1.927 | 1.939 | 1.946 | 5.82 | 6.32 | 6.40 | 6.50 | 3.04 | 3.28 | 3.30 | 3.34 |
| Granite | 1.703 | 1.727 | 1.739 | 1.756 | 5.33 | 5.89 | 6.00 | 6.11 | 3.13 | 3.41 | 3.45 | 3.48 |
| Gneiss | 1.642 | 1.736 | 1.756 | 1.788 | 4.86 | 6.11 | 6.25 | 6.40 | 2.96 | 3.52 | 3.56 | 3.58 |
| White marble | 1.5 | 1.930 | 1.958 | 1.964 | 3.72 | 6.37 | 6.52 | 6.60 | 2.48 | 3.30 | 3.33 | 3.36 |

Table B.2: Typical range of the velocity ratio $V_P/V_S$ for common lithologies. (Domenico, 1984; J. H. Schï¿½n, 1996)

| Lithology | $V_P/V_S$ |
|---|---|
| Dolomites | 1.78 - 1.84 |
| Limestones | 1.84 - 1.99 |
| Sandstones | 1.59 - 1.76 |
| Shales | 1.70 - 3.00 |

Table B.3: Crustal minerals, composition and abundances. After Anderson (2007)

| Mineral | Composition | Range of Crustal Abundances (vol. pct.) |
|---|---|---|
| Plagioclase | | 31–41 |
|   Anorthite | $Ca(Al_2Si_2)O_8$ | |
|   Albite | $Na(Al, Si_3)O_8$ | |
| Orthoclase | | 9–21 |
|   K-Feldspar | $K(Al, Si_3)O_8$ | |
| Quartz | $SiO_2$ | 12–24 |
| Amphibole | $NaCa_2(Mg, Fe, Al)_5[(Al, Si)_4O_11]_2(OH)_2$ | 0–6 |
| Biotite | $K(Mg, Fe^{2+})_3(Al, Si_3)O_{10}(OH, F)_2$ | 4–11 |
| Muscovite | $KAl_2(Al, Si_3)O_{10}(OH)$ | 0–8 |
| Chlorite | $(Mg, Fe^{2+})_5Al(Al, Si_3)O_{10}(OH)_8$ | 0–3 |
| Pyroxene | | |
|   Hypersthene | $(Mg, Fe^{2+})SiO_9$ | 0–11 |
|   Augite | $Ca(Mg, Fe^{2+})(SiO_3)_2$ | |
| Olivine | $(Mg, Fe^{2+})_2SiO_4$ | 0–3 |
| Oxides | | $\approx 2$ |
|   Sphene | $CaTiSiO_5$ | |
|   Allanite | $(Ce, Ca, Y)(Al, Fe)_3(SiO_4)_3(OH)$ | |
|   Apatite | $Ca_5(PO_4, CO_3)_3(F, OH, Cl)$ | |
|   Magnetite | $FeFe_2O_4$ | |
|   Ilmenite | $FeTiO_2$ | |

Table B.4: Average crustal abundance, density and seismic velocities of major crustal minerals. After Anderson (2007)

| Mineral | Volume % | $\rho$ g/cm$^3$ | $V_P$ km/s | $V_S$ km/s | $V_P/V_S$ |
|---|---|---|---|---|---|
| Quartz | 12 | 2.65 | 6.05 | 4.09 | 1.479 |
| K-feldspar | 12 | 2.57 | 5.88 | 3.05 | 1.928 |
| Plagioclase | 39 | 2.64 | 6.30 | 3.44 | 1.831 |
| Micas | 5 | 2.8 | 5.6 | 2.9 | 1.931 |
| Amphiboles | 5 | 3.2 | 7.0 | 3.8 | 1.842 |
| Pyroxene | 11 | 3.3 | 7.8 | 4.6 | 1.696 |
| Olivine | 3 | 3.3 | 8.4 | 4.9 | 1.714 |

Table B.5: Densities and elastic-wave velocities of upper-mantle rocks. From Anderson (2007)

| Rock | $\rho$ g/cm$^3$ | $V_P$ km/s | $V_S$ km/s | $V_P/V_S$ |
|---|---|---|---|---|
| Garnet Iherzolite | 3.53 | 8.29 | 4.83 | 1.72 |
|  | 3.47 | 8.19 | 4.72 | 1.74 |
|  | 3.46 | 8.34 | 4.81 | 1.73 |
|  | 3.31 | 8.30 | 4.87 | 1.70 |
| Dunite | 3.26 | 8.00 | 4.54 | 1.76 |
|  | 3.31 | 8.38 | 4.84 | 1.73 |
| Bronzitite | 3.29 | 7.89 | 4.59 | 1.72 |
|  | 3.29 | 7.83 | 4.66 | 1.68 |
| Eclogite | 3.46 | 8.61 | 4.77 | 1.81 |
|  | 3.61 | 8.43 | 4.69 | 1.80 |
|  | 3.60 | 8.42 | 4.86 | 1.73 |
|  | 3.55 | 8.22 | 4.75 | 1.73 |
|  | 3.52 | 8.29 | 4.49 | 1.85 |
|  | 3.47 | 8.22 | 4.63 | 1.78 |
| Jadeite | 3.20 | 8.28 | 4.82 | 1.72 |

After Clark (1966), Babuska (1972), Manghnani and Ramananotoandro (1974), Jordan (1979)

Table B.6: Anisotropy of upper-mantle rocks. From Anderson (2007)

| Mineralogy | Direction | $V_P$ km/s | $V_{S1}$ km/s | $V_{S2}$ km/s | $V_P/V_S$ |
|---|---|---|---|---|---|
| Peridotites | | | | | |
| 100 pct. ol | 1 | 8.70 | 5.00 | 4.85 | 1.74 - 1.79 |
| | 2 | 8.40 | 4.95 | 4.70 | 1.70 - 1.79 |
| | 3 | 8.20 | 4.95 | 4.72 | 1.66 - 1.74 |
| 70 pct. ol, | 1 | 8.40 | 4.90 | 4.77 | 1.71 - 1.76 |
| 30 pct. opx | 2 | 8.20 | 4.90 | 4.70 | 1.67 - 1.74 |
| | 3 | 8.10 | 4.90 | 4.72 | 1.65 - 1.72 |
| 100 pct. opx | 1 | 7.80 | 4.75 | 4.65 | 1.64 - 1.68 |
| | 2 | 7.75 | 4.75 | 4.65 | 1.63 - 1.67 |
| | 3 | 7.78 | 4.75 | 4.65 | 1.67 - 1.67 |
| Eclogites | | | | | |
| 51 pct. ga, | 1 | 8.476 | | 4.70 | 1.80 |
| 23 pct. cpx, | 2 | 8.429 | | 4.65 | 1.81 |
| 24 pct. opx | 3 | 8.375 | | 4.71 | 1.78 |
| 47 pct. ga, | 1 | 8.582 | | 4.91 | 1.75 |
| 45 pct. cpx | 2 | 8.379 | | 4.87 | 1.72 |
| | 3 | 8.300 | | 4.79 | 1.73 |
| 46 pct. ga, | 1 | 8.310 | | 4.77 | 1.74 |
| 37 pct. cpx | 2 | 8.270 | | 4.77 | 1.73 |
| | 3 | 8.110 | | 4.72 | 1.72 |

Manghnani and Ramananotoandro (1974), Christensen and Lundquist (1982).

# Bibliography

Abt, D. L. and K. M. Fischer, 2008. Resolving three-dimensional anisotropic structure with shear wave splitting tomography, *Geophys. J. Int.*, 173:859ï¿½886. doi: 10.1111/j.1365-246X.2008.03757.x. 133

Aki, K. and W. H. K. Lee, 1976. Determination of three dimensional velocity anomalies under a seismic array using first P arrival times from local earthquakes, 1. A homogeneous initial model, *J. Geophys. Res.*, 81:4381–4399. 7, 15, 19, 20, 23, 26

Aki, K. and P. G. Richards, 2002. *Quantitative Seismology, 2nd Ed.* Quantitative Seismology, 2nd Ed., by Keiiti Aki and Paul G. Richards. Published by University Science Books, ISBN 0-935702-96-2, 704pp, 2002. URL http://adsabs.harvard.edu/abs/2002quse.book.....A. 20

Anderson, D. L., 1984. The earth as planet: paradigms and paradoxes, *Science*, 223:347 – 355. 23, 180

Anderson, D. L., 1989. Theory of the Earth, *Boston: Blackwell Scientific Publications*. URL http://resolver.caltech.edu/CaltechBOOK:1989.001. 9

Anderson, D. L., 2007. *New Theory of the Earth*. Cambridge University Press. ISBN 978-0-521-84959-3. 177, 180, 181, 182, 240, 241, 242

Ando, M. and Y. Ishikawa, 1982. Observations of shear-wave polarisation anisotropy beneath Honshu, Japan: two masses with different polarizations in the upper mantle, *J. Phys. Earth*, 30:191ï¿½199. 44

Arlitt, R., 2001. *Teleseismic Body Wave Tomography across the Trans-European Suture Zone between Sweden and Denmark*. PhD thesis, ETH Zurich. 45, 46

Atkinson, K.E., 1976. A survey of numerical methods for the solution of Fredholm integral equations of the second kind. 26

Audoine, E.; M. K. Savage and K. Gledhill, 2000. Seismic anisotropy from local earthquakes in the transition region from a subduction to a strike-slip plate boundary, New Zealand, *JOURNAL OF GEOPHYSICAL RESEARCH*, 105(B4):8013–8034. URL http://www.agu.org/pubs/crossref/2000/1999JB900444.shtml. 44

Babuška, V. and J. Plomerová, 2000. Saxothuringian-Moldanubian Suture and Predisposition of Seismicity In The Western Bohemian Massif, *Studia Geophysica et Geodaetica*, 44(2):1573–1626. doi: 10.1023/A:1022123111966. URL http://www.springerlink.com/content/p4wl1008376214r1/. 42, 72, 221

Backus, G. E. and J. F. Gilbert, 1967. Numerical applications of a formalism for geophysical inverse problems, *Geophys. J. R. astr. SOC*, 13:247–276. 130

Backus, G.E. and F. Gilbert, 1968. The resolving power of gross earth data, *Geophys. J. R. astr. Soc.*, 16:169–205. 32

Bamford, D., 1973. Refraction data in western Germany - a time-term interpretation, *Z. Geophysik*, 39:907–927. 16, 42, 72, 221

Bamford, D. and S. Crampin, 1977. Seismic anisotropy - the state of the art, *Geophys. J. R. astr. Soe.*, 49:1–8. 40

Bass, J. D., 1995. *Mineral Physics and Crystallography: A Handbook of Physical Constants*, chapter Elasticity of Minerals, Glasses, and Melts, pages 45–64. American Geophysical Union, agu reference shelf 2 edition. 37

Bass, J. D.; D. J. Weidner; N. Hamaya; M. Ozima and S. Akimoto, 1984. Elasticity of the Olivine and Spinel Polymorphs of $Ni_2SiO_4$, *Phys Chem Minerals*, 10:261–272. 37

Bassin, C.; G. Laske and G. Masters, 2000. The Current Limits of Resolution for Surface Wave Tomography in North America, *EOS Trans AGU*, 81. URL http://mahi.ucsd.edu/Gabi/rem.html. 74

Beck, J.V. and K.J. Arnold, 1977. Parameter estimation science and engineering. 28

Berndt, M., 1999. Michael Berndt's Geologie-Paläontologie-Seite, *Website*. 9

Berthelsen, A., 1992. *The European Science Foundation: A Continent revealed, The European Geotraverse*, chapter 2. Mobile Europe. Derek Blundell, Roy Freeman and Stephan Mueller. 9, 14, 153, 185, 227

Bina, C. R., 1998. Lower Mantle Mineralogy and the Geophysical Perspective, *Reviews in Mineralogy, Mineralogical Society of America, Washington, DC*, 38:205–239. URL http://xoanon.earth.northwestern.edu/public/craig/publish/rm98.html. 37

Boness, N. L. and M. D. Zoback, 2004. Stress-induced seismic velocity anisotropy and physical properties in the SAFOD Pilot Hole in Parkfield, CA, *GEOPHYSICAL RESEARCH LETTERS*, 31. doi: 10.1029/2003GL019020. 9

Budweg, M., 2003. Der obere Mantel in der Eifel-Region untersucht mit der Receiver Function Methode. URL http://edoc.gfz-potsdam.de/gfz/3663. 77

Bystricky, M.; K. Kunze; L. Burlini and J.-P. Burg, 2000. High Shear Strain of Olivine Aggregates: Rheological and Seismic Consequences, *Science*, 290:1564 – 1567. URL www.sciencemag.org. 37

Cerveny, V., 2001. *Seismic Ray Theory*. Cambridge University Press. 23

Christensen, N. I. and S. Lundquist, 1982. Pyroxene orientation within the upper mantle, *Bull. Geol. Soc. Am.*, 93:279–88. 180

Clark, S. P. Jr., 1966. *Handbook of Physical Constants*, volume 97. Geol. Soc. Amer. Mem. 180

Courant, R. and D. Hilbert, 1962. *Methods of mathematical physics*. Springer, New York, Heidelberg. 26

Crampin, S., 1986. Anisotropy and transverse isotropy, *Geophysical Prospecting*, 34(1):94–99. doi: 10.1111/j.1365-2478.1986.tb00454.x. URL http://dx.doi.org/10.1111/j.1365-2478.1986.tb00454.x. 42

Domenico, S. N., 1984. Rock lithology and porosity determination from shear and compressional wave velocity, *Geophysics*, 49:1188 ff. doi: DOI:10.1190/1.1441748. 180, 240

Dziewonski, A. and D.L. Anderson, 1981. Preliminary reference Earth model, *Phys. Earth Planet. Inter.*, 25:297–356. doi: doi:10.1016/0031-9201(81)90046-7. 70

Eberhart-Phillips, D., 1995. Examination of seismicity in the central Alpine Fault region, South Island, New Zealand, *New Zealand Journal of Geology and Geophysics*, 38:571–578. doi: 0028-8306/95/ 3804-0571. URL http://www.rsnz.org/publish/nzjgg/1995/89.pdf. 7

Eberhart-Phillips, D. and M. Reyners, 1997. Continental subduction and three-dimensional crustal structure: The northern South Island, New Zealand, *J. Geophys. Res*, 102:11,843–11,862. doi: 10.1029/96JB03555. 98

Edel, J.-B.; K. Fuchs; C. Gelbke and C. Prodehl, 1975. Deep structure of the southern Rhinegraben area from seismic refraction investigations, *J. Geophys.*, 41:333–356. 46

Eisbacher, G., 1991. *Einführung in die Tektonik*. Enke. 39

Enderle, U., 1998. Signaturen in refraktionsseismischen Daten als Abbild geodynamischer Prozesse, *University of Karlsruhe, Geophysical Institute*. 46, 75, 76, 78, 92, 94, 231

Enderle, U.; J. Mechie; S. Sobolev and K. Fuchs, 1996a. Seismic anisotropy within the uppermost mantle of southern Germany, *Geophys. J. Int.*, 125:747 – 767. 9, 15, 16, 42, 46, 64, 72, 74, 75, 78, 92, 221

Enderle, U.; J. Mechie; S. Sobolev and K. Fuchs, 1996b. Seismic anisotropy within the uppermost mantle of southern Germany, *Geophys. J. Int.*, 125:747 – 767. 64, 72, 78, 223

Enderle, U.; K. Schuster; C. Prodehl; A. Schulze and J. Bribach, 1998. The refraction seismic experiment GRANU95 in the Saxothuringian belt, southeastern Germany, *Geophysical Journal International*, 133:245–259. 46

Forsyth, D.W., 1975. The early structural evolution and anisotropy of the oceanic upper mantle, *Geophys. J. R. astr. Soc.*, 43:103–162. 42

Fouch, Matthew J. and St. Rondenay, 2006. Seismic anisotropy beneath stable continental interiors, *Physics of the Earth and Planetary Interiors*, 158:292–320. 39

Franzke, H. J.; W. Werner and H.-U. Wetzel, 2003. Die Anwendung von Satellitenbilddaten zur tektonischen Analyse des Schwarzwalds und des angrenzenden Oberrheingrabens, *Jh. Landesamt f. Geologie, Rohstoffe und Bergbau Baden-Wü$\frac{1}{2}$rttemberg*, 39:25–54. 165

Fredholm, E.I., 1903. Sur une classe d'equations fonctionnelles, *Acta Mathematica*, 27:365–390. 26

Fuchs, K., 1975. Seismische Anisotropie des oberen Erdmantels und Intraplatten-Tektonik, *International Journal of Earth Sciences*, 64(1):700–716. URL http://www.springerlink.com/content/q46j43383863vx6t/. 39, 64, 221, 223

Fuchs, K., 1983. Recently formed elastic anisotropy and petrological models for the continental subcrustal lithosphere in southern Germany, *Physics of The Earth and Planetary Interiors*, 31: 93–118. 39, 42, 64, 72

Gajewski, D. and C. Prodehl, 1985a. Crustal structure beneath the Swabian Jura, SW Germany, from seismic refraction investigation, *J. Geophys.*, 56:69–80. 46

Gajewski, D. and C. Prodehl, 1985b. Test the citation Gajewski-1987, *J. Geophys.*, 56:69–80. 46

Gajewski, D.; R. Stangl; K. Fuchs and K.-J. Sandmeier, 1990. A new constraint on the composition of the topmost continental mantle - anomalously different depth increases of P and S velocity, *Geophys. J. Int.*, 103:479–507. 46

Gatzemeier, A. Elektrische Anisotropie durch ausgerichtete Olivinkristalle im oberen Mantel in Mitteleuropa: magnetotellurische Array-Messungen und ein Ansatz zum Vergleich mit seismischer Anisotropie, 2001. 37

Gledhill, K. R., 1991. Evidence for shallow and pervasive seismic anisotropy in the Wellington region, New Zealand, JOURNAL OF GEOPHYSICAL RESEARCH, 96:21,503–21,516. URL http://www.agu.org/pubs/crossref/1991/91JB02049.shtml. 9

Granet, M., 1998. Des Images du Système Lithosphère-Astenospère sous la France: L'apport de la Tomographie Télésismique de l'Anisotropie sismique, CNFGG - Rapport Quadriennal 1995 - 1998, pages 71 – 100. URL http://comp1.geol.unibas.ch/groups/3_2/Granet/s2granet.pdf. 23

Hartman, G.; M.Henger and A. Schick, 1990. DATA CATALOGUE OF EARTHQUAKES IN GERMANY AND ADJACENT AREAS, Seismological Data Catalogue. URL http://www.seismologie.bgr.de. 46

Haslinger, F. and E. Kissling, 2001. Investigating effects of 3-d ray tracing methods in local earthquake tomography, Phys. Earth Planet. Inter., 123:103–114. 103

Hawkins, K.; H. Kat; R. Leggott and G. Williams, 11 - 15 June 2001. ADDRESSING ANISOTROPY IN PRESTACK DEPTH MIGRATION : A SOUTHERN NORTH SEA CASE STUDY. In EAGE 63rd Conference and Technical Exhibition, 11 - 15 June 2001. 9

Hearn, T. M., 1996. Anisotropic Pn tomography in the western United States, Journal of Geophysical Research, 101:8403 – 8414. 23

Hearn, T. M. and R.W. Clayton, 1986a. Lateral velocity variations in southern California, 1, Results for the upper crust from $P_g$ waves, Bull. Seism. Am., 76:495 – 509. 16

Hearn, T. M. and R.W. Clayton, 1986b. Lateral velocity variations in southern California, 2, Results for the lower crust from $P_n$ waves, Bull. Seism. Am., 76:511 – 519. 16

Henger, M. and G. Leydecker, 1975. DATA CATALOGUE OF EARTHQUAKES IN GERMANY AND ADJACENT AREAS, Seismological Data Catalogue. URL http://www.seismologie.bgr.de. 46

Henningsen, D. and G. Katzung, 2002. Einführung in die Geologie Deutschlands. Spektrum Akademischer Verlag. 12

Hess, H. H., 1964. Seismic anisotropy of the upper mantle under oceans, Nature, 203:629–631. URL http://www.nature.com/nature/journal/v203/n4945/abs/203629a0.html. 42

Hoerl, H.A. and R.W. Kennard, 1970. Ridge regression; Biased estimation for nonorthogonal problems, Technometrics, 12:55–82. 29

Hoffmann, N.; H. Stiewe and G. Pasternak, 1996. Struktur und Genese der Mohorovicic-Diskontinuität (Moho) im Norddeutschen Becken - ein Ergebnis langzeitregistrierter Steilwinkelseismik, Z. angew. Geologie, 42:138 – 148. 12

Husen, S., 1999. Local Earthquake Tomography of a Convergent Margin, North Chile. PhD thesis, Mathematisch-Naturwissenschaftliche Fakultï¿½t der Christian-Albrechts-Universitï¿½t zu Kiel. 103

Husen, S. and E. Kissling, 2001. Local earthquake tomography between rays and waves: fat ray tomography, Physics of the earth and Planetary Interiors, 123:129–149. 34, 103

Husen, S.; E. Kissling; N. Deichmann; S. Wiemer; D. Giardini and M. Baer, 2003. Probabalistic earthquake location in complex three-dimensional velocity models: Application to Switzerland, Journal of Geophysical Research, 108(B2):5–49. 103

Husen, S.; E. Kissling; E. Flueh and G. Asch, 1999. Accurate Hypocentre Determination in the Seismogenic Zone of the Subducting Nazca Plate in Northern Chile Using a Combined On-/Offshore Network, *Geophys. J. Int.*, 138:687–701. 45

Illies, J. H., 1977. Ancient and recent rifting in the Rhinegraben, *GEOLOGIE EN MIJNBOUW*, 56(4):329–350. 165

J. H. Schï¿½n, , 1996. *Handbook of Geophysical Exploration Seismic Exploration*, volume 18, chapter Physical properties of rocks. K. Helbig and S. Treitel, Pergamon, New York. 240

Jordan, T. H., 1979. Mineralogies, densities and seismic velocities of garnet lherzolites and their geophysical implications. in the mantle sample, *Boyd, F. R. and Meyer, H. O. A. Washington, DC, American Geophysical Union*, pages 1–14. 180

Judenherc, S.; M. Granet and N. Boumbar, 1999. Two dimensional anisotropic tomography of lithosphere beneath France using regional arrival times., *J. Geophys. Res.*, 104:13 201 ï¿½ 13 215. 16

Kaminski, E. and N. M. Ribe, 2002. Timescales for the evolution of seismic anisotropy in mantle flow, *Geochemistry, Geophysics, Geosystems: G3*, 3(1). doi: 10.1029/2001GC000222. Published by AGU and the Geochemical Society AN ELECTRONIC JOURNAL OF THE EARTH SCIENCES. 39

Kaneshima, S. and P. G. Silver, 1992. A search for source side mantle anisotropy, *Geophys. Res. Lett.*, 19:1049–1052. 44

Kaschenz, J., 2006. *Regularisierung unter Berücksichtigung von Residuentoleranzen*. Dr.-ing., TU Berlin. URL http://www.gfz-potsdam.de/bib/zbstr.htm. 26

Kawasaki, I., 1989. Seismic anisotropy in the Earth, *The Encyclopedia of Solid Earth Geophysics*, pages 994–1005. 40

Keen, C. E. and D. L. Barrett, 1971. A measurement of seismic anisotropy in the Northeast Pacific, *Can. d. Earth Sci.*, 8:1056–64. 42

Kennett, B.L.N., 1991. IASPEI 1991 Seismological Tables, *Research School of Earth Sciences, Australian National University*. 70

Kennett, B.L.N., 2005. Seismological Tables: ak135, *Research School of Earth Sciences, The Australian National University, Canberra ACT 0200 Australia*. 70

Kennett, B.L.N. and E.R. Engdahl, 1991. Traveltimes for global earthquake location and phase identification, *Geophysical Journal International*, 105:429–465. 70

Kind, R.; G.L. Kosarev; L. I. Makeyeva and L. P. Vinnik, 1985. Observations of laterally inhomogeneous anisotropy in the continental lithosphere, 318:358–361. 44

Kissling, E., November 1988. Geotomography with local earthquake data, *Reviews of Geophysics*, 26: 659–698. 7, 26, 45

Kissling, E.; W. L. Ellsworth; D. Eberhart-Phillips and U. Kradolfer, 1994. Initial reference models in local earthquake tomography, *JOURNAL OF GEOPHYSICAL RESEARCH*, 99:19,635–19,646. 8

Kissling, E.; S. Husen and F. Haslinger, 2001. Model parameterization in seismic tomography: A choice of consequences, *Phys. Earth Planet. Int.*, 123:89–101. 103

Koch, M., 1983a. *Die Bestimmung lateraler Geschwindigkeitsinhomogenitäten aus der linearen und nichtlinearen Inversion tele- und lokalseismischer Laufzeiten - Anwendung auf die seismische Zone Vrancea, Rumänien*. PhD thesis, Geophysikalisches Institut, Uni Karlsruhe. 7, 9, 19, 24, 26

Koch, M., 1983b. Die Bestimmung lateraler Geschwindigkeitsinhomogenitäten aus der linearen und nichtlinearen Inversion tele- und lokalseismischer Laufzeiten - Anwendung auf die seismische Zone Vrancea, Rumänien, *Geophysikalisches Institut, Uni Karlsruhe*, University of Karlsruhe. 8, 133

Koch, M., 1985a. A theoretical and numerical study on the determination of the 3D structure of the lithosphere by linear and nonlinear inversion of teleseismic travel times, *Journal of Geophysics*, 56: 160 – 173. 7, 8, 9, 24, 45, 219, 220

Koch, M., 1985b. Nonlinear inversion of local seismic travel times for the simultaneous determination of 3D velocity structure and hypocenters - application to the seismic zone Vrancea, *Geophys. J. R. Astr. Soc*, 80:73–93. 24, 25, 26, 45, 102, 103, 174, 220

Koch, M., 1989. *Geophysical Inversion*, chapter Optimal Regularization of the Linear Seismic Inverse Problem, pages 183 – 244. SIAM Society for Industrial and applied Mathematics. 19, 24, 26, 31, 32, 33, 102, 103, 174

Koch, M., 1992. Bootstrap inversion for vertical and lateral variations of the S wave structure and the $v_p/v_s$-ratio from shallow earthquakes in the Rhinegraben seismic zone, Germany, *TECTONOPHYSICS*, 1:91–115. doi: 10.1029/1992-000001. 98, 131, 134, 177

Koch, M., 1993a. Simultaneous Inversion for 3D Crustal Structure and Hypocenters including direct, refracted and reflected phases. I. Development, Validation and optimal Regularization of the Method, *Geophys. J. Int.*, 112:385 –412. URL http://www.uni-kassel.de/fb14/geohydraulik/koch/paper/1993/SSH_Rhinegraben_I.pdf. 16, 25, 26, 28, 32, 69, 70, 78, 79, 102, 103, 131, 174, 220

Koch, M., 1993b. Simultaneous Inversion for 3D Crustal Structure and Hypocenters including direct, refracted and reflected phases. II. Application to the northern Rhinegraben/RhenishMassif, Germany, *Geophys. J. Int.*, 112:412–428. URL http://www.uni-kassel.de/fb14/geohydraulik/koch/paper/1993/SSH_Rhinegraben_II.pdf. 16, 32, 79, 131

Koch, M., 1993c. Simultaneous Inversion for 3D Crustal Structure and Hypocenters including direct, refracted and reflected phases. III. Application to the southern Rhinegraben, Germany, *Geophys. J. Int.*, 112:429–447. URL http://www.uni-kassel.de/fb14/geohydraulik/koch/paper/1993/SSH_Rhinegraben_III.pdf. 16, 32, 74, 79

Koch, M. and J. Kalata, 1992. Simultaneous Determination of Hypocenter Location and Crustal Structure in the Virginia Seismic Region, *Journal of Geophysical Research*, 97:17–481 – 17–502. 7, 9, 16, 31, 74, 78, 223

Konrad, H. J. and A. E. M. Nairn, 1972. The palaeomagnetism of the permian rocks of the black forest, germany, *Geophys. J. R. astr. SOC*, 27:369–382. 165

Krumbiegel, G., 1981. *Die Entwicklungsgeschichte der Erde*, chapter Historische Geologie (Erdgeschichte), page 281 ff. VEB F.A. Brockhaus Verlag, Leipzig. 9

Kummerow, J., 2004. Strukturuntersuchungen in den Ostalpen anhand des teleseismischen TRANSALP- Datensatzes. Master's thesis, FU Berlin. 15, 42, 43, 72, 231

Lahr, J.C., 1999. HYPOELLIPSE: A computer program for determining local earthquake hypocentral parameters, magnitude, and first motion pattern, (Y2K compliant version), *U.S. Geological Survey, Open-File Report 99-23*, page 112 p. URL http://greenwood.cr.usgs.gov/pub/open-file-reports/ofr-99-0023. 8, 15

Lay, T. and T. C. Wallace, 1995. MODERN GLOBAL SEISMOLOGY, 58. 19, 20, 237

Lee, W.H.K. and S.W. Stewart, 1981. Principles and application of microearthquake networks, *Academic Press, Orlando, Fl.* 25

Lee, W.H.K. and S.M. Valdes, 1985. HYPO71PC: A Personal Computer Version of the HYPO71 Earthquake Location Program, U. S. Geol. Surv. Open File Report 85-749, pages 1–43. 15

Levenberg, K., 1944. A method for the solution of certain nonlinear problems in least squares, Quart., Appl. Math., 2:164–168. 28

Little, T. A.; M. K. Savage and B. Tikoff, October 2002. Relationship between crustal finite strain and seismic anisotropy in the mantle, Pacific-Australia plate boundary zone, South Island, New Zealand, *Geophysical Journal International*, 151:106–116. doi: 10.1046/j.1365-246X.2002.01730.x. 39

Lombardi, D.; J. Braunmiller; E. Kissling and D. Giardini, 2008. Moho depth and Poisson's ratio in the Western-Central Alps from receiver functions, *Geophys. J. Int. (2008)*, 173:249–264. doi: {10.1111/j.1365-246X.2007.03706.x}. 181, 227

M. Powell, , 2005. *Lexikon der Geowissenschaften.* Spektrum Akademischer Verlag, Heidelberg. 12

Malvern, L.E., 1969. *Introduction to the Mechanics of a Continous Medium.* Prentice-Hall, Englewood Cliffs, NJ. 39

Manghnani, M. H. and C. S. P. Ramananotoandro, 1974. Compressional and shear wave velocities in granulite facies rocks and eclogites to 10 kbar, *J. Geophys. Res.*, 79:5427–46. 180

Marquardt, D.W., 1963. An algorithm of least squares estimation of nonlinear parameters, *J. Soc. Indust. Appl. Math.*, 11:431ï¿½441. 28

Mechie, J., 2005. Mohodepth by refraction profiling, *personal communication.* 46, 223

Meissner, Rolf; Wolfgang Rabbel and Hartmut Kern, 2006. Seismic lamination and anisotropy of the lower continental crust, *Tectonophysics*, 416:81–99. 231

Monna, S.; L. Filippi; L. Beranzoli and P. Favali, 2003. Rock properties of the upper-crust in Central Apennines (Italy) derived from high-resolution 3-D tomography, *GEOPHYSICAL RESEARCH LETTERS*, 30(7). doi: {10.1029/2002GL016780}. 98

Montanger, J.-P. and T. Tanimoto, 1990. Global anisotropy in the upper mantle inferred from the regionalization of phase velocities, *J. Geophys. Res.*, 95:4797–4819. 42

Mooney, W.; G. Laske and G. Masters, 1998. Crust 5.1: a global crustal model at 5x5 degrees, 103: 727–747. URL http://mahi.ucsd.edu/Gabi/rem.dir/crust/crust2.html. 74

Muench, T. W. Lokalbebentomografie der Vrancea-Zone, 2000. 92

Nataf, H.-C.; I. Nakanishi and D. L. Anderson, 1984. Anisotropy and shear velocity heterogeneities in the upper mantle, *Geophys. Res. Lett*, 11:109–112. 42

N.Draper, and H. Smith, 1981. Applied regression analysis, *John Wiley & Sons, New York.* 174

Nelson, S. A., 2006. Earth materials. URL http://www.tulane.edu/~sanelson/eens211/index.html. 37

Nicolas, A. and N. I. Christensen, 1987. Formation of anisotropy in upper mantle peridotites: A review, in Composition Structure and Dynamics of the Lithosphere-Asthenosphere System, vol. 16: 111 – 123. 39

Okaya, D.; N. Christensen; D. Stanley and T. Stern, 1995. Crustal anisotropy in the vicinity of the Alpine Fault Zone, South Island, New Zealand, *JOURNAL OF GEOPHYSICAL RESEARCH*, 38:579–583. doi: 0028-8306/95/3804-0579. URL http://www.rsnz.org/publish/nzjgg/1995/91.php. 9, 42

Orfanidis, Sophocles J., 2008. *Electromagnetic Waves and Antennas*. 44

Plomerová, J.; R. C. Liebermann and V. Babuška, 1998. Geodynamics of Lithosphere and Earth's Mantle: Seismic Anisotropy as a Record of the Past and Present Dynamic Processes, *Pure and Applied Geophysics*, 151:213–219. doi: 0033-4553/98/040213-07. 9

Plomerova, Jaroslava, January 1997. Seismic anisotropy in tomographic Studies of the upper mantle beneath Southern Europe, *Annali Di Geofisica*, XL(1):111 – 122. 9

Pratt, R. G.; F. Gao; C. Zelt and A Levander, 2002. A comparison of ray-based and waveform tomography: implications for migration, *EAGE 64th Conference and Exhibition ï¿½ Florence, Italy, 27 - 30 May*. 23, 33

Pujol, J., 1988. Comments on the Joint Determination of Hypocenters and Station Corrections, *Bulletin of the Seismological Society of America*, 78:1179 – 1189. 23

Pujol, J., 1992. Joint hypocentral location in media with lateral velocity variations and interpretation of the station corrections, *Physics of the Earth and Planetary Interiors*, 75:7 – 24. 26

Pulford, A.; M. Savage and T. Stern, 2003. Absent anisotropy: The paradox of the southern alps orogen, *GEOPHYSICAL RESEARCH LETTERS*, 30(20). URL http://www.agu.org/pubs/crossref/2003/2003GL017758.shtml. 44

Rabbel, W.; S. Siegesmund; T. Weiss; M. Pohl and T. Bohlen, 1998. Shear wave anisotropy of laminated lower crust beneath urach (sw germany): a comparison with xenoliths and with exposed lower crustal sections, *Tectonophysics*, 298:337–356. 231

Raitt, R.W.; G. G. Shor; T. J. G. Francis and G. B. Morris, June 1969. Anisotropy of the Pacific Upper Mantle, *J. Geophysical Research*, 74:3095–3109. URL http://adsabs.harvard.edu/abs/1969JGR....74.3095R. 42

Ribe, N.M., 1989a. A continuum theory for lattice preferred orientation, *Geophys. J*, 97:199 – 207. 39

Ribe, N.M., 1989b. Seismic anisotropy and mantle flow, *J. Geophys. Res.*, 94:4213–4223. 39

Ribe, N.M., 1992. On the relation between seismic anisotropy and finite strain., *J. Geophys. Res.*, 97: 8737–8747. 39

Ribe, N.M. and Y. Yu, 1991. A theory for plastic deformation and textural evolution of olivine polycrystals, *J. Geophys. Res.*, 96:8325–8336. 39

Ritter, J.R.R.; U. Achauer and U.R. Christensen, 2000. The Teleseismic tomography experiment in the Eifel Region, Central Europe: Design and first results, *Seismological Res. Lett.*, 71(4). 15

Ritter, J.R.R.; M. Jordan; U.R. Christensen and U. Achauer, 2001. A mantle plume below the Eifel volcanic fields, Germany, *Earth and Planetary Science Letters*, 186:7–14. 15

Savage, M.K., 1999. Seismic anisotropy and mantle deformation: What have we learned from shear wave splitting?, *Rev. Geophys.*, 37:65–106. 39, 42, 44

Schlittenhardt, J., 1999. Regional velocity models for Germany: A contribution to the systematic travel-time calibration of the International Monitoring System, *Proceedings of the 21st DoD/DOE Seismic Research Symposium: Technologies for Monitoring The Comprehensive Nuclear-Test-Ban Treaty. Las Vegas, Nevada*, I:263–273. 16, 46, 47, 221

Seber, G.A.F. and C.J. Wild, 1989. Nonlinear regression, *John Wiley, New York*. 102

Sherburn, S.; R. S. White and M. Chadwick, 2006. Three-dimensional tomographic imaging of the taranaki volcanoes, new zealand, *Geophys. J. Int.*, 166:957–969. 177

Silver., P. G. and C. Bina, 1993. An anomaly in the amplitude ratio of SKKS/SKS in the range 100-108o from portable teleseismic data, *Geophys. Res. Lett.*, 20:1135–1138. 44

Silver, P.G., 1996. Seismic anisotropy beneath the continents: Probing the depths of geology, *Annu. Rev. Earth Pl. Sc.*, 24:385–432. URL http://arjournals.annualreviews.org/doi/pdf/10.1146/annurev.earth.24.1.385. 40, 220

Silver, P.G. and W.W. Chan, 1988. Implications for continental structure and evolution from seismic anisotropy, *Nature*, 335:34–39. 44

Silver, P.G. and W.W.J. Chan, 1991. Shear-wave splitting and subcontinental mantle deformation, *J. geophys. Res.*, 96(16):429–454. 44

Song, L.-P. and A. G. Every, 2000. Approximate formulae for acoustic wave group slownesses in weakly orthorhombic media, *J. Phys. D: Appl. Phys.*, 33(2519):L81–L85. 40

Song, L. P.; A. G. Every and C. Wright, 2001a. Linearized approximations for phase velocities of elastic waves in weakly anisotropic media, *Journal of Physics D: Applied Physics*, 34:2052–2062. URL http://www.iop.org/EJ/article/0022-3727/34/13/316/d11316.ps.gz. 9, 16, 39, 40, 42, 46, 64, 72, 74, 78, 80, 92, 94, 223

Song, L. P.; M. Koch; K. Koch and J. Schlittenhardt, 2001b. Isotropic and anisotropic $P_n$ velocity inversion of regional earthquake travel-times underneath Germany, *Geophys. J. Int.*, pages 795 – 800. 9, 39, 72, 74, 78, 80, 221

Song, L. P.; M. Koch; K. Koch and J. Schlittenhardt, 2004. 2-D anisotropic $P_n$-velocity tomography underneath Germany using regional travel-times, *Geophys. J. Int.*, 157:645–663. 9, 16, 30, 46, 72, 92, 94, 96, 102, 152, 162, 165, 174, 223, 231

Song, L.P. and M. Koch, 2002. Anisotropic reference media and the possible linearized approximations for phase velocities of qS waves in weakly anisotropic media, *JOURNAL OF PHYSICS D: APPLIED PHYSICS*, 35:3007–3014. 9, 80, 221

Spakman, W., 1988. Upper mantle delay time tomography, *Ph.D Thesis, University of Utrecht; Netherlands*, page pp 200. 23

Spetzler, and R. Snieder, MAY-JUNE 2004. The Fresnel volume and transmitted waves, *GEOPHYSICS*, 69(3):653–663. doi: 10.1190/1.1759451. 33, 34

Stanley, S. M., 1999. Earth System History, *New York: W.H. Freeman and Company*. 11

Stöffler, D.; N. A. Artemieva and E. Pierazzo, December 2002. Modeling the Ries-Steinheim impact event and the formation of the moldavite strewn field, *Meteoritics and Planetary Science*, 37:1893–1907. URL http://cdsads.u-strasbg.fr/abs/2002M%26PS...37.1893S. 165

Studer, J. A.; J. Laue and M. G. Koller, 2007. *Bodendynamik*. Springer-Verlag Berlin Heidelberg. 19, 20

Su, L. and J. Park, November 1994. Anisotropy and the splitting of PS waves, *Physics of the Earth and Planetary Interiors*, 86:263–276. doi: {10.1016/0031-9201(94)90125-2}. 37

Tanimoto, T. and D.L. Anderson, 1985. Lateral heterogeneity and azimuthal anisotropy of the upper mantle: Love and Rayleigh waves 100-250 sec, *J. Geophys. Res.*, 90:1842–1858. 42

Tarantola, A., 1987. Inverse problem theory, methods for data fitting and model parameter estimation, *Elsevier; Amsterdam; New York*. 7, 133

Tarantola, A. and B. Valette, 1982a. Generalized nonlinear inverse problems solved using the least squares criterion, *Rev. Geophys. Space Phys.*, 20:219–232. 7

Tarantola, A. and B. Valette, 1982b. Inverse problems = Quest for information, *J. Geophys. Res.*, 50: 159 – 170. 7, 26, 45

Thurber, C. H., 1981. *Earth structure and earthquake locations in the Coyote Lake area, central California*. PhD thesis, Massachusetts Institute of Technology. Dept. of Earth and Planetary Sciences. 8, 19

Thurber, C. H., 1983. Earthquake locations and three-dimensional structure in the Coyote Lake area; Central California, *J. Geophys. Res.*, 88:8226–8236. 8

Thurber, C. H. and K. Aki, 1987. Three-Dimensional Seismic Imaging, *Ann. Rev. Earth Planet. Sci*, 15:115–139. 7, 8, 19, 20, 26

Trampert, J., 1998. Global seismic tomography: the inverse problem and beyond, *Inverse Problems*, 14(3):371–385. URL http://stacks.iop.org/0266-5611/14/371. 23

Tsoulis, D., September 2004. Spherical harmonic analysis of the CRUST 2.0 global crustal model, *Journal of Geodesy*, 78:7–11. doi: 10.1007/s00190-003-0360-3. URL http://adsabs.harvard.edu/abs/2004JGeod..78....7T. 74

Twoomey, S., 1977. Introduction to the mathematics of inversion in remote sensing and indirect measurements, *Elsevier; Amsterdam; New York;*. 23

V. Babuska, , 1972. Elasticity and anisotropy of dunite and bronzitite, *J. Geophys. Res.*, 77:6955–65. 180

Vavryčuka, Václav; Pavla Hrubcovaá; Milan Brož and Jiří Maék, 2004. Azimuthal variation of $p_g$ velocity in the moldanubian, czech republic: observations based on a multi-azimuthal common-shot experiment, *Tectonophysics*, 387:189–203. 231

Vinnik, L.P.; V. Farra and B. Romanowicz, 1989. Azimuthal anisotropy in the earth from observations of SKS at GEOSCOPE and NARS broadband stations, *Bull. seis. Soc. Am.*, 79:1542–1558. 44

Vinnik, L.P.; G.L. Kosarev and L.I. Makeyeva, 1984. Anisotropy of the lithosphere according to the observations of SKS and SKKS waves, *Dokl. Akad. Nauk. SSSR.*, 278:1335. 44

Volarovich, P. and V. A. Budnikov, 1978/79. Velocities of Elastic Waves and Vp/Vs Ratios in Dry and Water-Saturated Rock Samples at High Pressure, *Pageoph*, 117. 240

Walck, M.C., 1998. Three-dimensional Vp/Vs variations for the Coso region, California., *J. Geophys. Res.*, 93:2047–2052. 98

Walter, R., 1995. *Geologie von Mitteleuropa*. Schweizerbartsche Verlagsbuchhandlung, Stuttgart. 12

Weber, M.; G. Rümpker and D. Gajewski, 2007. Theory of Elastic Waves, *Scientific Technical Report STR 07/03.* doi: 10.2312/GFZ.b103-07037. URL http://bib.gfz-potsdam.de/pub/str0703/0703.htm. 23, 39, 98

Wenk, H.-R.; K. Bennett; G.R. Canova and A. Molinari, 1991. Modeling plastic deformation of peridotite with the self-consistent theory, *J. Geophys. Res.*, 96 (B5):8337–8349. 39

Wiechert, E., 1910. Bestimmung des Weges von Erdbebenwellen. I., *Theoretische. Phys. Z.*, 11:294–304. 7

Wielandt, E., 1987. On the validity of the ray approximation for interpreting delay times. In Nolet, G., editor, *Seismic Tomography*, pages 85–98. D. Reidel, Norwell, Mass., 1987. 23

Williamson, P. R. and M. Worthington, 1993. Resolution limits in ray tomography due to wave behavior: Numerical experiments, *Geophysics*, 58:727 – 735. doi: 10.1190/1.1443457. URL http://adsabs.harvard.edu/abs/1993Geop...58..727W. 23, 173

Group, EUGEMIworking , 1989. The European Geotraverse seismic refraction experiment of 1986 from Genova, Italy, to Kiel, Germany, 176:43–57. 46

Wu, H. and J. M. Lees, April 1999. Cartesian parametrization of anisotropic traveltime tomography, *Geophysical Journal International*, 137:64–80. 9

Zhang, S.; Si. Karato; J. Fitz Gerald; U. H. Faul and Y. Zhou, 2000. Simple shear deformation of olivine aggregates, *Tectonophysics*, 144:133–153. doi: 10.1126/science.287.5455.933c. 39

Zhang, S. and S. Karato, 1995. Lattice preferred orientation of olivine aggregates deformed in simple shear., *Nature 375*, pages 774–777. URL http://www.nature.com/nature/journal/v375/n6534/abs/375774a0.html. 39

# I want morebooks!

Buy your books fast and straightforward online - at one of world's fastest growing online book stores! Environmentally sound due to Print-on-Demand technologies.

Buy your books online at
**www.morebooks.shop**

Kaufen Sie Ihre Bücher schnell und unkompliziert online – auf einer der am schnellsten wachsenden Buchhandelsplattformen weltweit! Dank Print-On-Demand umwelt- und ressourcenschonend produziert.

Bücher schneller online kaufen
**www.morebooks.shop**

KS OmniScriptum Publishing
Brivibas gatve 197
LV-1039 Riga, Latvia
Telefax: +371 686 204 55

info@omniscriptum.com
www.omniscriptum.com

Printed by Books on Demand GmbH, Norderstedt / Germany